RADIATIVE TRANSFER IN STELLAR AND PLANETARY ATMOSPHERES

Radiative transfer is essential for obtaining information from the spectra of astrophysical objects. This volume provides an overview of the physical and mathematical background of radiative transfer and its applications to stellar and planetary atmospheres. It covers the phenomenology and physics of early-type and late-type stars, as well as ultra-cool dwarf stars and extrasolar planets. Importantly, it provides a bridge between classical radiative transfer and stellar atmosphere modelling and novel approaches, from both theoretical and computational standpoints. With new fields of application and a dramatic improvement in both observational and computational facilities, it also discusses the future outlook for the field. Chapters are written by eminent researchers from across the astronomical disciplines where radiative transfer is employed. Using the most recent observations, this is a go-to resource for graduate students and researchers in astrophysics.

Lucio Crivellari is Associate Researcher at the Instituto de Astrofísica de Canarias, La Laguna, Spain and at the INAF-Osservatorio Astronomico di Trieste, Italy. He has previously held positions as an ESA External Fellow at the Observatoire de Paris-Meudon (1982–1983) and European Union Senior Fellow Human Capital and Mobility at the Instituto de Astrofísica de Canarias (1994–1995). His research focuses on radiative transfer and stellar atmosphere theory.

Sergio Simón-Díaz is Staff Researcher at the Instituto de Astrofísica de Canarias. He is mainly involved in the observation and analysis of massive stars, and he has authored more than a hundred refereed papers. At present, he is leading the IACOB project, a long-term observational project aimed at providing an overview of the main physical properties of Galactic massive O- and B-type stars to be used to improve our current understanding of the current theories of stellar atmospheres, winds, interiors and evolution of massive stars.

María Jesús Arévalo is Associate Professor at Universidad de la Laguna and researcher at the IAC. She was also the Head of Graduate Studies Division of the IAC in 2016–2017. Her research focuses on binary stars.

Canary Islands Winter School of Astrophysics

Volume XXIX
Radiative Transfer in Stellar and Planetary Atmospheres

Series Editor
Rafael Rebolo *Instituto de Astrofísica de Canarias*

Previous Volumes in This Series

I. Solar Physics
II. Physical and Observational Cosmology
III. Star Formation in Stellar Systems
IV. Infrared Astronomy
V. The Formation of Galaxies
VI. The Structure of the Sun
VII. Instrumentation for Large Telescopes: A Course for Astronomers
VIII. Stellar Astrophysics for the Local Group: A First Step to the Universe
IX. Astrophysics with Large Databases in the Internet Age
X. Globular Clusters
XI. Galaxies at High Redshift
XII. Astrophysical Spectropolarimetry
XIII. Cosmochemistry: The Melting Pot of Elements
XIV. Dark Matter and Dark Energy in the Universe
XV. Payload and Mission Definition in Space Sciences
XVI. Extrasolar Planets
XVII. 3D Spectroscopy in Astronomy
XVIII. The Emission-Line Universe
XIX. The Cosmic Microwave Background: From Quantum Fluctuations to the Present Universe
XX. Local Group Cosmology
XXI. Accretion Processes in Astrophysics
XXII. Asteroseismology
XXIII. Secular Evolution of Galaxies
XXIV. Astrophysical Applications of Gravitational Lensing
XXV. Cosmic Magnetic Fields
XXVI. Bayesian Astrophysics
XXVII. High Time-Resolution Astrophysics
XXVIII. Solar System Exploration

RADIATIVE TRANSFER IN STELLAR AND PLANETARY ATMOSPHERES

Edited by

LUCIO CRIVELLARI

Instituto de Astrofísica de Canarias, Tenerife

SERGIO SIMÓN-DÍAZ

Instituto de Astrofísica de Canarias, Tenerife

MARÍA JESÚS ARÉVALO

Universidad de la Laguna, Tenerife

CAMBRIDGE
UNIVERSITY PRESS

CAMBRIDGE
UNIVERSITY PRESS

University Printing House, Cambridge CB2 8BS, United Kingdom

One Liberty Plaza, 20th Floor, New York, NY 10006, USA

477 Williamstown Road, Port Melbourne, VIC 3207, Australia

314–321, 3rd Floor, Plot 3, Splendor Forum, Jasola District Centre, New Delhi – 110025, India

79 Anson Road, #06–04/06, Singapore 079906

Cambridge University Press is part of the University of Cambridge.

It furthers the University's mission by disseminating knowledge in the pursuit of education, learning, and research at the highest international levels of excellence.

www.cambridge.org
Information on this title: www.cambridge.org/9781108499538
DOI: 10.1017/9781108583572

First published 2020

Printed and bound in Great Britain by Clays Ltd, Elcograf S.p.A.

A catalogue record for this publication is available from the British Library.

ISBN 978-1-108-49953-8 Hardback

Contents

Contributors

OLGA ATANACKOVIĆ Faculty of Mathematics, University of Belgrade, Serbia

TIMOTHY C. BEERS University of Notre Dame, USA

MARIA BERGEMANN Max Planck Institute for Astronomy, Heidelberg, Germany

MATS CARLSSON Institute of Theoretical Astrophysics, University of Oslo, Norway

LUCIO CRIVELLARI Instituto de Astrofísica de Canarias, La Laguna, Spain, and National Institute for Astrophysics (INAF) – Osservatorio Astronomico di Trieste, Italy

CAMILLA JUUL HANSEN Max Planck Institute for Astronomy, Heidelberg, Germany

ARTEMO HERRERO Instituto de Astrofísica de Canarias, La Laguna, Spain

MARK S. MARLEY National Aeronautics and Space Administration (NASA) Ames Research Center, Space Science and Astrobiology Division, USA

JOACHIM PULS Universitaetssternwarte der Ludwig Maximilian University (LMU), Munich, Germany

Preface

The transport of radiant energy plays an important role in physics and engineering. However, it is of paramount importance in astronomy: the development of theories of radiative transfer (RT) since the beginning of the last century has made possible our present knowledge of the basic physics working in the observable Universe. A. Schuster and K. Schwartzschild can be considered the founding fathers of modern radiative transfer. With his seminal paper *Radiation through a Foggy Atmosphere*, published in 1905, the former set the cornerstone of a theory that the former developed in the following years. In his paper *On the Equilibrium of the Sun's Atmosphere* (1906), Schwartzschild, by studying the problem of the temperature distribution in the solar atmosphere, reached the conclusion that the latter is in radiative equilibrium and derived a formula for the observed phenomenon of limb darkening. Later in 1914, Schwartzschild published another fundamental paper, *Diffusion and Absorption in the Sun's Atmosphere*, in which he derived the basic equations of RT and gave their formal solution in terms of an integral equation that Milne and Hopf developed later on. Moreover, this paper paved the way to what was later called local thermodynamic equilibrium. (The aforementioned three papers can be found in D. H. Menzel's *Selected Papers on the Transfer of Radiation*, published by Dover in 1966.) The situation in the 1930s and 1950s was given authoritative reviews by E. A. Milne in his *Thermodynamics of the Stars* (1930, *Handbuch der Astrophysik*, Vol. 3) and by D. Barbier in his *Théorie Générale des Atmosphères Stellaires* (1958, *Handbuch der Astrophysik*, Vol. 50).

The availability of large computers for the numerical solution of the RT equation and the computation of stellar atmosphere models marked a turn in the 1960s. Since then, effective numerical techniques for both aims have been steadily developed with impressive results. Nowadays the amazing performance of modern computers makes it possible to compute "all-singing, all-dancing" stellar atmosphere models that include many physical processes and consider time variability and three-dimensional (3D) geometry. However, during the conference *Stellar Atmosphere beyond Classical Models*, held in Trieste in 1990, V. V. Ivanov made a keen distinction between the industry of radiative transfer, based merely on numerical technology, and ART (analytical radiative transfer). Behind his joke there was the concern that the ever-wider application of overwhelming numerical simulations may hide the underlying physics. Twenty-seven years later, at a moment when radiative transfer is living a transition from old to new fields of application and a dramatic improvement both in observational and computational facilities is on the horizon, the fear that the new practitioners of radiative transfer may lack the necessary physical and mathematical background motivated the choice of devoting the XXIX IAC Winter School to the applications of radiative transfer to stellar and planetary atmospheres.

The present volume is the outcome of the School, and it has been the main aim of the editors that the book should preserve its spirit and purpose. Hence their ambitious aim to publish not just the compilation of the lectures delivered during the School, but a useful reference text for beginners in the theory and practice of stellar atmospheres. The material is organized in two parts. The first is intended as introductory to the second and presents the basis, both physical and mathematical, of radiative transfer. The authors of the first three contributions (L. Crivellari, A. Herrero and O. Atanaković) worked in close collaboration in order to complement homogeneously the content of their chapters. The first one addresses the physics of RT from a general point of view, while the second deals with specific issues of RT in stellar atmospheres and spectral line formation. Chapter 3 contains an exhaustive review of numerical methods for the solution of the RT equation, including the standard ones widely in use since the 1960s and their progressive improvements as well as novel algorithms, probably less well known until now. A review by M. Carlsson on the stellar atmosphere codes nowadays available concludes the first part. In the second, J. Puls, M. Bergemman and M. S. Marley discuss the phenomenology and physics of early-type stars, late-type stars and ultracool dwarfs and extrasolar planets. The purpose of the second part is to present a comprehensive although necessarily brief overview of the state of the art and future outlook vis-à-vis the next release of the Gaia mission data, which will provide precise information on distances, proper motion and spectral distribution for more than a billion Galaxy stars as well as radial velocities for many millions of the brightest stars, and the launch of the NASA James Webb Space Telescope.

The PowerPoint files with the original content of the lectures delivered can be found on the website of the School (www.iac.es/winterschool/2017).

Lucio Crivellari
Sergio Simón Díaz

Acknowledgements

As organizers, we are pleased to thank the lecturers for the level and quality of their contributions, both to the School and this book, to say nothing of the students, whose motivated participation showed us that radiative transfer is still alive. We are grateful to Cristina García Vargas, the direction secretary of the Spanish National University for Education at a Distance (UNED), which housed the event, and acknowledge the assistance of the UNED staff before and during the School. Needless to say we are highly indebted to all the colleagues of the *Instituto de Astrofsica de Canarias* who contributed to the success of the School. Special thanks go to Terry Mahoney for his thorough revision of some chapters, and to Gabriel Perez and Miguel Briganti for their diligent help, the former with the poster of the School and the figures of the book, the latter with the group photo. Carlos Allende and Sergio Simón-Díaz organized the tutorials, held in parallel with the lectures, and their outstanding job deserves praise. We acknowledge the hospitality of the University of La Laguna for making available the facilities for running the tutorials. We thank also the director of the *Museo de la Ciencia y del Cosmos* (*Museos de Canarias*, La Laguna) for allowing the use of the conference hall for the public lecture *'La vida de una estrella en un arco iris'*, delivered by Toñi Varela, whose contribution is warmly acknowledged. Last but not least, we express our gratitude to Lourdes Gonzalez, the veritable pillar of the IAC Winter Schools: the Institute has been most fortunate to rely on her dedicated and steadfast work during the thirty years since the first School in 1989.

Lucio Crivellari
Sergio Simón Díaz
María Jesús Arévalo

1. The Physical Grounds of Radiative Transfer

LUCIO CRIVELLARI

Abstract

Radiative transfer, i.e., the transport of radiant energy through a medium, can be described in several alternative ways, either at macroscopic or microscopic level. In order to set a common physical background for the applications of radiative transfer to stellar and planetary atmospheres, presented in the second part of this book, a macroscopic representation of the radiation field derived from radiometry, a microscopic picture based on the kinetics of photons and the transport of radiant energy in terms of Maxwell's electromagnetic theory are discussed.

1.1 Introduction

The aim of this chapter is to define a common physical framework, established by radiative transfer, for the ensuing discussion of the phenomenology of early-type, late-type and planetary atmospheres. Radiative transfer (RT), i.e., the transport of radiant energy, is a key phenomenon in astrophysics and plays a central role in this School.

Radiant energy may be defined as energy in form of electromagnetic waves that propagate either in a vacuum or through a material medium. In the latter case, the radiation field can exchange both energy and momentum with matter. In order to account for the transport of radiant energy, three different equivalent representations can be introduced: (a) the electrodynamic picture according to Maxwell's equations and the Poynting vector (waves); (b) the description in terms of a continuous stream of energy in given directions (rays); and (c) the corpuscular picture, revisited after Einstein's hypothesis of light quantum (flow of photons). The single underlying concept is that of radiant energy in motion, which is a transport process. This gives rise in a natural way to the concept of radiant flux, namely, the amount of radiant energy flowing per unit time across a given surface.

Newton's corpuscular hypothesis and the consequent light ray model (Newton, 1704) superseded Huygens's earlier wave theory (Huygens, 1690); Young's interference experiments in 1801 subsequently vindicated Huygens's theory. In 1873, Maxwell achieved the unification of optics and electrodynamics in terms of electromagnetic waves (Maxwell, 1891), later confirmed experimentally by Hertz. Einstein's interpretation of the photoelectric effect set the ball rolling again. Nowadays, quantum physics postulates the dual wave–particle nature of electromagnetic radiation so that its transport is ascribed to either of two different kinds of carriers: electromagnetic waves and photons. Only quantum electrodynamics, where the electromagnetic field is represented by operators and the radiation field by quantized harmonic oscillators, can the properties of photons be rigorously accounted for. Nevertheless, in our exposition we will assume a semiclassical point of view, according to which photons are considered localized particlelike entities with well-defined energy and momentum.

The solution of the Maxwell's equations implies the transport of energy through the surrounding medium by means of electromagnetic waves that propagate with a finite speed characteristic of the specific medium. An alternative phenomenological picture may be drawn from radiometric concepts, based on the single physical idea of radiant flux. In order to develop suitable mathematical tools for this representation, the first necessary step is to define, on a sound physical basis, the entity transported. Two alternative (but correlated) approaches are possible in terms of either macroscopic or microscopic quantities.[1]

After introducing in Section 1.2 the stellar atmosphere physical system and a brief mention of rays and geometrical optics in Section 1.3, we will recall the basic concept of radiometry (Section 1.4), prior to an operational definition of the macroscopic specific intensity of the radiation field that is given in Section 1.5. An alternative microscopic picture will be introduced in Section 1.6 in order to derive in Section 1.7 the RT equation as a kinetic (Boltzmann's) equation for the photons. Macroscopic RT coefficients are defined operationally in Section 1.8, which will allow us to analyse in Section 1.9 the structure of the source and sink terms in the RT equation, and to define the source function through which the specific RT equations are coupled. The statistical interpretation of radiative transfer is considered in Section 1.10. The transport of radiant energy is described as a fluid dynamics–like process in Section 1.11, which will serve as an introduction to Section 1.13, where the macroscopic and the electrodynamic pictures (Section 1.12) of the radiation field are compared.

1.2 The Stellar Atmosphere Physical System

1.2.1 *Definition of a Star*

A star may be defined as a gravitationally bound open concentration of matter and energy. From the point of view of thermodynamics, which is essentially concerned with the flow and balance of energy and matter, physical systems may be classified as:

open, when both matter and energy fluxes are present;
closed, when only energy fluxes occur; and
isolated, in which neither matter nor energy fluxes are present

The mere fact that we see the stars is clear-cut evidence that they emit radiant energy over a vast wavelength range. Moreover, the long-standing detection of the so-called solar and stellar winds has revealed important mass-loss phenomena, hence the justification of the previous definition.

The observed evidence of fluxes implies the existence of gradients in the physical properties inside the star and hence of transport phenomena that tend to establish equilibrium conditions. The energy generated by thermonuclear reactions in the inner core is carried through the stellar layers by two modes of transport: radiative and convective. Their relative weight depends on the point-by-point thermodynamic conditions inside the star. However, the primary observational fact that stars emit radiation is clear-cut proof that the former must always be present. Radiation pressure, acting outwards, is antagonist to the gravitational force. The balance between the force due to the total pressure (gas plus radiation) and gravity keeps the structure of the star stationary over very long periods of its life. All this characterizes a system that is out of equilibrium where irreversible processes are taking place. Since the system is not far from equilibrium, however, the transport phenomena are governed by linear phenomenological laws. Thus, to a first approximation, a linear nonequilibrium approach may be employed.

[1] Macroscopic quantities are suggested directly by experience, without any preliminary knowledge of their intrinsic nature. In contrast, a microscopic picture requires previous hypotheses concerning the entity to be represented.

1.2.2 *Qualitative Definition of a Stellar Atmosphere*

The outermost layers of a star, i.e., its atmosphere, constitute a boundary through which photons can escape in the form of radiation into the surrounding interstellar medium. We will therefore define the stellar atmosphere to be this frontier region, where the emergent electromagnetic spectrum forms. The radiative flux observed is the signature of gradients inside the atmosphere, which are in turn the consequence of departure from equilibrium conditions. Moreover, the evidence of mass loss from observations of solar and stellar winds shows that matter particles can also escape outwards, a proof of departure from mechanical equilibrium and hence the transport of matter. On the other hand, departure from radiative equilibrium in the atmosphere of late-type stars gives rise to convective transport, which plays a fundamental role in the energy balance, as well as in the generation and advection of the magnetic fields observed in these stars.

The *stellar atmosphere physical system* consists of two components, matter and a radiation field that permeates and interacts with the former. At the macroscopic level, its structure will be described at any point by the values of the thermodynamic variables T, P and ρ, as well as the velocity of the (ideal) matter elements on the one hand, and by the local properties of the radiation field on the other. As in any physical system, the structure of the atmosphere is shaped by the mutual interactions of the two components, which are determined by the physical relations among the variables, the internal energy of the system and the constraints imposed. Such relations will be expressed by conservation, state and transport equations. We note that the structure of a stellar atmosphere is mainly determined by the physical properties of the stellar interior, most importantly by the temperature gradient of the whole star. This gradient is responsible for the outward radiative flux. The outer layers do not quantitatively alter the outgoing flux. Owing to their low density, they cannot absorb and store a large amount of energy, neither is the energy produced inside them comparable to the energy generated in the stellar interior.

The physical conditions inside a stellar atmosphere are governed essentially by the gravitational field generated by the star and the outward radiation flux from the interior. In order to remain in a steady state, the configuration of the atmosphere must be such that radiation can flows outwards. Therefore, two external parameters – the gravitational acceleration at the surface and the total radiation flux (i.e., the bolometric luminosity of the star) – together with an internal parameter, namely the chemical composition, determine the state of the stellar atmosphere physical system.

1.2.3 *The Observational Side*

From an observational point of view, the spectral features of the emergent radiation, namely the qualitative properties of this flux, are determined by interactions between matter and radiation in the outer layers. These processes are responsible for a redistribution in frequency of the radiant energy. The emergent spectrum therefore reflects the physical state of the stellar atmosphere, where, by definition, it forms. This statement constitutes the key to the diagnostics of the physical properties of stars: from the theoretical modelling of stellar atmospheres, we try to predict the characteristics of the emergent spectrum via the computation of the radiative flux carried through the outer layers; we then compare successively the computed with the observed features.

Looking forwards to the lectures that follow, we mention just two examples:

1. The supersonic velocity fields in early-type stars are revealed by line shifts and P Cyg profiles, which provide diagnostic tools to infer the mechanical structure of the expanding outer layers of these stars. Moreover, velocity fields bear upon radiative transfer in spectral lines via the Doppler effect.

2. The observation of asymmetries in the profiles of late-type giant stars suggests a departure from hydrostatic equilibrium (or steady state) in their atmospheres.

1.3 Rays and Geometrical Optics

1.3.1 *Light Rays*

Against the background of the seventeenth-century atomistic view of natural philosophy Newton, going beyond the experimental evidence of his exhaustive investigations, put forward the conjecture that the geometrical behaviour of reflection and refraction could be explained only if light were made of corpuscles. The propagation of radiant energy as a flow of particles entails the ray concept. In his *Opticks*,[2] Newton states: 'The least Light, or part of Light, which may be stopp'd alone without the rest of the Light, or propagated alone, or do suffer any thing alone, which the rest of the Light doth not or suffers not, I call a Ray of Light.' The geometrical nature of the light rays is affirmed in the further definition: 'Mathematicians usually consider the Rays of Light to be Lines reaching from the luminous Body to the Body illuminated.'

To some extent, however, a ray can be realized physically by allowing light from a distant source to pass through a small circular aperture of radius r pierced into a screen. When $r \to 0$, the tube of light emerging from the screen shrinks to a curve. Therefore, we may assume that the radiation field consists of an ensemble of rays, characterized by their direction and the amount of energy they carry. Hence, the transport of radiant energy from one point of the medium to another can be formulated in terms of the creation, propagation and destruction of rays.[3]

1.3.2 *Geometrical Optics*

The ordinary properties of light, such as rectilinear propagation and reflection, or refraction at the interface between two material media, can be understood simply by knowing how light travels, without enquiring into its nature. In contrast with physical optics, which takes into account phenomena such as interference, diffraction and polarization, the foregoing simplified approach constitutes what is called geometrical (or ray) optics, an idealized model of light propagation in terms of rays. Because the wavelengths considered are very small in comparison with the spatial variation of any property of the medium through which the waves propagate, the value of the wavelength can be formally allowed to tend to zero. Hence, the laws of propagation, which determine the trajectories of the rays, have an essentially geometrical character.

In order to show that, under the necessary condition that the medium be isotropic, the rays are always perpendicular to the light wavefronts, it is necessary to make reference here to the Hamilton–Jacobi theory. As is well known, in this formulation of mechanics the motion of a particle (or a system of particles) is represented as a wave. The solution, S, of the equation $H + \partial S/\partial t = 0$, where H is the Hamiltonian of the system, is called Hamilton's principal function. For conservative systems, the motion of S in time is similar to the propagation of a wavefront in the configuration space.[4] In the specific instance of light waves, the corresponding scalar equation is satisfied by a plane wave if the refraction index, n, is constant (otherwise, variation of the latter will distort and bend the wave). When n does not vary greatly over distances of the order of the wavelength, which is the

[2] The first book of *Opticks*, Defs I and II, Part I.

[3] This, for example, is the point of view of Planck (1906).

[4] For a discussion of the motion of S in the configuration space, see, for example, ch. VIII of Brillouin (1938).

assumption of geometrical optics, the original scalar wave equation can be approximated by the eikonal equation for the unknown function $L(\mathbf{r})$, called the optical path length. Surfaces of constant L, determined by the solution of the eikonal equation, are surfaces of constant optical phase and thus define the wavefronts. The ray trajectories are everywhere perpendicular to the wavefronts because they are also determined by the eikonal equation. In this context, it is well founded to interpret the Poynting vector of electrodynamics as a ray vector.

1.4 Radiometric Concepts

Radiometry deals with the measurement of the production and propagation of observable electromagnetic fields. As thoroughly discussed in chapter II of Preisendorfer (1965), the radiometric concepts, based on the single physical idea of radiant flux, can be defined by means of the idealized results of real experiments. This is an operational approach, whose essential outcomes we are going to summarize here in the language of the mathematical theory of measure. We recall that the radiant flux is defined as the amount of radiant energy flowing per unit time across a given surface. Following Preisendorfer, the former can be operationally defined in terms of the readings of a standard meter, which senses and records the radiant flux within a given set F of frequencies incident on a collecting surface A from a set D of directions. We will denote by $\Phi(t; F, A, D)$ the reading at time t of the meter specified by the foregoing parameters.

1.4.1 *Monochromatic Radiant Flux*

The range of frequencies recorded by the radiant flux meter can be selected by means of a suitable filter. If the meter is adjusted so that it records simultaneously the frequencies in two disjoint sets F_1 and F_2 (i.e., $F_1 \cap F_2 = \Phi$) in a first experiment and successively the separate responses to F_1 and F_2 under the same conditions, two far-reaching properties of the function can be formalized in mathematical terms. Firstly, it turns out that

$$\Phi(t; F_1, A, D) + \Phi(t; F_2, A, D) = \Phi(t; F_1 \cup F_2, A, D). \tag{1.1}$$

This is the statement of the *linear additivity* of Φ in frequency. Then, if we denote by $m(F)$ the measure[5] of the set F, it holds that $\Phi = 0$ when $m(F) = 0$. This second property of Φ expresses its *absolute continuity* with respect to frequency.

To produce a finite amount of radiant energy, the frequency band must also be finite. We can, however, define a *monochromatic* radiant flux Φ_ν by taking the limit

$$\Phi_\nu(t; A, D) = \lim_{\Delta\nu \to 0} \frac{\Phi(t; A, D, \Delta\nu)}{\Delta\nu}. \tag{1.2}$$

where $\Delta\nu \equiv m(F)$. Because of the preceding properties, the existence of this limit is a reasonable assumption that can be rigorously proved if we recur to theory of measure. By definition, Φ_ν is the partial derivative of Φ with respect to ν. Hence *monochromatic* implies per unit frequency range.

1.4.2 *Geometrical Properties of the Radiant Flux*

Again by means of hypothetical experiments with a radiant flux meter,[6] based however on real ones, the following geometrical properties of Φ_ν can be deduced:

[5] The theory of measure assigns a real number to any subset of a set, i.e., its *measure* to be interpreted as its size.

[6] Fully described in ch. II, sec. 9 of Preisendorfer (1965).

(i) *Directional additivity.* For any pair of almost disjoint[7] sets such that $D_1 \cup D_2 = D$, from the results of three successive readings $\Phi_\nu(t; A, D_1)$, $\Phi_\nu(t; A, D_2)$ and $\Phi_\nu(t; A, D$, it holds that

$$\Phi_\nu(t; A, D_1) + \Phi_\nu(t; A, D_2) = \Phi_\nu(t; A, D_1 \cup D_2), \tag{1.3}$$

provided that the irradiation conditions are the same at the time of the three separate measurements.

(ii) *Absolute directional continuity.* If $m(D) = 0$, then $\Phi_\nu = 0$.

The properties of *surface additivity* (iii) and *absolute surface continuity* (iv) are deduced likewise.

Properties (i) through (iv) characterize the monochromatic radiant flux and constitute the essence of the phenomenological foundation of radiative transfer. They justify the treatment of radiative transfer as a process essentially *linear* in character. In particular, different rays are necessarily independent. While this excludes interference among the rays, it brings about at the same time an intrinsic difficulty in view of the representation of the radiation field: in order to specify completely its state at any point, a single quantity (either scalar or vectorial) is not enough. It shall be necessary to take into account *all* the rays – virtually an infinite set – passing through the point in question. It is worthwhile to remark that properties (i) and (ii) show where the macroscopic representation to be introduced in the next section contrasts with the electrodynamic one: diffraction of electromagnetic waves yields counterexamples of (i); interference disproves (ii).

1.5 Macroscopic Picture: The Specific Intensity of the Radiation Field

The energy carried on along a given direction is the fundamental physical observable in radiative transfer. We are going to introduce now a *local* and *directional* macroscopic quantity, namely *the specific intensity of the radiation field*, which makes possible a description of both the local properties of the radiation field and its propagation through a medium, suitable for the mathematical treatment of radiative transfer. Such a macroscopic representation is consistent with the corpuscular picture of the flow of photons, as will be shown at the end of Section 1.6.

1.5.1 *Operational Definition of the Specific Intensity*

We measure the amount of radiant energy $\Delta E_\nu(\mathbf{n})$ that flows during a time interval Δt through an oriented surface $\mathbf{k}\,\Delta\sigma$ around a point P_1 individuated by its position vector $\mathbf{r}$, inside the solid angle $\Delta\Omega$ around the direction of propagation $\mathbf{n}$ and vertex at P_1, with frequency in the range $(\nu, \nu + \Delta\nu)$. We must choose the surface element $\mathbf{k}\,\Delta\sigma$ such that at each of its points P' the value of the radiation field can be considered constant. In such a way the amount of energy that enters into any solid angle $\Delta\Omega$ with vertex at P' will be the same. Analogously, the energy spectral distribution will not vary noticeably within the frequency band $\Delta\nu$. (The geometrical layout is sketched in Figure 1.1.)

According to the experimental laws of radiometry, the extensive quantity $\Delta\ E_\nu(\mathbf{n})$ results proportional to each single element of the process of measurement, that is,

$$\Delta\ E_\nu(\mathbf{n}) \propto \mathbf{n} \cdot \mathbf{k}\ \Delta\sigma\,\Delta\Omega\,\Delta\nu\,\Delta t. \tag{1.4}$$

[7] That is, D_1 and D_2 have in common only a set of directions of zero measure.

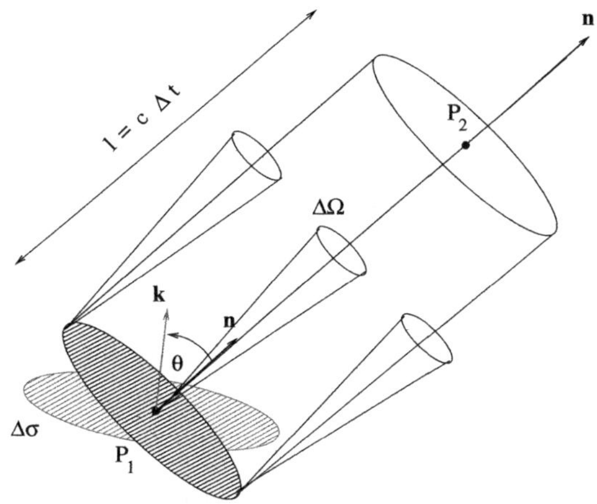

FIGURE 1.1. The geometrical elements of the measure of $\Delta E_\nu(\mathbf{n})$. During the time interval Δt, the photons that flow through $\mathbf{k}\ \sigma$ fill the volume $\Delta V = (\mathbf{n}\cdot\mathbf{k})\ \Delta\sigma\ c\ \Delta t$. By hypothesis, the properties of the radiation field are the same at each point of the surface $\mathbf{k}\Delta\sigma$, hence within each of the three solid angles drawn.

As a consequence of the foregoing radiometric properties and the absolute continuity in frequency, the limit

$$(\mathbf{n}\cdot\mathbf{k})^{-1} \lim_{\Delta\sigma\Delta\Omega\Delta\nu\Delta t\to 0} \Delta E_\nu(\mathbf{n}) \equiv I(\mathbf{r}, t; \mathbf{n}, \nu) \tag{1.5}$$

exists and takes on a finite value. This coefficient of proportionality between the measured value of the physical magnitude and the product of the values of the geometrical, spectral and time elements of the process of measurement is by definition the *specific intensity of the radiation field.* In such a way we have defined operationally the amount of specific energy carried on along a ray.

The dimension of I is $(ML^2T^{-2})\cdot L^{-2}\cdot T\cdot T^{-1} = MT^{-2}$, as it follows from (1.5). That is to say, I has the dimension of an energy flux per unit time and unit frequency band; in other words, a *monochromatic power flux.* In the centimetre-gram-second (*c.g.s.*) system, the units of I are $erg\ cm^{-2}\ st^{-1}\ hz^{-1}s^{-1}$. Dimensional analysis shows that the specific intensity has to be identified with the *radiance*, as defined in radiometry.

For further use, it will be convenient to write explicitly the infinitesimal amount of radiant energy $dE_\nu(\mathbf{n})$ as a function of the specific intensity $I(\mathbf{r}, t; \mathbf{n}, \nu)$ and the geometrical, spectral and temporal parameters involved, that is,

$$dE_\nu(\mathbf{n}) = I(\mathbf{r}, t : \mathbf{n}, \nu)\ \mathbf{n}\cdot\mathbf{k}\ d\sigma\ d\Omega\ d\nu\ dt. \tag{1.6}$$

1.5.2 *Moments of the Specific Intensity*

The straight average of the specific intensity over all solid angles defines the mean intensity of the radiation field, namely

$$J(\mathbf{r}, t; \nu) = \frac{1}{4\pi} \oint I(\mathbf{r}, t : \mathbf{n}, \nu)\ d\Omega. \tag{1.7}$$

The mean intensity J has the same dimension of the specific intensity I, but its units in the c.g.s. system are $erg\ cm^{-2}\ hz^{-1}\ s^{-1}$.

The mean intensity, defined in this way, is the zero-order moment with respect to $\mathbf{n}$ of the specific intensity. Two successive moments are defined likewise:

(1) The first-order moment

$$\mathbf{F}_\nu(\mathbf{r},t) \equiv \oint I(\mathbf{r},t:\mathbf{n},\nu)\mathbf{n}\; d\Omega. \tag{1.8}$$

(2) The second-order moment[8]:

$$\boldsymbol{T}_\nu \equiv \frac{1}{c}\oint I(\mathbf{r},t;\mathbf{n},\nu)\mathbf{nn}\; d\Omega. \tag{1.9}$$

The first-order moment is a vector to be identified with the *flux of radiation*; the second-order moment is a tensor that we will identify in the following with the *radiation pressure tensor*. (See Sections 1.13.2 and 1.13.3.)

1.5.3 *Energy Density of the Radiation Field*

Let us compute now the specific energy density of the radiation field in the frequency range $(\nu, \nu + d\nu)$. The photons that carry on the amount of energy $dE_\nu(\mathbf{n})$ specified by (1.6) fill during the time interval dt the volume $dV = \mathbf{n}\cdot\mathbf{k}\; d\sigma c dt$. (See Figure 1.1.) We define the *specific energy density* as

$$U(\mathbf{r},t;\mathbf{n},\nu)\; d\Omega\; d\nu \equiv \frac{dE_\nu(\mathbf{n})}{dV} = \frac{1}{c}\; I(\mathbf{r},t;\mathbf{n},\nu)\; d\Omega\; d\nu. \tag{1.10}$$

By integration of (1.10) over all the solid angles, it follow

$$\frac{d\nu}{c}\oint I(\mathbf{r},t;\mathbf{n},\nu)\; d\Omega = \frac{4\pi}{c}\; J(\mathbf{r},t;\nu)\; d\nu. \tag{1.11}$$

Hence we may define the *monochromatic energy density* (i.e., the energy density per unit frequency band) as

$$u_\nu(\mathbf{r},t) \equiv \frac{4\pi}{c}\; J_\nu(\mathbf{r},t). \tag{1.12}$$

The dimension of u_ν is $(ML^2\,T^{-2})\,L^{-3}T^{-1}$ and its c.g.s. units are *erg* cm^{-3} hz^{-1}. The integration of u_ν over the whole frequency range yields the *bolometric energy density* $u(\mathbf{r},t)$ that accounts for the localization of energy within the radiation field.

1.6 Microscopic Picture: The Corpuscular Model

It is needless to recall that, for a physical system constituted by very, very many particles, it is unfeasible to describe their individual properties. The only way out is a statistical approach, based on the average behaviour of a large number of them. In order to study the flows of energy and momentum inside a given system, information is required about the spatial and velocity distribution of the constituting particles. The simplest function that contains the necessary information is a *distribution function* defined on the phase space so that $f(\mathbf{r},\mathbf{v},t)\; d^3\mathbf{r}\; d^3\mathbf{v}$ is the expected number of particles at time t inside the infinitesimal volume $d^3\mathbf{r}$ around $\mathbf{r}$ with velocities in the infinitesimal volume $d^3\mathbf{v}$ around $\mathbf{v}$. The usual assumption is that $d^3\mathbf{r}$ is small compared with the spatial variation of any macroscopic propriety of the system, but large enough to contain a statistically significant number of particles.

[8] We have rather to employ here the dyadic notation, introduced by J. W. Gibbs in 1884. Although relatively obsolete nowadays, it is employed in continuum mechanics and electromagnetism.

The corpuscular nature of light suggests a picture in terms of a *flow of particles*: photons with energy $h\nu$ that carry on a momentum $\mathbf{p} = (h\nu/c)\ \mathbf{n}$ in their flight along a given direction specified by the unit vector $\mathbf{n}$. We are going to introduce two photon distribution functions, namely $f(\mathbf{r}, t; \mathbf{n}, \nu)$ and $F(\mathbf{r}, \mathbf{p}, t)$. The former, characterized by the pair of parameters $(\mathbf{n}, \nu)$ is consistent with the macroscopic description of the transport of radiant energy in terms of the specific intensity of the radiation field. The latter has the standard form of a distribution function, whose variables are $\mathbf{r}$, $\mathbf{p}$ and t.

1.6.1 *First Photon Distribution Function*

For the sake of a direct comparison with the specific intensity of the radiation field $I(\mathbf{r}, t; \mathbf{n}, \nu)$, we will define a distribution function f such that $f(\mathbf{r}, t; \mathbf{n}, \nu)\ d\Omega d\nu$ yields the number of photons per unit volume, at location r and time t in the range $(\nu, \nu + d\nu)$ that propagate along the direction $\mathbf{n}$ with speed c within the solid angle $d\Omega$ around $\mathbf{n}$. The dimension of f is $L^{-3}\ T$.

The specific photons $(\mathbf{n}, \nu)$ crossing $\mathbf{k} \cdot d\sigma$ during the time interval dt fill a volume $dV = \mathbf{n} \cdot \mathbf{k}\ d\sigma\ c\ dt$, hence $f(\mathbf{r}, t; \mathbf{n}, \nu)\ \mathbf{n} \cdot \mathbf{k}\ d\sigma\ c\ dt\ d\Omega d\nu$ will be their number. The corresponding energy is

$$dE_\nu(\mathbf{n} = h\nu\, c\, f(\mathbf{r}, t; \mathbf{n}, \nu)\, \mathbf{n} \cdot \mathbf{k}\ d\sigma\ d\Omega d\nu dt. \tag{1.13}$$

By direct comparison with (1.6), it follows that

$$I(\mathbf{r}, t; \mathbf{n}, \nu) = h\nu\ c\ f(\mathbf{r}, t; \mathbf{n}, \nu). \tag{1.14}$$

Equation (1.14) gives the quantitative link between the macroscopic description and the corpuscular picture of the radiation field.

1.6.2 *Second Photon Distribution Function*

In order to formulate the radiative transfer equation as a transport equation like in the kinetic theory of gases, we will make use of a second photon distribution function, defined so that $F(\mathbf{r}, \mathbf{p}, t)\ d^3\mathbf{r}\ d^3\mathbf{p}$ gives the number of photons per unit volume at point $\mathbf{r}$ and time t with momentum in the range $(\mathbf{p}, \mathbf{p} + d\mathbf{p})$. The dimension of F is $M^{-3}\ L^{-6}\ T^3$. The momentum $\mathbf{p}$ is linked to the frequency ν and the direction of propagation $\mathbf{n}$ of the photons according to the relation

$$\mathbf{p} = \frac{h\nu}{c}\ \mathbf{n}. \tag{1.15}$$

That is to say $\mathbf{p} = \mathbf{p}(\mathbf{n}, \nu)$ can be expressed as a function of the pair of parameters $\mathbf{n} = \mathbf{n}(\vartheta, \varphi)$ and ν. The infinitesimal volume in the R^3 momentum space is

$$d^3\ \mathbf{p} = dp_x\ dp_y\ dp_z = p^2\ sin\ \vartheta\ dp\ d\vartheta\ d\varphi. \tag{1.16}$$

From (1.15), it follows that $p = (h\nu)/c$ and $dp = (h/c)d\nu$, hence

$$d^3p = \frac{h^3\nu^2}{c^3}\ sin\ \vartheta\ d\vartheta\ d\varphi\ d\nu. \tag{1.17}$$

In the R^3 *direction* $\times$ *frequency* parameter space, the infinitesimal volume is given by

$$d\Omega\ d\nu = sin\ \vartheta\ d\vartheta\ d\varphi\ d\nu, \tag{1.18}$$

where $d\Omega$ is the solid angle around $\mathbf{n}$. According to the previous definitions it follows that

$$f(\mathbf{r},t;\mathbf{n},\nu)\ d^3\mathbf{r}\ d\Omega\ d\nu = F(\mathbf{r},\mathbf{p},t)\ d^3\mathbf{r}\ d^3\mathbf{p}. \tag{1.19}$$

Hence, taking into account (1.17) and (1.19), it follows straightforwardly that

$$f(\mathbf{r},t;\mathbf{n},\nu) = \frac{h^3\nu^2}{c^3}\ F(\mathbf{r},\mathbf{p},t). \tag{1.20}$$

Equations (1.14) and (1.20) proves that the specific intensity $I(\mathbf{r},t;\mathbf{n},\nu)$ is proportional to the distribution function $F(\mathbf{r},\mathbf{p},t)$. This result shows the correspondence between the macroscopic description of the radiation field and the corpuscular picture.

1.7 The Radiative Transfer Equation as a Kinetic Equation for Photons

In the kinetic theory of gases the distribution function $F(\mathbf{r},\mathbf{p},t)$ is assumed to vary with time because the particles constantly enter and leave a given volume of the phase space. The change is due to the streaming of the particles and binary collisions among them. This process can be expressed in words as

Total rate of change = Source terms − Sink terms.

The translation of this statement into mathematical language leads to the Boltzmann transport equation:[9]

$$\begin{aligned}\frac{d}{dt}\ F(\mathbf{r},\mathbf{p},t) &= \left(\frac{\partial}{\partial t} + \mathbf{v}\cdot\nabla_{\mathbf{r}} + \frac{\mathbf{F}_{ext}}{m}\cdot\nabla_{\mathbf{v}}\right)\ F(\mathbf{r},\mathbf{p},t)\\ &= \left[\frac{\partial}{\partial t}\ F(\mathbf{r},\mathbf{p},t)\right]_{coll},\end{aligned} \tag{1.21}$$

where $\nabla_{\mathbf{r}}$ and $\nabla_{\mathbf{v}}$ denote the gradient with respect to $\mathbf{r}$ and $\mathbf{v}$, respectively, and $\mathbf{F}_{ext}$ the external force acting on a particle of mass m. The right-hand side (RHS) is the Boltzmann collisional operator, which accounts for the kinetic balance between gains (sources) and losses (sinks).

It follows from (1.14) and (1.20) that the specific intensity $I(\mathbf{r},t;\mathbf{n},\nu)$ is proportional to the photon distribution function $F(\mathbf{r},\mathbf{p},t)$. Because there are not external forces acting on the photons and $\mathbf{v} = c\mathbf{n}$, we can write the transport equation for the distribution function $I(\mathbf{r},t;\mathbf{n},\nu)$ as

$$\frac{1}{c}\frac{\partial I(\mathbf{r},t;\mathbf{n},\nu)}{\partial t} + \mathbf{n}\cdot\nabla I(\mathbf{r},t;\mathbf{n},\nu) = \frac{1}{c}\left[\frac{\delta I_\nu}{\delta t}\right]^+ - \frac{1}{c}\left[\frac{\delta I_\nu}{\delta t}\right]^-. \tag{1.22}$$

The dimension of the terms in (1.22) is $(ML^2T^{-2})\cdot L^{-3}$, that is, an *energy density*.

Equation (1.22) is the mathematical formulation of a directional problem: the transport of a *local* and *scalar* quantity along a given direction $\mathbf{n}$. As $\delta l = c\delta t$ is a path length along $\mathbf{n}$, the left-hand side (LHS) of (1.22) gives the variation per unit length of the specific intensity that propagates in the direction $\mathbf{n}$. This variation must be equal to the difference

[9] The derivation of the Boltzmann equation is nicely sketched in section 2-2 of Mihalas (1978). Among the text books on kinetic theory, Huang (1963) covers in full details the physics of transport processes.

between the creation and destruction of the specific intensity along a unit length path in the direction $\mathbf{n}$. This quantity is formally denoted in the RHS of (1.22) by two square brackets that will be specified in Section 1.9. Equation (1.22) is the standard form of the RT equation. The *macroscopic* properties of the medium through which radiation propagates, like its steady state, homogeneity and isotropy, together with its geometry, shape any particular form of the LHS of (1.22). On the other hand, the interactions between matter and radiation at a microscopic level as well as the thermodynamic state of the medium determine the functional form of the RHS.

The term $\mathbf{n} \cdot \nabla I(\mathbf{r}, t; \mathbf{n}, \nu)$ in (1.22) can be immediately recognized as the *directional derivative* $dI(\mathbf{r}, t; \mathbf{n}, \nu)/ds$ of the specific intensity along the straight line (*characteristic*), whose direction is specified by the unit vector $\mathbf{n}$. Then (1.22) can be recast into the form

$$\frac{1}{c}\frac{\partial I(s,t:\mathbf{n},\nu)}{\partial t} + \frac{dI(s,t;\mathbf{n}\nu)}{ds} = \left[\frac{\delta I_\nu}{\delta s}\right]^+ - \left[\frac{\delta I_\nu}{\delta s}\right]^-, \tag{1.23}$$

which is a first-order ordinary differential equation whenever the source and sink terms in the RHS do not include the unknown $I(s, t; \mathbf{n}, \nu)$. Here s is the abscissa measured along the characteristic; ds and δs denote path elements along the latter.

There is a specific equation (1.23) for each direction $\mathbf{n}$. As already remarked at the end of Section 1.4, the complete specification of the radiation field requires a virtually infinite set of rays, hence of specific intensities. In practice, however, the choice of a proper finite number of direction is enough for an adequate representation. Therefore, the radiation field will be determined by the solution of a *finite system of differential equations* like (1.23), one for each monochromatic specific intensity, labelled by the parameter $\mathbf{n}$.

Different kinds of objects (stellar and planetary atmospheres, circumstellar envelopes, etc.) are characterized by different physical properties. The evidence of phenomena like departure from hydrodynamical and/or thermodynamic equilibrium, convective instability or the presence of magnetic fields compels one, case by case, to design accordingly the form of the total derivative of the specific intensity and to include all the relevant physical information into the source and sink terms. The former is a problem of mathematics; the latter implies a suitable representation of the physics to be taken into account, which will be achieved by means of the macroscopic RT coefficients that will be defined in Section 1.8.

1.8 The RT Macroscopic Transport Coefficients

On the microscopic scale, matter cannot be assumed to be homogeneous. However, consistently with the macroscopic picture introduced for the radiation field, we may consider homogeneous volume elements that emit and absorb radiant energy so that all the physical information on the atomic level can be incorporated into the treatment of radiative transfer by means of a limited number of *macroscopic transport coefficients*.

1.8.1 *The Emission Coefficient*

Let us start by introducing a macroscopic *thermal* emission coefficient that accounts for the energy emitted by a volume element ΔV into a solid angle $\Delta\Omega$ along a direction $\mathbf{n}$ during a time interval Δt with frequency in the range $(\nu, \nu + \Delta\nu)$ at the expense of the internal energy of the bulk of matter contained inside ΔV. For the sake of simplicity, we assume, as already said, that the element be physically homogeneous and isotropic so that the radiation emitted will be the same in all directions.

As in the case of the specific intensity, the emission coefficient can be introduced as the coefficient of proportionality between a measurable physical quantity, namely the amount of energy emitted, and the set of geometric and spectral parameters that characterize the thermal emission process. That is to say,

$$\Delta E_\nu^{th} \propto \Delta V \; \Delta\Omega \; \Delta\nu \; \delta t. \tag{1.24}$$

The limit

$$\lim_{\Delta V \Delta\Omega \Delta\nu \Delta t \to 0} \frac{\Delta E_\nu^{th}}{\Delta V \Delta\Omega \Delta\nu \Delta t} \equiv \eta_\nu^{th}, \tag{1.25}$$

which exists and is finite because of the usual considerations, defines the *thermal emission* coefficient. Its dimension is $(ML^2T^{-2}) \cdot L^{-3}$, i.e., an *energy density.*

1.8.2 *The Absorption, Scattering and Extinction Coefficients*

We now describe the decrease in specific intensity along a path δl in the direction $\mathbf{n}$, due to the absorption of the radiation field by a matter element. In the case of a weak electromagnetic field and a diluted medium, the physics of the removal process is *linear.* The decrease will therefore be proportional to the incoming beam of radiation. For a small enough path δl, the ratio $\delta I/I$ will be proportional to the path itself.

In the case of true absorption, we may then write

$$\delta I(\mathbf{n}) \propto I(\mathbf{n}) \; \delta l, \tag{1.26}$$

hence

$$\frac{\delta I(\mathbf{n})}{I(\mathbf{n})} = a_\nu(\mathbf{n}) \; \delta l, \tag{1.27}$$

where the proportionality coefficient $a_\nu(\mathbf{n})$ is the *absorption coefficient*, giving the fraction of monochromatic radiant energy removed from the incident beam along a unit path and converted into the internal energy of the material.[10]

The foregoing considerations hold true also for the loss of specific intensity due to scattering, which allows us to introduce likewise the *scattering coefficient* $\sigma_\nu(\mathbf{n})$ that gives the fraction of energy diverted from the beam propagating along $\mathbf{n}$ into a different direction. The *extinction coefficient*, defined as

$$\chi_\nu(\mathbf{n}) \equiv a_\nu(\mathbf{n}) + \sigma_\nu(\mathbf{n}), \tag{1.28}$$

accounts for the global effect of the removal of photons. Dimensional analysis shows that $[\chi_\nu] = [a_\nu] = [\sigma_\nu] = L^{-1}$.

1.8.3 *The Structure of the Absorption and Scattering Coefficients*

Looking at the *microscopic* (atomic) structure of the foregoing macroscopic coefficients, we may consider the latter as the product of two factors. The first one accounts for the probability that a photon is either absorbed or scattered by a single particle. This probability is measured by a proper, in general frequency-dependent, cross section defined in the normal way as the ratio between the number of photons either absorbed or scattered and the flux of incoming photons. The second factor expresses the number of particles

[10] Equation (1.27) is a statement of the more general Beer–Lambert law that relates the attenuation of light to the properties of the medium through which it propagates.

per unit volume capable of absorbing or scattering the photons of the incident beam. Thus, we can write[11]

$$a_\nu = a_P(\nu)\ n_{P_a} \tag{1.29-a}$$

and

$$\sigma_\nu = \sigma_P(\nu)\ n_{P_s}, \tag{1.29-b}$$

where $a_P(\nu)$ and $\sigma_P(\nu)$ are the cross sections for absorption and scattering respectively, and n_{P_a} and n_{P_s} are the corresponding particle density numbers. The former factor, whose dimensions are L^2, is an atomic property of the particles; the latter depends on the thermodynamic state of the medium. In the case that the hypothesis of thermodynamic equilibrium may be assumed (at least locally), the relative number of atoms and ions in successive stages of ionization is given by Saha's ionization equation, and their distribution over the quantum level allowed by the Boltzmann excitation equation. In the general case, however, the occupation numbers of bound and free states must be computed by solving the kinetic equations of statistical equilibrium.

1.9 The Source and Sink Terms of the RT Equation

We now cast into an explicit form the source and sinks terms of the RT equation, only formally indicated in (1.22) and (1.23), by means of the foregoing transport coefficients.

1.9.1 *The Source Term*

The term $[\delta I_\nu/\delta l]^+$ denotes the local increase in specific intensity with frequency ν travelling the path δl along $\mathbf{n}$ due to the interaction between matter and the radiation field. Such an increase may be either a *true creation* of photons or the result of a *scattering process*. We write the global term as the sum of the emission and scattering contributions; that is,

$$\left[\frac{\delta I_\nu}{\delta l}\right]^+_{\rm tot} = \left[\frac{\delta I_\nu}{\delta l}\right]^+_{\rm e} + \left[\frac{\delta I_\nu}{\delta l}\right]^+_{\rm s}. \tag{1.30}$$

Let us first evaluate the energy $\delta E_\nu^{\rm th}(\mathbf{n})$ created by *thermal emission* per unit time interval and unit frequency band within the volume δV of unit base perpendicular to $\mathbf{n}$ and height equal to the distance δl between two successive points P_1 and P_2 travelled by the photons during the unit time interval considered. According to the definition of $\eta_\nu^{\rm th}$ given implicitly by (1.25), the amount of energy injected along $\mathbf{n}$ is

$$\delta E_\nu^{\rm th}(\mathbf{n}) = \eta_\nu^{\rm th}\ \delta V\ \Delta\Omega = \eta_\nu^{\rm th}\ \delta l\ \Delta\Omega. \tag{1.31}$$

On the other hand, by the definition of specific intensity, the variation of energy between the points P_1 and P_2 will be equal to

$$\delta E_\nu(\mathbf{n}) = [I_\nu(P_1) - I_\nu(P_2)]\ (\mathbf{n}\cdot\mathbf{k})\ \Delta S\ \Delta\Omega\ \Delta\nu\ \Delta t, \tag{1.32}$$

which in the present instance reduces to

$$\delta E_\nu^{\rm th}(\mathbf{n}) = \delta I_\nu^{\rm e}\ \Delta\Omega. \tag{1.33}$$

[11] So far we have explicitly taken into account the possible dependence of the absorption and scattering coefficients in the direction $\mathbf{n}$. For the sake of simplicity, in the following we assume (justifiably in most cases) that they be isotropic.

From (1.31) and (1.33), it follows that

$$\lim_{\delta l \to 0} \frac{\delta I_\nu^{\mathrm{e}}}{\delta l} \equiv \left[\frac{\delta I_\nu}{\delta l}\right]_{\mathrm{e}}^{+} \equiv \eta_\nu^{\mathrm{th}}. \tag{1.34}$$

Equation (1.34) specifies the thermal contribution to the source term. Under the assumption of (at least local) thermodynamic equilibrium (TE), the thermal contribution is given by the Kirchhoff's law $\eta_\nu^{th} = a_\nu B_\nu(T)$.

The energy carried by the beam propagating in direction $\mathbf{n}$ is also increased by the contribution of all those photons that are diverted from their original direction $\mathbf{n}'$ into the direction $\mathbf{n}$. According to Section 1.8.2, the fraction of energy lost by scattering along $\mathbf{n}'$ is equal to

$$\delta I_\nu^{\mathrm{s}}(\mathbf{n}') = \sigma_\nu \; I_\nu(\mathbf{n}') \; \delta l. \tag{1.35}$$

If we now denote by $p(\mathbf{n}', \mathbf{n})$ the joint probability that a photon will be diverted from $\mathbf{n}'$ into $\mathbf{n}$, the total contribution to the beam propagating in direction $\mathbf{n}$ due to losses from all directions $\mathbf{n}'$ will be

$$\delta I_\nu^{s}(\mathbf{n}) = \sigma_\nu \; \delta l \oint p(\mathbf{n}', \mathbf{n}) \; I_\nu(\mathbf{n}') \; d\Omega', \tag{1.36}$$

and it follows straightforwardly that

$$\lim_{\delta l \to 0} \frac{\delta I_\nu^{\mathrm{s}}}{\delta l} \equiv \sigma_\nu \oint p(\mathbf{n}', \mathbf{n}) \; I_\nu(\mathbf{n}') \; d\Omega' \equiv \eta_\nu^{\mathrm{s}}. \tag{1.37}$$

In the simplest case of isotropic scattering, it holds that $p(\mathbf{n}', \mathbf{n}) = 1/4\pi$ so that the contribution due to scattering reduces to

$$\left[\frac{\delta I_\nu}{\delta l}\right]_{\mathrm{s}}^{+} = \sigma_\nu \; \frac{1}{4\pi} \oint I_\nu(\mathbf{n}') \; d\Omega' = \sigma_\nu \; J_\nu. \tag{1.38}$$

1.9.2 *The Sink Term*

The term $[\delta I_\nu / \delta l]^-$ denotes the decrease in specific intensity along a path δl in the direction $\mathbf{n}$ due to the interaction of the radiation field with matter. According to the previous definitions of the absorption and scattering coefficients, the total amount of specific intensity removed from the beam per unit path and unit time interval along the direction $\mathbf{n}$ will be the sum

$$\left[\frac{\delta I_\nu}{\delta l}\right]_{\mathrm{tot}}^{-} = \left[\frac{\delta I_\nu}{\delta l}\right]_{\mathrm{a}}^{-} + \left[\frac{\delta I_\nu}{\delta l}\right]_{\mathrm{s}}^{-} = (a_\nu + \sigma_\nu) \; I_\nu(\mathbf{n}). \tag{1.39}$$

Dimensional analysis shows straightforwardly that the source and sink terms have the required dimensions of energy per unit volume, i.e., $(ML^2T^{-2}) \cdot L^{-3}$.

1.9.3 *Monochromatic Optical Depth and the Source Function*

Let us define in the normal way the monochromatic optical depth by means of the differential relations

$$d\tau_\nu \equiv -\chi_\nu \; ds. \tag{1.40}$$

Optical depth is the natural variable of radiative transfer.

The source function is defined as the ratio of the total emissivity to the extinction coefficient, namely

$$S_\nu \equiv \frac{\eta_\nu}{\chi_\nu}. \tag{1.41}$$

By taking now into account (1.28), (1.34) and (1.37), the source function can be written as

$$S_\nu = \frac{\eta_\nu^{\mathrm{th}} + \eta_\nu^{\mathrm{s}}}{a_\nu + \sigma_\nu} = \frac{a_\nu B_\nu(T) + \sigma_\nu \oint p(\mathbf{n}', \mathbf{n})\, I_\nu(\mathbf{n}')\, d\Omega'}{a_\nu + \sigma_\nu}. \tag{1.42}$$

For a steady-state case, the partial derivative with respect to time in the RT equation vanishes, and we can therefore recast (1.23) in the form

$$\frac{dI_\nu(s; \mathbf{n})}{ds} = -\chi_\nu(s)\; I_\nu(s; \mathbf{n}) + \eta_\nu(s), \tag{1.43}$$

where $\eta_\nu(s)$ is the total source term and $\chi_\nu(s)$ the extinction coefficient. From (1.43) it follows that the time-independent RT equation in terms of the monochromatic optical depth reads

$$\frac{dI_\nu(\tau_\nu; \mathbf{n})}{d\tau_\nu} = I_\nu(\tau_\nu; \mathbf{n}) - S_\nu(\tau_\nu). \tag{1.44}$$

1.9.4 *Coupling of the RT Equations*

At the end of Section 1.7 we stated that a suitable representation of the radiation field may be obtained from the solution of a system of independent, first-order differential equations whenever the source and sink terms do not depend on the specific intensity (i.e., the unknowns of the problem). In general, however, the source function of each specific RT equation may contain a term that includes the full set of specific intensities, either explicitly or implicitly. According to the physics of the particular problem considered, this term may be either the same for all the RT equations frequency- and/or direction-dependent. The coupling of the specific intensities inside the source functions has an important consequence for the mathematical structure of the system of RT equations. In the case that the latter are *uncoupled*, the solution of the system depends on a set of independent initial conditions, each assigned at one point in each specific direction of propagation. In contrast, the solution of a system of *coupled equations* for the whole set of specific intensities, those travelling inwards and those travelling outwards, requires the assignment of boundary conditions on the two limiting surfaces of the geometric structure of the atmosphere. This constitutes a typical *two-point boundary problem.*

Different physical instances lead to a similar structure of the source function: a *local* term independent of radiative transfer and a *nonlocal* one that accounts for the RT process. Two illustrative examples are now considered.

(A) *The emission and extinction of radiation are governed by true absorption and scattering.*

Global information on the physics of the two processes is included in the macroscopic RT parameters a_ν and σ_ν, respectively. Scattering may be pictured as an almost elastic binary collision between a photon and a scattering centre (scatterer) that is not endowed with an internal structure (e.g., Thomson or Rayleigh scattering by free electrons). The photon incident from a given direction is almost instantaneously diverted into a new direction with nearly the same frequency.

In the case of coherent isotropic scattering, the source function is

$$S_\nu = \frac{a_\nu B_\nu(T) + \sigma_\nu J_\nu}{a_\nu + \sigma_\nu}, \tag{1.45}$$

which can be recast as

$$S_\nu = \varepsilon_\nu B_\nu(T) + (1 - \varepsilon_\nu) J_\nu \tag{1.46}$$

by defining the parameter

$$\varepsilon_\nu \equiv \frac{a_\nu}{a_\nu + \sigma_\nu}. \tag{1.47}$$

The source function given by (1.46) is a weighted average of the thermal emission contribution $B_\nu(T)$, a *local* property of matter assumed in local thermodynamic equilibrium (LTE), and the *non-local* term J_ν brought about by radiative transfer. The latter couples all the specific monochromatic RT equations as it includes all their solutions. The weight is the branching parameter ε_ν. In the literature, the quantity $1 - \varepsilon_\nu$ is called sometimes the *single-scattering albedo.*

A probabilistic interpretation of the parameter as given by (1.47) can be given in terms of the photons' mean-free-path l_ν, which is by definition the average length that a photon of frequency ν travels along a ray between the point where it is emitted and that of the successive encounter with a matter particle. It is determined by the extinction coefficient defined by (1.28); that is, $l_\nu = 1/(a_\nu + \sigma_\nu)$. Clearly, ε_ν is a measure of the probability that a photon be converted into thermal energy along a mean free path. According to the definition of a_ν and σ_ν, the fraction of energy subtracted from the beam by absorption processes over the total amount of energy removed is equal to $a_\nu/(a_\nu + \sigma_\nu)$. This ratio is clearly proportional to the ratio between the number of absorptions and that of total extinction processes along the ray, hence its probabilistic interpretation.

In the following, we discuss two limiting instances of case A.

(i) *Pure (coherent isotropic) scattering*

In this case $a_\nu = 0$, hence $\varepsilon_\nu = 0$ and $S_\nu(\tau_\nu) = J_\nu(\tau_\nu)$. It holds, therefore, that

$$S_\nu(\tau_\nu) = J_\nu(\tau_\nu) = \frac{1}{4\pi} \oint I_\nu(\tau_\nu; \mathbf{n})\, d\Omega. \tag{1.48}$$

In the lack of absorption, the source function is totally decoupled from the local thermodynamic state of the material and depends on the radiation field only. Its value at any given point will therefore be set by the physical conditions at all the surrounding points because of the radiative transfer process. Equation (1.48) shows clearly that the source function of any specific RT equation for the mode $(\mathbf{n}, \nu)$ includes all the specific intensities $I_\nu(\tau_\nu; \mathbf{n}')$ that constitute the solution of the global problem. We note that the coupling of the specific intensities is *linear.* This is a direct consequence of the radiometric directional additivity formulated in Section 1.4.2. The only agent responsible for the coupling, and hence of the value of each monochromatic source function, is the radiation field itself; namely, an *explicit internal* mechanism for radiative transfer.

(ii) *Pure absorption*

Here, it holds that $\sigma_\nu = 0$, and consequently that the source function is the Planck function at the local temperature of matter, namely $S_\nu = B_\nu(T)$. Under the hypothesis that the stellar atmosphere is in radiative equilibrium, the amount of energy subtracted from the radiation field by absorption processes must be exactly compensated by thermal emission. That is,

$$\int_0^\infty a_\nu \; J_\nu \; d\nu = \int_0^\infty a_\nu \; B_\nu(T) \; d\nu. \tag{1.49}$$

Equation (1.49) is a constraint imposed on the RT process by the physical conditions assumed for matter, i.e., it an *external* coupling mechanism. The value of each specific source function is fixed by the temperature that fulfils the RE condition given by (1.49).[12]

(B) *The Two-Level Atom*

When the scatterer is an atom, with its ladder of internal energy levels, the photon can be either scattered by a bound electron without alteration of the atomic structure or absorbed with the consequent photoexcitation of an electron to a higher energy level. In the latter case, if the atom is not deexcited by an inelastic collision with another particle, it decays to a lower level by a radiative process in which a new photon is emitted in another direction at a frequency corresponding to the quantum jump, nearly the same as that of the incident photon only in the case that the atom goes back to its initial state. This *scattering-like* process is a two-step quantum process whose finite duration is determined by the lifetime of the excited level.

The important case study of the two-level atom without continuum is perhaps the best example of the two-step process described earlier. The atomic model considered here consists of only two energy levels (a lower level l and an upper level u), between which both radiative and collisional transitions occur. In spite of its crude simplicity, this model provides an illustrative paradigm for the formation of spectral lines, in particular the resonance lines that arise from the ground level. In the following, we are going to focus on those aspects of the problem that are relevant to the present exposition[13]. For the particular instance under consideration, the only sources of absorption and emission are from the atomic populations of the levels. The radiative transitions will be described at a microscopic level by Einstein's phenomenological theory (Einstein, 1917), in terms of his coefficients A_{ul}, B_{ul} and B_{lu} for spontaneous emission, stimulated emission and absorption respectively.[14] We stress that scattering is not considered by Einstein's theory: the photons of the radiation field can only be absorbed or emitted. The bridge between the microscopic description and that in terms of the macroscopic RT coefficients η_ν and χ_ν is given by the relations introduced by Füchtbauer (1920) and Ladenburg (1921).

Following Füchtbauer, we write the extinction coefficient as

$$\chi_{lu} = \frac{h\nu}{4\pi} \left(n_l \; B_{lu} - n_u B_{ul} \right), \tag{1.50}$$

where n_l and n_u denote the level population densities (occupation numbers). As is customary, stimulated emission is included in the extinction coefficient because, as in the case of absorption, the rate of this process is proportional to the specific intensity.

[12] Here the coupling mechanism is expressed by a *single* law of conservation. In the more complex instance of multiline transfer, for example, the coupling is through the *system of statistical equilibrium equations* for the level populations.

[13] A clear statement of the two-level atom problem and the detailed derivation of the relevant formulae can be found in section 11-2 of Mihalas (1978).

[14] Actually, we employ here the radiative coefficients B_{ul} and B_{lu} as defined by Milne (1924), which differ from Einstein's original B_u^l and B_l^u by the dimensional factor $4\pi/c$.

Likewise, the emission coefficient will be

$$\eta_{ul} = h\nu/4\pi n_u A_{ul}. \tag{1.51}$$

Hence, the source function takes the form

$$S_L = \frac{n_u A_{ul}}{n_l B_{lu} - n_u B_{ul}} = \frac{2h\nu^3}{c^2}\left(\frac{n_l}{n_u}\frac{g_u}{g_l} - 1\right)^{-1}. \tag{1.52}$$

(The relations among the Einstein–Milne coefficients have been taken into account here; g_l and g_u denote the statistical weights of the levels.)

The ratio n_l/n_u is straightforwardly obtained from the single statistical equilibrium equation

$$n_l\left(B_{lu}\ J_\varphi + C_{lu}\right) = n_u\ \left(A_{ul} + B_{ul}\ J_\varphi + C_{ul}\right), \tag{1.53}$$

which accounts for the balance among all the processes, both radiative and collisional, that populate and depopulate the two levels. C_{lu} and C_{ul} denote the collisional rates, which depend on the temperature and density of the medium. The term J_φ is the integral over the frequencies of the monochromatic mean intensity weighted by the line profile, namely

$$J_\varphi \equiv \int_0^\infty \varphi_\nu\ J_\nu\ d\nu. \tag{1.54}$$

By algebraic manipulation, it follows that

$$S_L = \frac{J_\varphi + \varepsilon'\ B(T)}{1+\varepsilon'}, \tag{1.55}$$

where $B(T)$ is the Planck function at the nominal frequency of the line. The parameter ε' is defined as

$$\varepsilon' \equiv \frac{C_{ul}\left(1 - \exp\{-h\nu/kT\}\right)}{A_{ul}}, \tag{1.56}$$

where $1 - \exp\{-h\nu/kT\}$ is the correction factor for stimulated emission. The latter is of the order of unity in the case of resonance lines, for which it holds that $h\nu >> kT$.

The parameter ε' expresses the ratio between the fraction of photons destroyed by collisional deexcitation over that of those generated by spontaneous emission; in other words, it gives the relative weight of collisional over radiative processes. Because inelastic collisions tend to restore LTE conditions, altered by the escape of photons through the open upper boundary surface of the stellar atmosphere, ε' may be read as a measure of the degree of departure of the matter from LTE.

If we now define

$$\varepsilon \equiv \frac{\varepsilon'}{1+\varepsilon'} = \frac{C_{ul}\left(1 - \exp\{-h\nu/kT\}\right)}{A_{ul} + C_{ul}\left(1 - \exp\{-h\nu/kT\}\right)}, \tag{1.57}$$

the source function can be recast in the form

$$S_L = \varepsilon B(T) + (1-\varepsilon)\ J_\varphi. \tag{1.58}$$

Equation (1.58) has the same structure as (1.46): the sum of the *local* thermal contribution $\varepsilon B(T)$ and the *nonlocal* term $(1-\varepsilon)J_\varphi$, which includes all the specific intensities.

Its physical meaning, however, is different. The second term does not account for a true scattering process. In the two-level atom model a photon is removed from the radiation beam by photoexcitation of the lower level and successively a new photon is emitted by spontaneous emission, if the upper level is not depopulated by an inelastic collision. Although different on physical grounds, the effect of the diffusion term $(1-\varepsilon)J_\varphi$ in this two-step process is similar to that brought about by true scattering. The branching parameter ε (according to its definition the ratio of the collisional deexcitation rate to the total deexcitation ones) may be interpreted as the photon destruction probability in the scatteringlike process; $(1-\varepsilon)$ is the probability of the complementary event, in which a photon is returned to the radiation field.

1.10 Statistical Interpretation of Radiative Transfer

Each photon originating at a given point in a medium has a definite probability of leaving it, either directly or after a series of successive scattering (or scattering-like) processes. This escape probability, called quantum exit by Sobolev,[15] is independent of the origin of the photons and depends only on the optical properties of the medium through which they propagate. Knowing the escape probability $P_e(\tau)$ at each depth point makes it is possible to compute the emergent radiation for any kind of source within the medium because it is easy to set the correspondence between the escape probability and the microscopic description of the radiation field in terms of the first photon distribution function introduced in Section 1.6.1, and consequently with the specific intensity of the radiation field of the macroscopic picture. We will derive a linear integral equation for $P_e(\tau)$, whose mathematical structure is akin to that of the Schwartzschild–Milne formal solution for the specific intensity of the radiation field, integrated over all directions, in the paradigm instance where the source function includes a thermal emission component and a coherent isotropic scattering term. Hence, a statistical interpretation of radiative transfer can be applied.

1.10.1 *One-Dimensional Finite Medium*

In order to introduce the basic definitions for a statement of the general problem, let us start with the case of a one-dimensional homogeneous medium of finite optical thickness τ_0, delimited by the boundary points at $\tau = 0$ and $\tau = \tau_0$. We call incoming photons those propagating inwards along increasing optical depths from 0 to τ_0 and outgoing photons those travelling the other way. As an illustrative instance, let us consider here isotropic scattering: at each point, a photon propagating along either of the two directions has the same constant probability $\lambda/2$ of being scattered either forwards or backwards. The scattering parameter λ expresses the ratio of the scattered radiation to the total amount of energy lost. It is clearly identifiable with the single-scattering albedo $1-\varepsilon_\nu$ introduced in case (A) of Section 1.9.4.[16]

Our aim is to find the escape probability through the boundary at $\tau = 0$ for a photon absorbed at optical depth τ and successively scattered forwards. Let us first consider those photons that do not undergo scattering along their path from τ to $\tau = 0$. If we denote by $f(\tau)$ the number of photons per unit volume at τ, of which the fraction $\lambda/2$ is scattered forwards, according to the general attenuation law the fraction

[15] See Sobolev (1963).

[16] Although for the sake of a simpler notation we drop the subscript ν, all the quantities considered in the following are considered monochromatic.

$f^{out}(0) = \lambda/2\ f(\tau)\ \exp\{-\tau\}$ will leave undisturbed the medium. By the definition of probability, we can therefore define the probability of *direct escape*,

$$p_d(\tau) = (\lambda/2)\ e^{-\tau}, \tag{1.59}$$

as the ratio $f^{out}(0)/f(\tau)$. In spite of the simplification brought about by the stationary 1-D case and our choice to drop explicit reference to the frequency, $f^{\rm out}(0)$ and $f(\tau)$ should be easily recognized as the values at $\tau = 0$ and τ of the corresponding distribution function.

Let us now take into account the possibility that a photon undergoes scattering during its flight and consider separately the two cases in which the point of interaction is either downstream or upstream. In the first case, a fraction of the outgoing photons, originating at τ, experiences forward scattering at a depth point $\tau' < \tau$. The probability that one of these photons reaches τ' undisturbed from τ is $\exp\{-(\tau-\tau')\}$; the probability of its travelling along a further infinitesimal path $d\tau'$ without interaction is $\exp\{-d\tau'\} \approx 1 - d\tau'$. Hence, the probability of its undergoing a scattering event is that of the complementary event, i.e., $d\tau'$. Therefore, if we denote by $p_e(\tau')$ the hitherto unknown escape probability at τ', the probability of quantum exit for the photons originating at τ after single scattering between τ' and $\tau' + d\tau'$ will be the product of the foregoing probabilities, namely

$$p_s(\tau, \tau')\ d\tau' = (\lambda/2)\ e^{-(\tau-\tau')}\ p_e(\tau')\ d\tau'. \tag{1.60}$$

Likewise, the quantum exit probability that a fraction of incoming photons, originating at τ, undergoes backward scattering at a depth point $\tau' > \tau$ will be the same as in (1.60) but for the argument $(\tau' - \tau)$ of the exponential. In order to find the probability of quantum exit after a series of scattering processes, we must integrate the foregoing probability over the entire optical path from $\tau = 0$ to $\tau = \tau_0$. Eventually, the total escape probability at a given depth point τ will be given by $p_d(\tau)$ plus the integral over all depths of the single-scattering probability distribution, that is,

$$p_e(\tau) = p_d(\tau) + (\lambda/2) \int_0^{\tau_0} p_e(\tau')\ e^{|\tau-\tau'|}\ d\tau'. \tag{1.61}$$

The solution of this linear integral equation yields the required escape probability through the boundary at $\tau = 0$ for a photon originating at τ.

1.10.2 *The Case of a 3-D Semi-Infinite Atmosphere*

The previous results are easily generalized to the case of a 3-D semi-infinite atmosphere constituted by homogeneous plane–parallel layers. We again consider isotropic scattering, this time in any direction $\mathbf{n} = \mathbf{n}(\vartheta, \varphi)$ from among the virtually infinite number of directions of propagation of the photons. The equal probability of scattering in any direction is now $\lambda/4\pi$. The geometry of the simplified case under study here requires only two of the three coordinates of a cylindrical reference frame for a complete treatment of the problem: the geometrical depth measured along the z-axis, specified by the unit vector $\mathbf{e}$ chosen perpendicular to the layers, and the direction of propagation in any plane of the sheaf that has the z-axis in common, individuated by the angle $\vartheta = arc\ cos(\mathbf{n}\cdot\mathbf{e})$. The azimuthal angle φ is cyclical because of the axial symmetry assumed. The parameter $\mu \equiv \cos\vartheta$ is therefore sufficient to characterize a direction. In each plane of the sheaf, any slant optical depth scale $\tau_{\mathbf{n}}$ measured along $\mathbf{n}$ can be transformed into the standard optical depth scale τ on the z-axis according to the geometric relation $\tau_{\mathbf{n}} = \tau/\mu = \tau \sec\vartheta$.

It is self-evident that the probability of directional escape is null for the inward directions, for which it holds that $0 > \mu \geq -1$. Only the photons propagating outwards, for which $0 < \mu \leq 1$, have a nonzero directional escape probability, which we now determine. As in the 1-D case, the probability that a photon, absorbed at τ, will be scattered into an infinitesimal solid angle $d\Omega$ about $\mathbf{n}$ is $(\lambda/4\pi)\sin\vartheta d\vartheta d\varphi$; after reaching undisturbed the point at slant optical depth $\tau_{\mathbf{n}} = \tau' \sec\vartheta$, the probability of undergoing a successive scattering along the path from $\tau'_{\mathbf{n}}$ to $\tau'_{\mathbf{n}} + d\tau'_{\mathbf{n}}$ is $\exp\{- \mid \tau - \tau' \sec\vartheta \mid\} \sec\vartheta \; d\tau'$. The integration of this probability over all outward directions defines the direction-independent probability distribution function $P_s(\tau, \tau')$, such that

$$
\begin{aligned}
P_s(\tau, \tau')\; d\tau' &= (\lambda/4\pi) \int_0^{2\pi} d\varphi \int_0^1 d\mu \; e^{-|\tau-\tau'|/\mu}\; d\tau'/\mu \\
&= (\lambda/2)\;\; E_1(\mid \tau - \tau' \mid)\; d\tau' \qquad (1.62)
\end{aligned}
$$

is the probability that a photon originating at τ is scattered once between τ' and $\tau' + d\tau'$. The total escape probability from τ after a series of successive scattering processes will therefore be given by adding the direct escape probability $p_d(\tau, \mu) = \exp\{-\tau/\mu\}$, integrated over all outward directions, to the product of $P_s(\tau, \tau')\; d\tau$ times the so far unknown directional escape probability $p_e(\tau')$, integrated over all outward directions. The final result is

$$
P(\tau) = (\lambda/2)\;\; E_2(\tau) + (\lambda/4\pi) \int_0^{\infty} \; E_1\,(\mid \tau - \tau' \mid)\;\; P_e(\tau')\; d\tau'. \qquad (1.63)
$$

Here $E_2(\tau)$ and $E_1(\mid \tau - \tau' \mid)$ are the second and the first exponential integral functions for the arguments τ and $\mid \tau - \tau' \mid$, respectively. Equation (1.63) is a linear integral equation of the second kind (Fredholm), whose kernel is the function $K(\tau, \tau') = (\lambda/2) E_1(\mid \tau - \tau' \mid)$. The function $g(\tau) \equiv (\lambda/2) E_2(\tau)$ is a known term.

1.10.3 *The Neumann Series Solution of the Integral Equation*

The general form of a Fredholm equation like (1.63) is

$$
\phi(s) = f(s) + \rho \int K(s,t)\; \phi(t)\; dt, \qquad (1.64)
$$

where $\phi(s)$ is the unknown function and $f(s)$ is the known term. In order to find a solution of (1.64), the following sketched iterative procedure can be employed.[17] Let us rewrite (1.64) inserting in place of $\phi(t)$ inside the integral the expression for $\phi(s)$ given by (1.64) itself, and repeat the procedure indefinitely. To illustrate the scheme, we write the first two steps explicitly.

In the first one, we have

$$
\phi(s) = f(s) + \rho \int K(s,t) \left[f(t) + \rho \int K(\tau, \tau')\; \phi(\tau')\; d\tau' \right]\; dt. \qquad (1.65)
$$

Splitting the integral of the sum into the sum of two integrals and inverting the order of integration in the double second integral, we obtain

$$
\phi(s) = f(s) + \rho \int K(s,t)\; f(t)\; dt + \rho^2 \int dt' \phi(t') \int \; dt\; K(s,t) K(t,t'). \qquad (1.66)
$$

[17] An exhaustive treatment can be found in chapter III of Courant and Hilbert (1953) .

For the sake of a compact notation, it is convenient to introduce the *iterated kernels*

$$K^{(2)}(s,t) \equiv \int K(s,\sigma)\, K(\sigma,t)\, d\sigma, \tag{1.67-a}$$

$$\begin{aligned} K^{(3)}(s,t) &\equiv \int K^{(2)}(s,\sigma)\, K(\sigma,t)\, d\sigma \\ &= \int\int K(s,\sigma_1)\, K(\sigma_1,\sigma_2)\, K(\sigma_2,t) d\sigma_1\, d\sigma_2, \end{aligned} \tag{1.67-b}$$

$$\cdots\cdots\cdots$$

$$\begin{aligned} K^{(n)}(s,t) &\equiv \int K^{(n-1)}(s,\sigma)\, K(\sigma,t)\, d\sigma \\ &= \int\cdots\int K(s,\sigma_1)\, K(\sigma_1\sigma_2),\ldots K(\sigma_{n-1},t)\, d\sigma_1\, d\sigma_2\, \ldots\, d\sigma_{n-1}. \end{aligned} \tag{1.67-c}$$

In the second step, with the aid of the iterated kernels, we can rewrite (1.66) in the form

$$\begin{aligned} \phi(s) =& f(s) + \rho \int K(s,t)\, f(t)\, dt + \rho^2 \int K^{(2)}(s,t)\, f(t)\, dt \\ &+ \rho^3 \int K^{(3)}(s,t)\, f(t)\, dt. \end{aligned} \tag{1.68}$$

The indefinite repetition of this procedure leads to the *Neumann series*

$$\phi(s) = f(s) + \sum_{n=1}^{\infty} \rho^n \int K^{(n)}(s,t)\, f(t)\, dt, \tag{1.69}$$

which yields the solution to (1.64), provided that the series converges uniformly.

In order to give now a stochastic interpretation of (1.63), let us write the successive terms of its Neumann series:

$$\begin{aligned} P_e(\tau) = g(\tau) &+ \int_0^\infty d\tau'\, (\lambda/2)\, E_1(|\,\tau-\tau'\,|)\, g(\tau') \\ &+ \int_0^\infty \int_0^\infty d\sigma\, d\tau'(\lambda/2) E_1(|\,\tau-\sigma\,|)(\lambda/2)\, E_1(|\,\sigma-\tau'\,|) \\ &+ \int_0^\infty \int_0^\infty \int_0^\infty d\sigma_1\, d\sigma_2\, d\tau' \\ &\times (\lambda/2)\, E_1(|\,\tau-\sigma_1\,|)\, (\lambda/2)\, E_1(|\,\sigma_1-\sigma_2\,|)\, (\lambda/2)\, E_1(|\,\sigma_2-\tau'\,|)\, g(\tau') + \cdots. \end{aligned} \tag{1.70}$$

According to (1.62), $(\lambda/2)\, E_1(|\, s - s'\, |)$ represents the probability distribution integrated over all the outward directions that a photon, after travelling undisturbed from s to s', will be scattered once in the interval $s', s' + ds'$. Therefore, the successive terms of the Neumann series can be interpreted as the probability that, during its flight from τ to τ', the photon is scattered once, twice and so on.

The probability $P(\tau - t)$ that a photon generated at τ suffers forward scattering at a point $0 < t < \tau$ is by definition $(1/2) \int_t^\tau E_1(\tau - t) dt'$. Figure 1.2 illustrates the behaviour of the normalized probability distribution $(1/2)\, E_1(\tau - t)$ and the probability $P\, (\tau - t)$. The shape of the former is the same for all values of τ and shows the rapid decrease of the exponential integral for an increasing argument. On the other hand, the probability of being scattered is greater than 0.8 after a path longer than the unit optical depth.

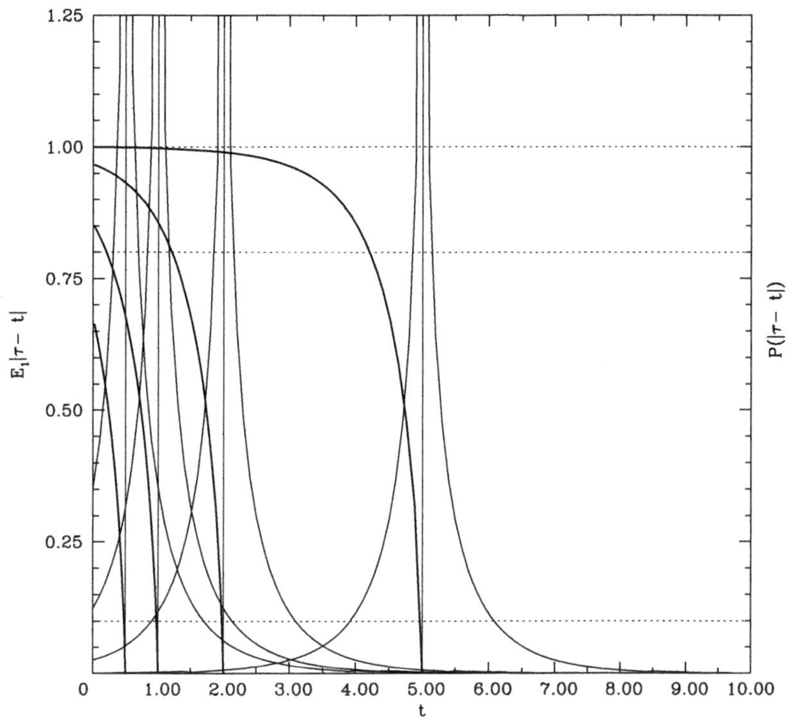

FIGURE 1.2. The behaviour of (1/2) $E_1(\tau - t)$, normalized to its area (thin line), and $P\,(\tau - t) =$ (1/2) $\int_t^\tau\, dt'$ at $\tau = 0.5, 1., 2., 5.$ (thick line).

1.11 The Transport of Radiant Energy as a Fluid Dynamical Process

1.11.1 *Analogy between Fluid Dynamics and Radiative Transfer*

Fluid dynamics considers fluid elements whose size is small enough compared with the spatial gradient of the macroscopic properties of the material medium, but so large that they contain a very great number of particles. In spite of the intrinsic corpuscular nature of the medium, these elements can, however, be considered homogeneous. The state of the system will then be determined by assigning the values of a suitable set of physical quantities to any fluid element. Moreover, the foregoing hypothesis justifies the process of taking the limit that assigns to each element a pointlike position in space. The state of a moving fluid will then be described by the local values of the fluid velocity distribution function and those of any pair of thermodynamic variables. The fluid elements carry mass, energy and momentum during their motion, whose trajectory and velocity are determined by the relevant equations supplemented by the proper boundary conditions. Trajectories such that their tangent at any point gives the direction of the velocity at that point are called streamlines.

In the macroscopic representation of the radiation field, the amount of specific radiant energy $\Delta E_\nu(\mathbf{n})$ transported along a ray $\mathbf{n}$, as defined in Section 1.4.1, takes the place of the fluid elements. This analogy, established on the grounds of radiometric principles, allows one to extend to the radiation field the hypotheses previously assumed for a fluid, along with the consequent properties. Accordingly, the equations of motion for fluids may be put into correspondence with the eikonal equation of geometric optics and streamlines with rays.

1.11.2 *Transport of Energy and Momentum*

Consistent with the foregoing standpoint, we may describe any transport process in terms of a generic *scalar* quantity $Q = qN_c$, where q is the quantity referring to an individual particle (carrier) and N_c the number of carriers, transported in a given direction $\mathbf{n}$ with constant velocity $\mathbf{v} = v\mathbf{n}$ and its associated *vector* quantity $Q\mathbf{v}$. The total rate of change of $Q(\mathbf{r}, t)$ is the total derivative

$$\begin{aligned} \frac{dQ(\mathbf{r},t)}{dt} &= \frac{\partial\, Q(\mathbf{r},t)}{\partial t} + \nabla\, Q(\mathbf{r},t) \cdot \mathbf{v} \\ &= \frac{\partial\, Q(\mathbf{r},t)}{\partial t} + \nabla \cdot [Q(\mathbf{r},t)\ \mathbf{v}]\,. \end{aligned} \tag{1.71}$$

This last identity holds true when $\mathbf{v}$ is constant. If the quantity Q is conserved, the *continuity equation*

$$\frac{\partial\, Q(\mathbf{r},t)}{\partial t} = -\nabla \cdot [Q(\mathbf{r},t)\ \mathbf{v}] \tag{1.72}$$

follows immediately from (1.71). By definition, the flux of $Q\mathbf{v}$ through an orientated surface $\mathbf{k}\ d\sigma$ is $Q\ \mathbf{v}\ (\mathbf{n} \cdot \mathbf{k})\ d\sigma$. Therefore, according to the divergence theorem, it holds that

$$\int\!\!\int\!\!\int_V \frac{\partial\, Q(\mathbf{r},t)}{\partial t}\, dV = -\int\!\!\int_\Sigma Q(\mathbf{r},t)\ \mathbf{v}\,(\mathbf{n}\cdot\mathbf{k})\ d\sigma. \tag{1.73}$$

That is to say, the integral over some volume V of the time derivative of the transported scalar quantity is equal to the flux of the associated vector quantity through the boundary surface Σ of V. Equation (1.73) has the customary form of a *conservation equation.*

In the case of radiative transfer, the carriers are the specific photons $(\mathbf{n}, \nu)$, whose number is given by their distribution function, that carry their own energy $h\nu$ and momentum $(h\nu/c)\mathbf{n}$ in the direction $\mathbf{n}$ with velocity $c\mathbf{n}$.

1.12 The Electrodynamic Picture

1.12.1 *Energy Balance of the Electromagnetic Field*

The physical foundation of Maxwell's theory is that any distribution of charges and currents generates an electromagnetic field that permeates the medium hosting them. The electromagnetic field is characterized by four fundamental vectors: namely, the electric field $\mathbf{E}$, the magnetic induction $\mathbf{B}$, the electric displacement $\mathbf{D}$ and the magnetic field intensity $\mathbf{H}$.

In the following, we adopt the conventional Gaussian system of units e.s.u., where it holds that $[\mathbf{E}] = [\mathbf{D}] = [\mathbf{B}] = [\mathbf{H}] = M^{1/2}L^{-1/2}T^{-1}$ and $\varepsilon = \mu = 1$ are dimensionless constants. Hence, Maxwell's four basic equations read

$$\nabla \cdot \mathbf{D} = 4\pi\, \rho \tag{1.74-a}$$

$$\nabla \cdot \mathbf{B} = 0; \tag{1.74-b}$$

$$\nabla \times \mathbf{E} + \frac{1}{c}\frac{\partial \mathbf{B}}{\partial t} = 0; \tag{1.74-c}$$

$$\nabla \times \mathbf{H} - \frac{1}{c}\frac{\partial \mathbf{D}}{\partial t} = \frac{4\pi}{c}\mathbf{J}. \tag{1.74-d}$$

These equations link the space and time variations of the aforementioned fundamental vectors with the current density $\mathbf{J} = \rho\mathbf{v}$, i.e., the charge density ρ times the velocity $\mathbf{v}$ of the charges. The dimensions of $\mathbf{J}$ are $M^{1/2}L^{-1/2}T^{-2}$.

If we form the scalar product of (1.74-c) by $\mathbf{H}$ and (1.74-d) by $\mathbf{E}$, and subtract the former from the latter, we obtain

$$\mathbf{E}\cdot(\nabla\times\mathbf{H}) - \mathbf{H}\cdot(\nabla\times\mathbf{E}) = \frac{1}{c}\left[\mathbf{H}\cdot\dot{\mathbf{B}} + \mathbf{E}\cdot\dot{\mathbf{D}} + 4\pi\,\mathbf{E}\cdot\dot{\mathbf{J}}\right]. \tag{1.75}$$

By introducing the Poynting vector, defined as

$$\mathbf{S} \equiv \frac{c}{4\pi}(\mathbf{E}\times\mathbf{H}), \tag{1.76}$$

and taking into account the identity

$$\mathbf{E}\cdot(\nabla\times\mathbf{H}) - \mathbf{H}\cdot(\nabla\times\mathbf{E}) = \nabla\cdot(\mathbf{H}\times\mathbf{E}), \tag{1.77}$$

we eventually get

$$\frac{1}{4\pi}\,\mathbf{H}\cdot\dot{\mathbf{B}} + \frac{1}{4\pi}\,\mathbf{E}\cdot\dot{\mathbf{D}} + \mathbf{E}\cdot\mathbf{J} + \nabla\cdot\mathbf{S} = 0. \tag{1.78}$$

Equation (1.78) is the statement of Poynting's theorem, which expresses the energy balance in the electromagnetic field.

From its definition, it follows that the dimensions of the Poynting vector are $MT^{-3} = (ML^2T^{-2})\,L^{-2}T^{-1}$, i.e., *power flux*. Each term in (1.78) has the dimensions $ML^{-1}T^{-3} = (ML^2T^{-2})\,L^{-3}T^{-1}$, i.e., *power density*.

1.12.2 *Localization and Transfer of Energy in the Electromagnetic Field*

Electromagnetic waves, which form the solution of Maxwell's equations, propagate through the medium with a *finite* speed characteristic of the medium itself. Accordingly, energy is *localized* in the electromagnetic field: a finite amount of energy must be contained in any finite volume of the medium. We may therefore define the electric and magnetic energy density as

$$W_{\text{elec}} \equiv 1/8\pi\;\mathbf{E}\cdot\mathbf{D} \tag{1.79-a}$$

and

$$W_{\text{mag}} \equiv 1/8\pi\;\mathbf{H}\cdot\mathbf{B} \tag{1.79-b}$$

It follows from (1.79-a) that

$$\dot{W}_{\text{elec}} = \frac{1}{8\pi}\,\mathbf{E}\cdot\dot{\mathbf{D}} + \frac{1}{8\pi}\,\dot{\mathbf{E}}\cdot\mathbf{D}. \tag{1.80}$$

It can be shown that $\mathbf{E}\cdot\dot{\mathbf{D}} = \dot{\mathbf{E}}\cdot\mathbf{D}$, hence $\dot{W}_{\text{elec}} = (1/4\pi)\,\mathbf{E}\cdot\dot{\mathbf{D}}$. The same holds true for the magnetic energy density.

If we define the Joule heat density as $W_J \equiv \mathbf{E}\cdot\mathbf{J}$, we can recast (1.78) into the form of an *equation of continuity*, namely

$$\dot{W} + \nabla\cdot\mathbf{S} = -W_J, \tag{1.81}$$

where $W \equiv W_{\text{elec}} + W_{\text{mag}}$. Thus, the Poynting theorem expresses the *energy balance* of the electromagnetic field: the LHS of (1.81) accounts for the exchange of energy between a volume element dV and the neighbouring volume elements; the Joule heat in the RHS represents a dissipative loss of energy. By integration over a given volume V and successive application of Gauss's theorem, (1.81) transforms into

$$\int\int_{\Sigma} \mathbf{S} \cdot \mathbf{n} \, d\sigma = -\int\int\int_{V} \left[\dot{W} + W_J\right] \, dV, \tag{1.82}$$

i.e., a *conservation law*: the energy flux per unit time across the boundary surface is equal to the sum of the time derivative of the energy density and the Joule heat dissipated inside the volume V bounded by Σ. Equation (1.82) is a statement of the possibility of the transport of energy by means of electromagnetic waves. Poynting's vector in particular paves the way for an electromagnetic interpretation of optical phenomena.

1.12.3 *Transport of Radiant Energy by an Electromagnetic Wave*

In the following, we consider the transport of radiant energy from the standpoint of the propagation of an electromagnetic wave. For simplicity, let us consider a monochromatic polarized plane wave propagating along the x-axis, whose direction is specified by the unit vector $\hat{\mathbf{x}}$. As is well known, the vectors $\hat{\mathbf{x}}$, $\mathbf{E}$ and $\mathbf{H}$ are mutually orthogonal. In this particular instance, only the components E_y of $\mathbf{E}$ and H_z of $\mathbf{H}$ differ from zero. The wave equation for E_y, derived from Maxwell's equations specialized for a vacuum, reads

$$\frac{1}{c^2}\frac{\partial^2 E_y}{\partial t^2} - \frac{\partial^2 E_y}{\partial x^2} = 0, \tag{1.83}$$

whose solution can be written as

$$E_y(x,t) = E_0 \cos\left(kx - \omega t\right), \tag{1.84}$$

where $k = \omega/c$ is the wave number. A similar equation is obtained for H_z.

Multiplication of (1.83) by $(1/4\pi k^2)$ $(\partial E_y/\partial t)$ yields

$$\frac{1}{4\pi k^2}\frac{\partial E_y}{\partial t}\left(\frac{1}{c^2}\frac{\partial^2 E_y}{\partial t^2} - \frac{\partial^2 E_y}{\partial x^2}\right) = 0. \tag{1.85}$$

By taking into account that

$$\frac{1}{c^2}\frac{\partial E_y}{\partial t}\frac{\partial^2 E_y}{\partial t^2} = \frac{1}{2}\frac{1}{c^2}\frac{\partial}{\partial t}\left(\frac{\partial E_y}{\partial t}\right)^2 \tag{1.86}$$

and

$$\frac{\partial}{\partial x}\left(\frac{\partial E_y}{\partial t}\frac{\partial E_y}{\partial x}\right) = \frac{1}{2}\frac{\partial}{\partial t}\left(\frac{\partial E_y}{\partial x}\right)^2 + \frac{\partial E_y}{\partial t}\frac{\partial^2 E_y}{\partial x^2}, \tag{1.87}$$

(1.85) becomes

$$\frac{\partial}{\partial t}\left\{\frac{1}{8\pi k^2}\left[\frac{1}{c^2}\left(\frac{\partial E_y}{\partial t}\right)^2 + \left(\frac{\partial E_y}{\partial x}\right)^2\right]\right\} - \frac{\partial}{\partial x}\left(\frac{1}{4\pi k^2}\frac{\partial E_y}{\partial t}\frac{\partial E_y}{\partial x}\right) = 0. \tag{1.88}$$

If we denote by e the term in curly brackets and f the opposite of that in round brackets, (1.88) can be rewritten as

$$\frac{\partial e}{\partial t} + \frac{\partial f}{\partial x} = 0. \tag{1.89}$$

In this way, we have derived an *equation of continuity* from the wave equation.

Dimensional analysis shows that $[e] = (ML^2T^{-2})L^{-3}$; in other words, e is an energy density. By replacing (1.84) in the definition of e, we obtain

$$e(t) = \frac{E_0^2}{4\pi} \sin^2(kx - \omega t). \tag{1.90}$$

In the specific case under consideration, as $H_z = E_y$, from equations (1.79-a) and (1.79-b), it follows that

$$W(t) = W_{\rm elec}(t) + W_{\rm mag}(t) = \frac{E_0^2}{4\pi} \cos^2(kx - \omega t). \tag{1.91}$$

By direct comparison of (1.90) and (1.91), it is immediately seen that e and W differ only by a phase factor equal to $\pi/2$,

On the other hand, the dimensions of f are $(ML^2T^{-2})L^{-3}(LT^{-1})$, i.e., the product of energy density times velocity, i.e., a *power flux* like the dimensions of the Poynting vector. By definition

$$f(t) = \frac{E_0^2}{4\pi}\frac{\omega}{k} \sin^2(kx - \omega t) = \frac{c}{4\pi}\, E_0^2 \, \sin^2\, (kx - \omega t). \tag{1.92}$$

The Poynting vector associated with the electromagnetic wave is

$$\mathbf{S}(t) = \frac{c}{4\pi}\, E^2(t)\, \hat{\mathbf{x}} = \frac{c}{4\pi}\, E_0^2 \, \cos^2\, (kx - \omega t)\, \hat{\mathbf{x}}. \tag{1.93}$$

Therefore, we identify $f\hat{\mathbf{x}}$ with $\mathbf{S}$ again but for the preceding phase factor.

It is straightforward to prove that, when averaging over the period T of the wave, it holds that $\langle e(t)\rangle_T = \langle W(t)\rangle_T$ and $\langle f(t)\rangle_T = \langle S(t)\rangle_T$.

1.13 Electrodynamics vs. the Macroscopic Picture of the Radiation Field

1.13.1 *Correspondence between Specific Intensity and Electric Field Strength*

Let us again consider a polarized monochromatic plane electromagnetic wave of frequency $\nu_0 = 1/T$ propagating in a vacuum in a given direction $\mathbf{n}_0 = \mathbf{n}_0(\vartheta_0, \varphi_0)$ that we take to be coincident with the x-axis of a Cartesian reference frame, specified by the unit vector $\hat{\mathbf{x}}$. The corresponding wave equation and its solution are given again by (1.83) and (1.84) respectively. Taking into account the considerations made at the beginning of Section 1.12.3, if we average over the period T of the wave the instantaneous energy density $W(t)$ of the corresponding electromagnetic field, from (1.79-a) and (1.79-b) it follows that

$$\langle W(t)\rangle_T = \frac{E_0^2}{8\pi}. \tag{1.94}$$

The monochromatic specific intensity corresponding to the preceding wave can be written as

$$I(\mathbf{r}; \vartheta, \varphi, \nu) = I_0(\mathbf{r})\; \delta(\vartheta - \vartheta_0)\; \delta(\varphi - \varphi_0)\; \delta(\nu - \nu_0), \tag{1.95}$$

where δ denotes the Dirac delta distribution function. For consistency, the dimensions of I_0 must be MT^{-3}, because the dimensions of any distribution function is the reciprocal of its argument, so that $[\delta(\nu-\nu_0)] = T$. By substituting (1.95) into (1.12), the integration over all frequencies yields the bolometric energy density

$$u(\mathbf{r},t) = \frac{1}{c}\int d\nu \int I(\mathbf{r},t;\mathbf{n},\nu)\; d\Omega = \frac{I_0(\mathbf{r})}{c}. \tag{1.96}$$

From a physical standpoint the time-averaged energy density of the electromagnetic field must be equal to the bolometric energy density of the radiation field. Hence, in a natural way, we can make the identification

$$I_0 = \frac{c}{8\pi}\, E_0^2, \tag{1.97}$$

which establishes a consistent relation between the electromagnetic description of the radiation field and the macroscopic picture in terms of the specific intensity. The dimension of I_0 is consistent with that of cE_0^2. Although the foregoing derivation applies strictly only to a plane monochromatic wave, this result may be generalized to fields with arbitrary angular and frequency distributions by taking into consideration a proper decomposition of the latter into elementary plane waves.

1.13.2 *The Monochromatic Flux*

The scalar quantity transported is here $h\nu\; f(\mathbf{r},t;\mathbf{n},\nu)$, i.e., the energy carried on by an individual photon (carrier) times the first photon distribution function that gives the number of carriers. The associated vector quantity will be

$$h\nu\; f(\mathbf{r},t;\mathbf{n},\nu)\; c\mathbf{n} = I(\mathbf{r},t;\mathbf{n},\nu)\; \mathbf{n}, \tag{1.98}$$

whose dimensions are $(ML^2T^{-2})L^{-2}$, i.e., those of an energy flux. This vector represents the transport along $\mathbf{n}$ of the *monochromatic*[18] energy $\Delta E_\nu(\mathbf{n})$ propagating per unit time into unit solid angle about $\mathbf{n}$ across the orientated unit surface, as defined in Section 1.5.1.

The number of specific photons $(\mathbf{n},\nu)$ that cross per unit time a unit area of an orientated surface $\mathbf{k}\cdot d\sigma$ is equal to the product of $f(\mathbf{r},t;\mathbf{n},\nu)$ times the volume $(\mathbf{k}\cdot\mathbf{n})\; c$. The corresponding net flux of monochromatic radiant energy will therefore be given by

$$\begin{aligned}\Phi(\mathbf{r},t;\nu) &= h\nu \oint f(\mathbf{r},t:\mathbf{n},\nu)\; (\mathbf{k}\cdot\mathbf{n})\; c d\Omega \\ &= \mathbf{k}\cdot \oint h\nu\; f(\mathbf{r},t;\mathbf{n},\nu)\; c\;\mathbf{n}\; d\Omega,\end{aligned} \tag{1.99}$$

i.e., the net flux of the vector quantity transported. It can be shown that $[\Phi] = (ML^2T^{-2})T^{-1}L^{-2}T$.

According to the definition of $\mathbf{F}_\nu(\mathbf{r},t)$ given by (1.8), it follows that

$$\Phi(\mathbf{r},t;\nu) = \mathbf{k}\cdot\mathbf{F}_\nu(\mathbf{r},t). \tag{1.100}$$

Hence $\mathbf{F}_\nu(\mathbf{r},t)$ is the *monochromatic power flux* of the radiation field, as proved by its dimensions.

In order to find the electromagnetic counterpart of $\mathbf{F}_\nu(\mathbf{r},t)$, let us consider once more the polarized monochromatic plane electromagnetic wave previously considered. According to the foregoing considerations, the associated Poynting vector, averaged over the period T, will be

[18] Namely per unit frequency range.

$$\langle \mathbf{S}(t) \rangle_T = \frac{c}{8\pi}\, E_0^2\, \mathbf{n}_0, \tag{1.101}$$

because $\mathbf{E}$, $\mathbf{H}$ and $\mathbf{n}_0$ are mutually orthogonal.

On the other hand, the corresponding monochromatic specific intensity is

$$I(\mathbf{r}, t; \mathbf{n}, \nu) = I_0(\mathbf{r})\ \delta(\mathbf{n} - \mathbf{n}_0)\ \delta(\nu - \nu_0). \tag{1.102}$$

By integrating over all frequencies and directions, it follows that

$$\int_0^\infty d\nu \oint d\Omega\ I(\mathbf{r}, t; \mathbf{n}, \nu)\ \mathbf{n} = I_0(\mathbf{r})\ \mathbf{n}_0 = \frac{c}{8\pi}\ E_0^2(\mathbf{r})\ \mathbf{n}_0. \tag{1.103}$$

This last equality comes from (1.97). This result, which may be be generalized for any electromagnetic wave, justifies the identification of the bolometric vector flux $\mathbf{F}(\mathbf{r}, t) \equiv \int_0^\infty \mathbf{F}_\nu(\mathbf{r}, t)\ d\nu$ with the Poynting vector.

1.13.3 *The Radiation Pressure Tensor*

Each specific photon $(\mathbf{n}, \nu)$ also carries a momentum equal to

$$\mathbf{p}(\mathbf{n}, \nu) = \begin{pmatrix} p_x \\ p_y \\ p_z \end{pmatrix} = \frac{h\nu}{c} \begin{pmatrix} n_x \\ n_y \\ n_z \end{pmatrix}, \tag{1.104}$$

where n_x, n_y and n_z are the direction cosine of $\mathbf{n}$ in a Cartesian frame. We consider the three scalar quantities p_x, p_y and p_z transported along $\mathbf{n}$, together with the vector quantities $p_x c\mathbf{n}$, $p_y c\mathbf{n}$ and $p_z c\mathbf{n}$. In place of the single vector equation (1.98), we have now the three vector equations

$$\frac{h\nu}{c}\ n_j\ f(\mathbf{r}, t; \mathbf{n}, \nu)\ c\mathbf{n} = \frac{1}{c}\ I(\mathbf{r}, t; \mathbf{n}, \nu)\ n_j\ \mathbf{n}, \tag{1.105}$$

where j is any of the three components. According to the semiclassical point of view adopted (cf. Section 1.1), $(h\nu/c)$ is a well-defined quantity; the spread in frequency is introduced by the distribution function $f(\mathbf{r}, t; \mathbf{n}, \nu)$.

The symmetric tensor $\boldsymbol{T}_\nu(\mathbf{r}, t) \equiv (1/c) \oint I(\mathbf{r}, t; \mathbf{n}, \nu)\ \mathbf{nn}\ d\Omega$ was defined in Section 1.5.2 by equation (1.9). In the dyadic notation, it holds that

$$\mathbf{n}\,\mathbf{n} = \begin{pmatrix} n_x n_x & n_x n_y & n_x n_z \\ n_y n_x & n_y n_y & n_y n_z \\ n_z n_x & n_z n_y & n_z n_z \end{pmatrix}. \tag{1.106}$$

Each component T_{ij} of $\boldsymbol{T}_\nu$ has dimensions $(MLT^{-1})L^{-2}$, i.e., those of momentum flux.

According to the rules of vector calculus, the product of the vector $\mathbf{k} = (k_x, k_y, k_z)$ times the tensor $\boldsymbol{T}_\nu$ is the vector

$$\mathbf{k} \cdot \boldsymbol{T}_\nu = \begin{pmatrix} \frac{1}{c} \oint I(\mathbf{r}, t; \mathbf{n}, \nu)\ n_x\ (\mathbf{k} \cdot \mathbf{n})\ d\Omega \\ \\ \frac{1}{c} \oint I(\mathbf{r}, t; \mathbf{n}, \nu)\ n_y\ (\mathbf{k} \cdot \mathbf{n})\ d\Omega \\ \\ \frac{1}{c} \oint I(\mathbf{r}, t; \mathbf{n}, \nu)\ n_z\ (\mathbf{k} \cdot \mathbf{n})\ d\Omega \end{pmatrix}. \tag{1.107}$$

The net flux of each component of $\mathbf{p}$ carried by the photons $(\mathbf{n}, \nu)$ across a unit area of the orientated surface $\mathbf{k} \cdot d\sigma$ per unit time is

$$\begin{aligned} &\oint \frac{h\nu}{c}\ f(\mathbf{r}, t : \mathbf{n}, \nu)\ n_j\ (\mathbf{k} \cdot \mathbf{n})\ c\ d\Omega \\ &= \oint \frac{1}{c}\ I(\mathbf{r}, t : \mathbf{n}, \nu)\ n_j\ (\mathbf{k} \cdot \mathbf{n})\ d\Omega = (\mathbf{k} \cdot \boldsymbol{T}_\nu)_j\,. \end{aligned} \tag{1.108}$$

In fluid dynamics, the pressure is defined as the net transport of the momentum of the fluid through a unit surface per unit time. Hence the identification of $\boldsymbol{T}_\nu$ as the *monochromatic radiation pressure tensor* is justified by the foregoing result, consistent with the dimensions of its components, which are those of a force per unit surface and unit frequency band as it holds that $(MLT^{-1})L^{-2} = (MLT^{-2})L^{-2}T$.

1.13.4 *Identification of Radiation Pressure with the Maxwell Stress Tensor*

Clearly, it holds that

$$(\mathbf{n} \cdot \boldsymbol{T}_\nu)_j = \frac{1}{c} \oint I(\mathbf{r}, t; \mathbf{n}, \nu)\ n_j\ d\Omega, \tag{1.109}$$

where the RHS of (1.109) is the jth component of $(1/c)\mathbf{F}_\nu$. It follows that the net transport of $\mathbf{p}$ through a unit surface perpendicular to $\mathbf{n}$ per unit time is

$$\mathbf{n} \cdot \oint \frac{h\nu}{c}\ f(\mathbf{r}, t; \mathbf{n}, \nu)\ c\ \mathbf{n}\ d\Omega = \mathbf{n} \cdot \frac{1}{c}\ \mathbf{F}_\nu. \tag{1.110}$$

The amount of momentum transported under the conditions previously considered fills the volume $V = c$. Therefore, the vector

$$\mathbf{G}_\nu(\mathbf{r}, t) \equiv \frac{1}{c^2}\ \mathbf{F}_\nu(\mathbf{r}, t) \tag{1.111}$$

represents the *monochromatic momentum density* associated with the radiation field, as confirmed by dimensional analysis: $[G_\nu] = (MT^{-2})\ L^{-2}T^2 = (MLT^{-1})L^{-3}T$. By integration of (1.111) over the whole frequency range, we obtain

$$\mathbf{G} \equiv \int_0^\infty \mathbf{G}_\nu\ d\nu = \frac{1}{c^2} \int_0^\infty \mathbf{F}_\nu\ d\nu = \frac{1}{c^2}\ \mathbf{S}, \tag{1.112}$$

where $\mathbf{S}$ is the Poynting vector, previously identified with the bolometric vector flux.

The net flux of each vector quantity $(G_\nu)_j\ c\ \mathbf{n}$ transported across a closed boundary surface Σ surrounding some volume V of space will be

$$\begin{aligned} (\Phi_G)_j &= \int\!\!\int_\Sigma (G_\nu)_j\ c\ \mathbf{n} \cdot \mathbf{n}\ d\sigma = \int\!\!\int_\Sigma c\,(G_\nu)_j\ d\sigma \\ &= \int\!\!\int_\Sigma d\sigma \oint \frac{h\nu}{c}\ f(\mathbf{r}, t; \mathbf{n}, \nu)\ c\ n_j\ d\Omega. \end{aligned} \tag{1.113}$$

On the other hand, the corresponding net flux of the vector $(\boldsymbol{T}_\nu)_j$ is

$$\begin{aligned} (\Phi_T)_j &= \int\!\!\int_\Sigma (\boldsymbol{T}_\nu)_j \cdot \mathbf{n}\ d\sigma \\ &= \int\!\!\int_\Sigma d\sigma \oint \frac{h\nu}{c}\ f(\mathbf{r}, t : \mathbf{n}, \nu)\ c\ n_j\ d\Omega = (\Phi_G)_j\,. \end{aligned} \tag{1.114}$$

By applying the divergence theorem, from (1.113) and (1.114) it follows that

$$\int\int_{\Sigma} (\boldsymbol{T}_\nu)_j \cdot \mathbf{n}\, d\sigma = \int\int\int_V \nabla \cdot (\boldsymbol{T}_\nu)_j \; dV \\ = \int\int\int_V \nabla \cdot \left[(\mathbf{G}_\nu)_j \; c\,\mathbf{n} \right] \; dV. \tag{1.115}$$

The identity $(\boldsymbol{T}_\nu)_j = (\mathbf{G}_\nu)_j c\mathbf{n}$ comes from (1.109) through (1.111) that link the monochromatic radiation pressure tensor with the monochromatic flux of the momentum density. On the same grounds, we can deduce for each component j of the monochromatic moment density the equation

$$\frac{\partial (\mathbf{G}_\nu)_j}{\partial t} = \frac{1}{c^2} \frac{\partial (\mathbf{F}_\nu)_j}{\partial t} = -\nabla \cdot \left[(\mathbf{G}_\nu)_j \; c\,\mathbf{n} \right], \tag{1.116}$$

which has the form of a continuity equation: the rate of change of any component j of the moment density is equal to the divergence of the corresponding vector quantity transported. Collecting the three components $(\mathbf{G}_\nu)_j$ and taking into account (1.115) and (1.116), we can eventually write

$$\frac{\partial \mathbf{G}_\nu}{\partial t} = -\nabla \cdot \boldsymbol{T}_\nu. \tag{1.117}$$

By introducing the *bolometric* radiation pressure tensor $\boldsymbol{T} \equiv \int_0^\infty \boldsymbol{T}_\nu \, d\nu$, we obtain

$$\frac{\partial \mathbf{G}}{\partial t} = -\nabla \cdot \boldsymbol{T}. \tag{1.118}$$

The terms in (1.118) have dimensions $\left(MLT^{-1}\right) L^{-3} T^{-1}$.

In the theory of electromagnetism,[19] the momentum density $\mathbf{G}_{\mathrm{em}}$ associated with the field is defined such that

$$\frac{\partial \mathbf{G}_{em}}{\partial t} = \nabla \cdot \boldsymbol{T}^M, \tag{1.119}$$

where $\boldsymbol{T}^M$ is the Maxwell stress tensor, whose components are

$$T_{ij}^M = \frac{1}{4\pi} \left[E_i E_j + H_i H_j - \delta_{ij} \left(\mathbf{E}^2 + \mathbf{H}^2 \right) \right], \tag{1.120}$$

where δ_{ij} denotes the Kronecker symbol. The dimensions of the terms in (1.119) are $\left(MLT^{-1}\right) L^{-3} T^{-1}$, the same as in (1.118). the same as in (1.118). Hence, the physical interpretation of the rate of change of $\boldsymbol{T}^M$: the rate of change of the momentum density is equal to the force density exerted by the Maxwell stresses.

Clearly, $\mathbf{G}_{\mathrm{em}}$ and the bolometric momentum density $\mathbf{G}$ are the same entity. The important physical meaning of (1.112) is that the field possesses a momentum, continuously distributed through the space wherever there is a net power flux $\mathbf{S}$. In this way, we go beyond the mere localization of the energy of the field and can learn how it propagates. Consequently, an electromagnetic wave carries momentum from a source of radiation to an absorber, and exerts pressure (i.e. radiation pressure) on the latter. On the other hand, when electromagnetic radiation is emitted by a body, it imparts a recoil on the latter whose magnitude is equal and opposite to the momentum carried.

[19] For details see, for example, chapter DI complemented by chapter AIII of Becker (1964) or section 31 of Sommerfeld (1964).

By comparing (1.118) with (1.119) it is immediately seen that the Maxwell tensor is to be identified with the bolometric radiation pressure tensor, taken with the negative sign. We have previously shown the correspondence between the specific intensity of the radiation field and the electric field strength, and that between the bolometric vector flux and the Poynting vector. These results prove the complete correspondence between Maxwell's electrodynamics and the continuous macroscopic description of the radiation field in terms of the specific intensity, which in turn is directly linked to the photon distribution functions introduced in the microscopic representation.

Acknowledgements

I owe due thanks to my mentor and friend Eduardo Simonneau, who helped me to broaden my knowledge on radiative transfer and the methods for the numerical solution of the RT equation during the years of our long-standing and fruitful collaboration.

REFERENCES

Becker, D. 1964. *Electromagnetic Fields and Interactions.* Dover

Brillouin, L. 1938. *Les Tenseurs en Mécanique et en Élasticité.* Original edition reprinted in 1987 by Éditions Jacques Gabay

Courant, R. and Hilbert, D. 1953. *Methods of Mathematical Physics*, Vol. I. John Wiley & Sons

Einstein, A. 1917. *Phys. Zeit.*, **18**, 121

Füchtbauer, Ch. 1920. *Phys. Zeit.*, **21**, 322

Huang, K. 1963. *Statistical Mechanics.* John Wiley & Sons

Huygens, Ch. 1690. *Traité de la Lumièere.* Pierre Van der Aa

Ladenburg, R. 1921. *Zeit. Phys.*, **4**, 451. English transl. in van der Waerden, B. L., 1967 *Sources of Quantum Mechanics*, Dover

Maxwell, J. C. 1891. *A Treatise on Electricity & Magnetism.* Clarendon Press

Mihalas, D. 1978. *Stellar Atmospheres*, 2nd edition. W. H. Freeman and Co.

Milne, E. A. 1924. *Phil. Mag. Series 6*, Vol. 47, **277**, 209

Newton, I. 1704. *Opticks.* Samuel Smith & Benjamin Walfords

Planck, M. 1906. *Vorlesungen uber die Theorie der Warmestrahlung*, Verlag Von Johann Ambrosius-Barth, English transl. Masius, M., 1914, in *The History of Modern Physics 1800–1950*, Thomash Publishers and American Institute of Physics

Preisendorfer, R. W. 1965. *Radiative Transfer on Discrete Spaces.* Pergamon Press

Sobolev, V. V. 1963. *A Treatise on Radiative Transfer.* van Nostrand

Sommerfeld, A. 1964. *Electrodynamics, Lectures on Theoretical Physics*, Vol. III. Academic Press

2. Fundamental Physical Aspects of Radiative Transfer

ARTEMIO HERRERO

Abstract

In this chapter, we discuss the fundamental physical processes that will allow us to reproduce the stellar spectra that we observe, together with the basic equations, approximations and techniques used to model them in atmosphere codes. The interaction between matter and radiation is described by the radiative transfer (RT) equation, whose solution gives the emergent stellar spectrum. The coupling of the RT equations with the statistical equilibrium equations that give the atomic populations (which under particular conditions can be computed by means of the local thermodynamic equilibrium [LTE] approximation) is discussed, as well as the role of the atomic properties. The structure equations (equation of state, momentum and energy conservation) that complete the set of equations required to compute a model atmosphere are also examined. Finally, the broadening mechanisms that change the appearance of the spectral line are presented.

2.1 Introduction

Almost all the information at our disposal about the visible Universe is in the form of photons. However, the properties of the photons we collect are not the same they had at the moment of their creation. Most of them were generated in the stellar cores by nuclear reactions, others during the process of star formation or in the last phases of stellar life, further photons are generated in accretion processes around massive and supermassive black holes. Along their path to us, they interact with matter and change their own original properties while altering those of the medium through which they propagate. Thus, by the analysis of their spectral characteristics, we can infer the state of the physical system from which they ultimately come.

The present chapter forms a whole with Chapters 1 and 3 of this book. The derivation of the RT equation, its particular form in plane-parallel and spherical geometry and its formal solution are exhaustively discussed in the latter. Here we will consider some specific issues, stressing the underlying physical content. The general physical bases of the transport of radiant energy, presented in Chapter 1, are particularized and deepened in this chapter, where the main focus is on the problem of nonlocal thermodynamic equilibrium (NLTE) line formation.

Among the numerous textbooks on the application of radiative transfer to stellar atmospheres, we suggest those by Mihalas (1978), Gray (2008) and Hubený and Mihalas (2015).

2.2 The Specific Intensity and Its Moments

In order to recover information from the radiation field, we must first define it on physical grounds. In the following, we will adopt the macroscopic representation introduced in Section 1.5.

Let dE_ν be the energy with frequency between ν and $\nu + d\nu$ crossing a surface dS at point $\mathbf{r}$ in direction $\mathbf{n}$ within a solid angle $d\Omega$ during the time interval dt. Then, as in Section 1.5.1, the specific intensity of the radiation field will be defined as

$$I(\mathbf{r}, t; \mathbf{n}, \nu) = \lim_{\Delta S, \Delta\Omega, \Delta t, \Delta\nu \to 0} \frac{\Delta E_\nu}{\Delta S cos\theta \Delta\Omega \Delta t \Delta\nu} = \frac{dE_\nu}{dS cos\theta d\Omega dt d\nu}. \tag{2.1}$$

This limit has a physical interpretation in terms of a light ray of given frequency travelling in a given direction. (See Section 1.3.) The specific intensity is measured in units of energy per unit area, time, spectral band and stereoradian (the corresponding c.g.s. units are $erg\ cm^{-2}s^{-1}Hz^{-1}str^{-1}$). The spectral band can be measured either in frequency or wavelength units. Care must be taken when converting from the former to the latter, as it holds that $d\nu = -(c/\lambda^2)\ d\lambda$. An important property of the specific intensity is its invariance along a path, if it propagates in a vacuum. In thermodynamical equilibrium (TE), the specific intensity is equal to the Planck distribution function.

The first three moments of $I(\mathbf{r}, t; \mathbf{n}, \nu)$ with respect to the direction $\mathbf{n}$ play an important role in radiative transfer. In the following, we will denote by $d\mathbf{n}$ the infinitesimal solid angle $d\Omega = sin\theta\ d\theta\ d\phi$, where θ is the polar and ϕ the azimuth angle.

- The zero-order moment is the *monochromatic mean intensity*, defined as

$$J_\nu(\mathbf{r}, t) \equiv \frac{1}{4\pi} \oint I_\nu(\mathbf{r}, t; \mathbf{n})\ d\Omega. \tag{2.2}$$

It is the average of the intensity over all directions around a point and has the same units as the specific intensity. Like the specific intensity, the mean intensity also is a scalar. However, the former is a *directional* quantity, as it depends on the direction of propagation of the photons. On the contrary, the latter is, by definition, independent of the direction.

If the radiation field is isotropic, we have

$$J_\nu = \frac{1}{4\pi}\ I_\nu \oint d\Omega = I_\nu. \tag{2.3}$$

In the case of azimuthal symmetry, (2.2) becomes

$$J_\nu = \frac{1}{2} \int_{-1}^{1} I_\nu\ d\mu, \tag{2.4}$$

where the customary notation $\mu \equiv cos\ \theta$ is adopted. It is clear from (2.4) that J_ν is the first (zero-order) μ-moment of the specific intensity. In TE, it holds that $J_\nu = I_\nu = B_\nu$. As shown in Section 1.5.3, the relation between J_ν and the monochromatic energy density u_ν (whose c.g.s. units are $erg\ cm^{-3}\ Hz^{-1}$) is given by

$$u_\nu = \frac{4\pi}{c}\ J_\nu \tag{2.5}$$

By integration over frequencies from zero to infinity, we get the bolometric relations among u, J and B, namely

$$(c/4\pi)\ u = J = B.$$

- The first-order moment, the *monochromatic radiative flux* $\mathcal{F}_\nu$, is the *vector* defined by

$$\mathcal{F}_\nu = \frac{1}{4\pi} \oint I_\nu(\mathbf{r}, t; \mathbf{n})\ \mathbf{n}\ d\Omega. \tag{2.6}$$

The scalar product $\mathcal{F}_\nu(\mathbf{r},t;\mathbf{n})\cdot d\mathbf{S}$ gives the net rate of energy flowing across the oriented surface $d\mathbf{S}$. The c.g.s. units of the flux are $erg\ cm^{-2}\ s^{-1}\ Hz^{-1}$.

In the case the radiation field is symmetric with respect to an axis, the net flux across an oriented surface perpendicular to that axis is zero. For the particular instance of a plane-parallel atmosphere, homogeneous in the x and y axes, only the component $\mathcal{F}_\nu(z)$ can be different from zero. Therefore, we can consider this flux as if it were a scalar and write

$$\mathcal{F}_\nu(z,t) = 2\pi \int_{-1}^{1} I_\nu(z,t;\mu)\ \mu\ d\mu \tag{2.7}$$

as the first μ-moment.

If we denote by $\mathcal{F}_\nu^+$ and $\mathcal{F}_\nu^-$, the quantities crossing the oriented surface $d\mathbf{S}$ outwards and inwards, respectively. the preceding flux can be split into the difference

$$\mathcal{F}_\nu = \mathcal{F}_\nu^+ - \mathcal{F}_\nu^- = 2\pi \int_0^1 I_\nu\ \mu\ d\mu - 2\pi \int_{-1}^{0} I_\nu\ \mu\ d\mu. \tag{2.8}$$

Note that for an isotropic radiation field the total (net) flux is zero.

The energy from a star, received by an observer at large distance, is directly related to the flux $\mathcal{F}_* = \int_0^\infty \mathcal{F}_\nu^+ d\nu$ emitted at the stellar surface.[1]

The law of Stefan–Boltzmann states that $u = a\ T^4$, where $a = (4/c)\ \sigma$ is the radiation density constant and $\sigma = 5.67 \cdot 10^{-5}\ erg\ cm^{-2}\ s^{-1}\ K^{-4}$ the Stefan–Boltzmann constant. From the foregoing bolometric relations, it follows that the flux emitted by an ideal black body, integrated over all the frequencies, is

$$\mathcal{F}_{bb}^+ = \pi\ B(T) = \sigma\ T^4. \tag{2.9}$$

On the basis of the preceding results, it is customary to define the effective temperature T_{eff} of a star as that of a black body that irradiates the same flux as the star.[2] A fundamental magnitude that characterizes a star is the total amount of energy leaving it per unit time (i.e., the emitted power), known as *stellar luminosity* and defined by

$$L = \oint \mathcal{F}_* dS = 4\pi\ R_*^2\ \sigma\ T_{\text{eff}}^4, \tag{2.10}$$

where R_* is the stellar radius and the integral is extended to the whole stellar surface. Because of the invariance of the specific intensity, in absence of sinks and sources of radiant energy along the path, the total amount of energy crossing the surface of a sphere of radius d centred at the star shall be equal to the luminosity of the latter. Therefore, the link between f_{obs}, the flux measured by an observer at the distance d from the star, and the flux at the stellar surface is given by

$$4\pi\ R_*^2 \mathcal{F}_* = 4\pi\ d^2 f_{obs} \tag{2.11}$$

[1] See, e.g., section 1-3 of Mihalas (1978).

[2] Note that black body curves never cross. A black body with higher temperature radiates at all wavelengths more energy per surface unit than another one at a lower temperature. Thus a hotter star will emit more energy per unit surface than a cooler one at any wavelength (as far as the emerging energy is fairly well approximated by a black body curve). Thus a hot white dwarf emits more energy per surface unit than a red supergiant even at those wavelengths where the flux of the red supergiant has its peak.

Hence we have the relation

$$\mathcal{F}_* = \sigma T_{\text{eff}}^4 = \left(\frac{d}{R_*}\right)^2 f_{obs}, \tag{2.12}$$

from which any of the three parameters T_{eff}, R_* and d can be determined, provided that the other two are known (a well as the alterations suffered by radiation during his travel from the star to the observer).

According to (2.8), for isotropic radiation the outward flux $\mathcal{F}_\nu^+$ is equal to $\pi\ I_\nu$. Therefore, it is convenient to define the *astrophysical flux*

$$F_\nu = \frac{\mathcal{F}_\nu}{\pi}, \tag{2.13}$$

so that the outward astrophysical flux is simply $F_\nu = I_\nu$.

In the practice of stellar atmospheres, it is customary to introduce the *Eddington flux*

$$H_\nu \equiv \frac{\mathcal{F}_\nu}{4\pi}. \tag{2.14}$$

In case of azimuthal symmetry, (2.14) becomes

$$H_\nu = \frac{1}{2}\int_{-1}^{+1} I_\nu\, \mu\, d\mu, \tag{2.15}$$

that is, the second μ-moment of the specific intensity. The relation among the preceding three fluxes[3] is

$$\mathcal{F}_\nu = \pi F_\nu = 4\pi H_\nu. \tag{2.16}$$

- The second-order moment of the specific intensity is the symmetric *tensor* of rank two

$$\boldsymbol{T}_\nu(\mathbf{r},t) \equiv \frac{1}{c}\oint I_\nu(\mathbf{r},t;\mathbf{n})\ \mathbf{n}\ \mathbf{n}\ d\Omega, \tag{2.17}$$

whose components are

$$T_{ij}(\mathbf{r},t;\nu) = \frac{1}{c}\oint I_\nu(\mathbf{r},t;\mathbf{n})\ n_i n_j\ d\Omega \quad (i,j = 1,2,3). \tag{2.18}$$

Dimensional analysis shows that the components T_{ij} have the dimension of the flux of a moment, namely $(MLT^{-1})\ L^{-2}$. In the c.g.s. system, they are measured in $erg\ cm^{-3}\ Hz^{-1}$.

In the case of an azimuthally invariant radiation field (plane-parallel or spherical geometry), $\boldsymbol{T}_\nu(\mathbf{r},t)$ becomes a diagonal tensor, whose three nonzero terms, $T_{ii}(\mathbf{r},t;\nu)$, are equal and depend on a single scalar position variable. For plane-parallel geometry, we have

$$\frac{c}{4\pi}\ \boldsymbol{T}_\nu(\mathbf{r},t) \to \frac{1}{2}\int_{-1}^{1} I_\nu(z,t;\mu)\ \mu^2\ d\mu \equiv K_\nu(z,t). \tag{2.19}$$

Equation (2.19) defines the second-order μ-moment.

[3] As the values of the numerical constants involved are small compared with those of the fluxes, a careful check is recommended when using the output of model atmosphere codes because it is not always clear which of the three fluxes has been taken into account.

According to Maxwell's theory, each photon of frequency ν travelling along $\mathbf{n}$ carries on a moment $\mathbf{p}_\nu(\mathbf{n})$. As shown in Section 1.13.3, the net flux per unit time of the jth component of the latter across a unit area of the oriented surface $\mathbf{k}dS$ is given by

$$\frac{1}{c} \oint I_\nu(\mathbf{n})\; n_j\; (\mathbf{k} \cdot \mathbf{n})\; d\Omega = (\mathbf{k} \cdot \boldsymbol{T}_\nu)_j\,. \tag{2.20}$$

(Cf. (1.108).)

In fluid mechanics, the symmetric cartesian tensor of components $\Pi_{ij} = P\delta_{ij} + \rho v_i v_j$ is the moment flux density. Each component accounts for the amount of moment that flows per unit time across a unit area perpendicular to the x_j-axis, and has therefore the dimension of a pressure. Hence the identification of $\boldsymbol{T}_\nu$ with the *monochromatic radiation pressure.*[4]

The average of the diagonal components of $\boldsymbol{T}_\nu$ define the *scalar* monochromatic mean radiation pressure

$$\bar{P}_\nu \equiv \frac{1}{3} \sum_{j=1}^{3} \frac{1}{c} \oint I_\nu(\mathbf{n})\; n_j\; n_j\; d\Omega = \frac{1}{3c} \oint d\Omega\; I_\nu \sum_{j=1}^{3}\; n_j^2 = \frac{1}{3c} \oint I_\nu\; d\Omega. \tag{2.21}$$

From the foregoing results, it follows that

$$\bar{P}_\nu = \frac{4\pi}{3c}\; J_\nu = \frac{1}{3}\; u_\nu \tag{2.22}$$

and

$$\bar{P}_R = \frac{4\pi}{3c}\; J = \frac{1}{3}\; u. \tag{2.23}$$

The preceding relations show the link between the average radiation pressure and the energy density of the radiation field. It must be noted that $\bar{P}_\nu$ (or $\bar{P}_R$) does not give the actual radiation pressure unless the radiation field is isotropic. Analogously to the hydrodynamical case of a fluid at rest, in which the velocities of the molecules are symmetric with respect to the direction and therefore only the diagonal components of the momentum flux density tensor are not zero and define the hydrostatic pressure, the radiation pressure is one-third of the energy density only for isotropic radiation. Otherwise, the normal pressure on a surface is given by a weighted mean of μ^2 instead of the factor 1/3.

2.3 The Radiative Transfer Equation

The way the properties of the radiation field change when travelling through and interacting with matter are described by the radiative transfer equation (RTE). The physical interpretation of this process at microscopic level in terms of the kinetics of photons is given in Section 1.7, where the RTE is introduced as a Boltzmann transport equation. We will consider its general form (akin to (1.22) therein), that is,

$$\frac{\partial}{\partial t} \frac{I_\nu(\mathbf{r}, t; \mathbf{n})}{c} + \nabla \cdot \left[\frac{I_\nu(\mathbf{r}, t; \mathbf{n})}{c}\; c\,\mathbf{n} \right] = \eta_\nu(\mathbf{r}) - \chi_\nu(\mathbf{r})\; J_\nu(\mathbf{r}, t), \tag{2.24}$$

where, like in Section 1.8, we define the emission and the extinction (absorption plus scattering) coefficients η_ν and χ_ν as

$$\eta_\nu = \lim_{\Delta V, \Delta\Omega, \Delta t, \Delta\nu \to 0} \frac{\Delta E_\nu}{\Delta V \Delta\Omega \Delta\nu \Delta t} \tag{2.25}$$

[4] A detailed introduction of the radiative pressure tensor and its relation with the Maxwell's stress tensor is contained in Sections 1.13.3 and 1.13.4.

and

$$\frac{dI_\nu}{I_\nu} = -\chi_\nu ds. \tag{2.26}$$

The nature of the emission and absorption coefficients is different: the former, more properly called emissivity, accounts for the amount of energy emitted by a volume into a solid angle per time interval and frequency band, whereas the latter gives the fraction of the specific intensity absorbed per unit length. Thus the first one has units of intensity per unit length, i.e., in the c.g.s. system $erg\ cm^{-3}\ s^{-1}\ str^{-1}\ Hz^{-1}$, the second one of cm^{-1}. The inverse of the macroscopic absorption coefficient is directly related to the mean free path of the photons. Note that both coefficients may be anisotropic, i.e., direction dependent.

Let's consider some particular cases[5] of the RTE to illustrate its behaviour.

• *Stationary atmosphere*

If the properties in the atmosphere are time independent, then $\partial I_\nu/\partial t = 0$ and the RTE is simply

$$\mathbf{n} \cdot \nabla I_\nu(\mathbf{r};\mathbf{n}) = \eta_\nu(\mathbf{r}) - \chi_\nu(\mathbf{r})\ I_\nu(\mathbf{r};\mathbf{n}). \tag{2.27}$$

Observations of the solar atmosphere (sunspots, flares, loops, etc.) show the limits of this approximation.

• *Plane-parallel geometry*

We assume that the atmosphere consists of parallel planes that extend infinitely with constant physical properties in the two directions defining the plane. There is radiation transfer only along the direction normal to the parallel planes. In this case, assuming that direction is along the x-axis (see Figure 2.1), it holds that $\partial I_\nu/\partial y = \partial I_\nu/\partial z = 0$ and the RTE takes the form

$$\frac{1}{c}\frac{\partial I_\nu(x,\mu)}{\partial t} + \mu\frac{\partial I_\nu(x,\mu)}{\partial x} = \eta_\nu(x) - \chi_\nu(x)\ I_\nu(x,\mu), \tag{2.28}$$

where the customary notation $\mu \equiv (\mathbf{e}_x \cdot \mathbf{n} = cos\ \vartheta)$ is adopted. Again, this is obviously an approximation. It is valid as long as the angle ϑ formed by the normal to the plane and

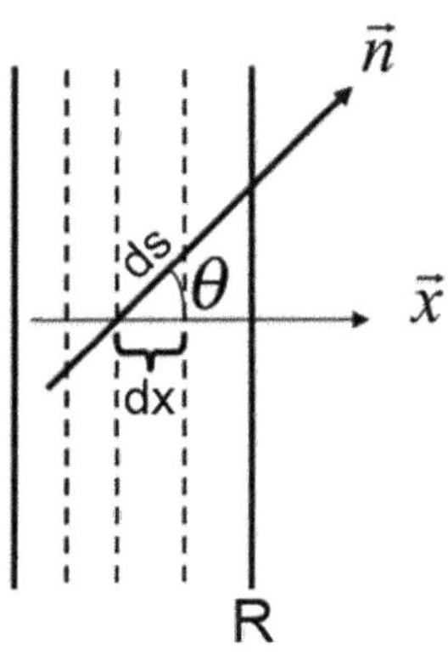

FIGURE 2.1. Plane-parallel geometry. The light ray propagates in a direction $\vec{n}$ that forms and angle θ with the direction $\vec{x}$ normal to the planes. The geometrical scale (i.e., the height in the atmosphere) is measured along $\vec{x}$.

[5] Cf. Section 3.2.

the direction of propagation of the radiation is significantly constant between the first and the last of the planes considered. In other words, the extension of the atmosphere is smaller than the stellar radius (by a factor $\approx 10^{-3}$). The main advantage of plane-parallel geometry is the reduction of the dimensions of the problem. Another important property of a plane-parallel atmosphere is that the total flux is constant through the atmosphere. Thus, when we characterize a plane-parallel atmosphere by a given effective temperature, the resulting total net flux σT_{eff}^4 is conserved throughout all the layers.

- *Plane-parallel and stationary atmosphere*

If we combine the two preceding conditions, the RTE becomes a first-order differential equation, namely

$$\mu \frac{dI_\nu(x,\mu)}{dx} = \eta_\nu(x) - \chi_\nu(x)\ I_\nu(x,\mu). \tag{2.29}$$

This is the particular equation that we are going to use in this chapter.

- *Spherical geometry*

When the stellar atmosphere extends significantly compared to the stellar radius, we must adopt spherical geometry. Using the expression of the divergence in spherical coordinates and assuming azimutal symmetry, we have

$$\frac{1}{c}\frac{\partial I_\nu(r,t;\mu)}{\partial t} + \mu \frac{\partial I_\nu(r,t;\mu)}{\partial r} + \frac{1-\mu^2}{r}\frac{\partial I_\nu(r,t;\mu)}{\partial \mu} = \eta_\nu(r) - \chi_\nu(r)\ I_\nu(r,t;\mu). \tag{2.30}$$

This will be the case, for example, when we deal with extended atmospheres or stellar winds, considered in Chapter 5. Compared with the previous case of a plane-parallel atmosphere, there are two important differences: firstly, the angle formed by the direction of propagation and the normal to the spherical surface is not constant; secondly, what is constant is now the product $r^2 \mathcal{F}_\nu(r,t)$ and not the net total flux.

2.4 Optical Depth

The solution of the differential equation (2.26) that defines the extinction coefficient χ_ν is

$$I_\nu(s) = I_\nu(0)\ e^{-\int_0^s \chi_\nu(') \ ds'}. \tag{2.31}$$

It gives the emergent intensity at a point P_s after an intensity $I_\nu(0)$ incident at point P_0 has travelled a distance s through a layer whose extinction coefficient is $\chi_\nu(s)$. The distance travelled is not significant to know the fraction of the emergent intensity. Photons may travel a large path without suffering any interaction with matter if the extinction coefficent is small, but may be absorbed or scattered along a very short distance if χ_ν is large. In order to describe properly the propagation of the radiation through a medium, the proper parameter is the monochromatic optical dept τ_ν, defined by the differential relation

$$d\tau_\nu \equiv -\chi_\nu \ ds. \tag{2.32}$$

Hence

$$\tau_\nu(s) = -\int_0^s \chi_\nu(s')\ ds', \tag{2.33}$$

so that

$$I_\nu(s) = I_\nu(0)\ e^{-\tau_\nu(s)}. \tag{2.34}$$

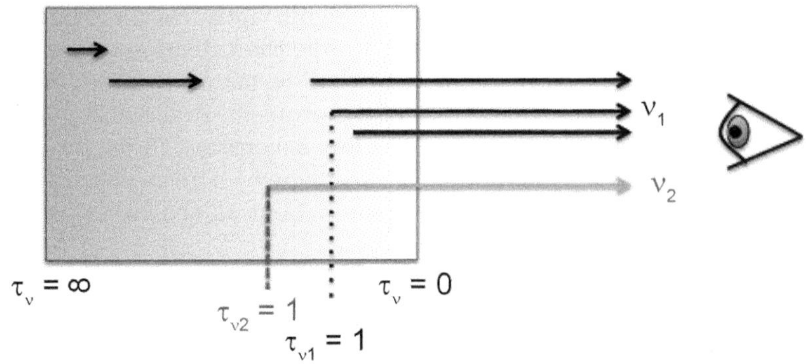

FIGURE 2.2. Photons' depth of formation.

From (2.34), it clearly results that for large values of τ_ν only a small fraction of the photons generated at a point P_0 will reach a successive point at P_s. As shown in Section 1.10, the exponential factor is closely related to the direct escape probability of a photon of frequency ν. The expected value of the optical depth from which this photon can escape through the surface of the atmosphere will be by definition

$$< \tau_\nu >= \frac{\int_0^\infty \tau_\nu \, e^{-\tau_\nu} \, d\tau_\nu}{\int_0^\infty e^{-\tau_\nu} \, d\tau_\nu}. \tag{2.35}$$

Integration by parts yields $\langle \tau_\nu \rangle = 1$, hence we can expect that in a semi-infinite stellar atmosphere emerging photons of frequency ν come in the average from depths where $\tau_\nu \approx 1$. It follows from the statistical interpretation of the RT process that photons generated at optical depth $\tau_\nu = 1$ carry on information of the physical conditions of the medium inside a volume of radius $\Delta\tau_\nu \approx 1$. (See Figure 1.2.)

By definition, the atmosphere of a star consists of the outer regions where the emergent flux forms. The radiation emitted covers a wide spectral range, so that an observer sees photons of different frequency coming from different atmospheric layers (the deeper those where the opacity is lesser), which sample different atmospheric layers.[6] A pictorial explanation is given in Figure 2.2, where a semi-infinite atmosphere extending from $\tau_\nu = 0$ to $\tau_\nu = \infty$ is sketched. Let us consider first photons of frequency ν_1, whose propagation is represented by black lines. Those emitted at great depth are trapped and eventually destroyed by absorption, so that they cannot escape. Only photons generated at a distance $\Delta\tau_\nu \approx 1$ from the outer surface have a definite chance of escaping and may reach the observer. For photons of a different frequency ν_2, whose propagation is drawn as a grey line, the absorption coefficient will be different and consequently the geometrical depth at which $\tau_{\nu_2} = 1$ will be different from that where $\tau_{\nu_1} = 1$.

2.5 The Source Function

The optical depth is the natural variable for radiative transfer. If we define in the usual way the source function[7] as

[6] A quantitative characterization of the outermost stellar layers in terms of the monochromatic optical depth is contained in Crivellari (2018).

[7] The source function has the units of the specific intensity.

$$S_\nu \equiv \frac{\eta_\nu}{\chi_\nu}, \tag{2.36}$$

the RTE (2.29) can be rewritten in terms of the independent variable τ_ν as

$$\mu \frac{dI_\nu(\tau_\nu, \mu)}{d\tau_\nu} = I_\nu(\tau_\nu, \mu) - S_\nu(\tau_\nu). \tag{2.37}$$

2.5.1 *Simple Cases of the Source Function*

The specific intensity can be computed by solving (2.37) if the source function $S_\nu(\tau_\nu)$ is known (together with the corresponding initial condition). Let us consider in the following a few simple cases of the source function.

• *TE*

In TE, the ratio between the emission and the absorption coefficients is given by Kirchhoff's law, in terms of an universal function of the temperature only: Planck's distribution function. It holds therefore that

$$\frac{\eta_\nu}{\chi_\nu} = S_\nu = B_\nu(T). \tag{2.38}$$

That is to say that the source function is equal to the Planck function at the temperature of the system, and the same is true for all its points.

• *LTE*

We know that (in general) inside a stellar atmosphere the temperature decreases outwards. Therefore, we cannot assume TE and a unique temperature for the system. Nevertheless, under particular conditions that will be examined later in Section 2.8, we can set at any point the *local* source function equal to the Planck function at the *local* value of the temperature, so that

$$S_\nu(\tau_\nu) = B_\nu[T(\tau_\nu)]. \tag{2.39}$$

The LTE source function is coupled to the local temperature $T(\tau_\nu)$. (More details can be found in Section 1.9.4.) If we know the temperature distribution inside the atmosphere, the RTE (2.37) is solved directly.

• *Pure coherent scattering*

Let us consider the instance where a photon is scattered by a particle (usually a free electron), as illustrated in the left panel of Figure 2.3. From the standpoint of the observer, a scattered photon is lost when removed from his line of sight. On the other hand, photons scattered from other directions into the observer's line of sight are added to the beam. The former is an extinction process, the latter is like an emission one.

If scattering is isotropic, the macroscopic scattering coefficient σ_ν is independent of the direction. As shown in Section 1.9.1, the emissivity due to scattering is

$$\eta_\nu^s = \sigma_\nu \oint p(\mathbf{n}', \mathbf{n})\, I_\nu(\mathbf{n}')\, d\mathbf{n}', \tag{2.40}$$

where $p(\mathbf{n}', \mathbf{n})$ is the joint probability that a photon be diverted from $\mathbf{n}'$ into $\mathbf{n}$. In the case of isotropic scattering, $p(\mathbf{n}', \mathbf{n}) = 1/4\pi$, so that $\eta_\nu^s = \sigma_\nu J_\nu$. It follows immediately that the source function for isotropic coherent scattering is $S_\nu = J_\nu$. It is completely decoupled from the local thermal conditions of the medium and depends only on the radiation field through its mean intensity, brought about by radiative transfer.

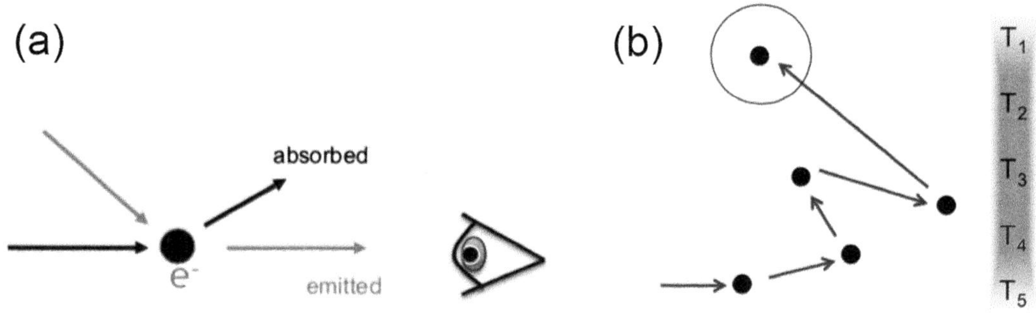

FIGURE 2.3. (a) Photons scattered coherently are removed from and diverted into the observer's line of sight. (b) The local temperature throughout the atmosphere is indicated in the scale in the right margin, increasing from T_1 to T_5. A photon, generated at the depth where the temperature is T_5, is repeatedly scattered without change in its frequency before reaching the point where the temperature is T_1. The thermal parameters in the neighbourhood of this point, where it may be eventually absorbed, are different from those of the region where the photon was generated.

A photon that reaches a given point after a series of coherent scatterings carries on information on the local conditions at the point where it was generated, which are different from those of the point where it may be eventually destroyed by absorption. The right panel of Figure 2.3 is a schematic picture of this process.

Coherent scattering decouples the radiation field and the source function from the local thermodynamic conditions of matter. On the other hand, if scattering is either noncoherent or nonisotropic, it will couple different frequencies and/or directions.

2.6 Formal Solution of the Radiative Transfer Equation

By definition, the *formal* solution of the RTE is its solution for a given source function. This issue is extensively treated in Section 3.4. Here we will just discuss its basic physical content from the standpoint of radiative transfer. Let us consider a finite slab of a plane-parallel atmosphere, delimited by two boundaries respectively at optical depths τ_1 and τ_2 ($\tau_1 < \tau_2$).[8] Depending on the value of the parameter μ, equation (2.37) will denote either the propagation along an outward direction $\mathbf{n}$ of a beam of radiation (ray) incident at angle $\vartheta = \mathbf{k} \cdot \mathbf{n} = arccos\ \mu$ onto the bottom boundary surface at τ_2, whose orientation is specified by the unit vector $\mathbf{k}$, or the propagation along an inward direction $\mathbf{n}$ of a beam incident onto the oriented upper boundary surface at τ_1.

From the mathematical point of view, (2.37) is an ordinary first-order differential equation with constant coefficients. We know from the general theory that its solution for the outward ($\mu > 1$) specific intensities at τ_1 is given by

$$I_\nu(\tau_1, \mu) = I_\nu(\tau_2, \mu) e^{-(\tau_2-\tau_1)/\mu} + \int_{\tau_1}^{\tau_2} S_\nu(t_\nu) e^{-(t_\nu-\tau_1)/\mu} \frac{dt_\nu}{\mu}, \tag{2.41}$$

The physical interpretation of the two terms in the RHS of (2.41) is quite straightforward . The first one accounts for the contribution brought about by the intensity incident at the initial point at τ_2, reduced by the exponential decay due to the optical depth between τ_1 and τ_2. The second term corresponds to the integration of the intensities created at each intermediate point between τ_1 and τ_2, attenuated by the optical depth

[8] For simplicity's sake, the explicit dependence of τ_1 and τ_2 on ν is dropped.

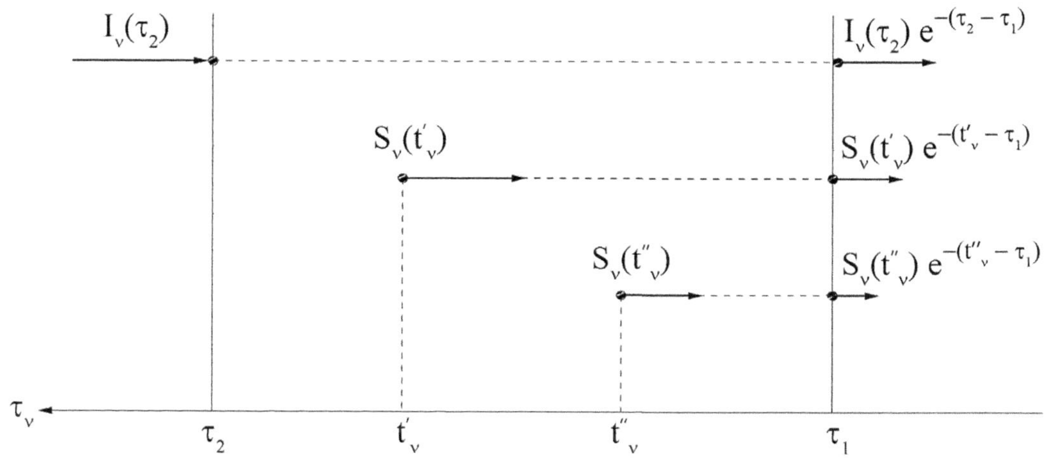

FIGURE 2.4. An homogeneous slab of matter, delimited by two plane-parallel boundaries at τ_2 and τ_1, is irradiated at τ_2 by a beam of parallel specific intensities $I_\nu(\tau_2$. (The explicit dependence of τ_1 and τ_2 is dropped for simplicity's sake.) According to (2.41) the emergent intensity at τ_1 is the incident intensity at τ_2, attenuated across the whole slab, plus the integral of all the contribution $S_\nu(t_\nu)$, emitted at each depth point t_ν, attenuated by the optical distance from t_ν to the boundary at τ_1.

between the creation point t_ν and the final point at τ_1. The propagation of a radiation beam perpendicular to an homogeneous slab of stellar matter from τ_2 to τ_1 is sketched in Figure 2.4.

In the case of a semi-infinite atmosphere, the limits of the slab are $\tau_1 = 0$ and $\tau_2 \to \infty$. Therefore, the emergent intensity at the stellar surface will be

$$I_\nu(0,\mu) = \int_0^\infty S_\nu(t_\nu)\; e^{-t_\nu/\mu}\frac{dt_\nu}{\mu}. \tag{2.42}$$

It follows from (2.42) that we can determine the radiation emerging from a star, provided we know the values of the source function throughout the atmosphere. In more general terms, (2.42) establishes a quantitative link between the observed stellar flux and the physical properties inside the atmosphere.

2.6.1 *The Schwarzschild–Milne Equations*

Consider a point inside the preceding semi-infinite plane-parallel atmosphere at optical depth τ_ν. It is expedient to define over the half range $0 < \mu \leq 1$ of the full range $-1 \leq \mu \leq 1$ the *outgoing* intensities $I_\nu^+(\tau_\nu,\mu) \equiv I_\nu(\tau_\nu,\mu > 0)$ and the *incoming* intensities $I_\nu^-(\tau_\nu,\mu) \equiv I_\nu(\tau_\nu,\mu < 0)$. Thus the specific intensities propagating outwards along **n** belong to the former family, those propagating inwards to the latter. If, as in most cases, there is no incident radiation onto the stellar surface, (2.41) for the incoming intensities ($\mu < 0$) will be replaced by

$$I_\nu(\tau_\nu,\mu) = \int_0^{\tau_\nu} S_\nu(t_\nu)\; e^{-(\tau_\nu - t_\nu)/(-\mu)}\frac{dt_\nu}{(-\mu)}. \tag{2.43}$$

Because the contribution of the outgoing intensities from the bottom boundary surface is null, due to the infinite value of the optical depth path, the outgoing intensities ($\mu > 0$) will be given by

$$I_\nu(\tau_\nu,\mu) = \int_{\tau_\nu}^{\infty} S_\nu(t_\nu)\ e^{-(t_\nu-\tau_\nu)/\mu}\frac{dt_\nu}{\mu}. \tag{2.44}$$

By taking into account (2.43) and (2.44), we can split the nth μ-moment of the specific intensity into the sum of two terms, namely

$$M_\nu^{(n)} \equiv \frac{1}{2}\int_{-1}^{1} I_\nu(\tau_\nu,\mu)\mu^n\ d\mu = \frac{1}{2}\int_{-1}^{0}\mu^n\left(-\frac{d\mu}{\mu}\right)\int_0^{\tau_\nu} S_\nu(t_\nu)\ e^{-(\tau_\nu-t_\nu)/(-\mu)}\ dt_\nu$$
$$+\frac{1}{2}\int_0^1 \mu^n\frac{d\mu}{\mu}\int_{\tau_\nu}^{\infty} S_\nu(t_\nu)\ e^{-(t_\nu-\tau_\nu)/\mu}\ dt_\nu. \tag{2.45}$$

If we make the substitution $w = 1/\mu$ in the term where $\mu < 0$ and $w = -1/\mu$ in the term where $\mu > 0$, we obtain

$$M_\nu^{(n)} = \int_{\tau_\nu}^{\infty} S(t_\nu)\ dt_\nu \int_1^{\infty}\frac{e^{-(t_\nu-\tau_\nu)w}}{w^{n+1}}\ dw + (-1)^n\int_0^{\tau_\nu} S(t_\nu)\ dt_\nu\int_1^{\infty}\frac{e^{-(\tau_\nu-t_\nu)w}}{w^{n+1}}\ dw$$
$$= \int_{\tau_\nu}^{\infty} S(t_\nu)\ E_{n+1}(t_\nu-\tau_\nu)\ dt_\nu + (-1)^n\int_0^{\tau_\nu} S(t_\nu)\ E_{n+1}(\tau_\nu-t_\nu)\ dt_\nu, \tag{2.46}$$

where we have made use of the definition of the exponential integral

$$E_n(x) \equiv \int_1^{\infty}\frac{e^{-x\ w}}{w^n}dw = \int_0^1 e^{-x/\mu}\ \mu^{n-1}\frac{d\mu}{\mu} \tag{2.47}$$

For $x \gg 1$, the asymptotical behaviour of the exponential integrals is

$$E_n(x) = \frac{e^{-x}}{x}\left[1-\frac{n}{x}+\frac{n(n+1)}{x^2}+\cdots\right] \approx \frac{e^{-x}}{x}. \tag{2.48}$$

By applying the formulae, we can write the moments in terms of exponential integrals.

For the mean intensity (i.e., $n = 0$), we have

$$J_\nu(\tau_\nu) = \frac{1}{2}\int_{-1}^{+1} I_\nu(\tau_\nu,\mu)\ d\mu = \frac{1}{2}\int_{\tau_\nu}^{\infty} S(t_\nu)\ E_1(t_\nu-\tau_\nu)\ dt_\nu + \frac{1}{2}\int_0^{\tau_\nu} S(t_\nu)\ E_1(\tau_\nu-t_\nu)\ dt_\nu$$
$$= \frac{1}{2}\int_0^{\infty} S(t_\nu)\ E_1(|t_\nu-\tau_\nu|)\ dt_\nu \tag{2.49}$$

Equation (2.49) shows that the mean intensity at any depth point τ_ν is the result of integrating the contribution of the source function at all points within the atmosphere. The exponential integral $E_1(|\ t_\nu - \tau_\nu\ |)$ plays here the role of the exponential decay factor. It follows from (2.49) that we may have either $J_\nu(\tau_\nu) > S_\nu(\tau_\nu)$ or $J_\nu(\tau_\nu) < S_\nu(\tau_\nu)$, depending on the behaviour of $S_\nu(\tau_\nu)$.

For the the first- and the second-order μ-moments, H_ν and K_ν, we will have

$$H_\nu(\tau_\nu) = \frac{1}{2}\left[\int_{\tau_\nu}^{\infty} S(t_\nu)E_2(t_\nu-\tau_\nu)\ dt_\nu - \int_0^{\tau_\nu} S(t_\nu)E_2(\tau_\nu-t_\nu)\ dt_\nu\right] \tag{2.50}$$

and

$$K_\nu(\tau_\nu) = \frac{1}{2}\int_0^{\infty} S(t_\nu)\ E_3(|t_\nu-\tau_\nu|)\ dt_\nu \tag{2.51}$$

The foregoing equations can be written in operator form. For instance, in the case of the mean intensity $J_\nu(\tau_\nu)$ we can define the so-called Λ-operator

$$\Lambda_{\tau_\nu}\left[f(t)\right] \equiv \frac{1}{2}\int_0^\infty f(t)\; E_1(|t_\nu-\tau_\nu|)\; dt_\nu \tag{2.52}$$

and rewrite (2.49) as

$$\Lambda_{\tau_\nu}\left[S_\nu(t_\nu)\right] = \frac{1}{2}\int_0^\infty S_\nu(t_\nu)\; E_1(|t_\nu-\tau_\nu|)\; dt_\nu = J_\nu(\tau_\nu). \tag{2.53}$$

Analogous operators can be introduced for the successive moments H_ν and K_ν.[9]

2.6.2 *At the Bottom of the Atmosphere: The Diffusion Approximation*

Irrespectively of the method employed for the solution of the RTE, whenever the specific intensity is not given at the bottom it shall be necessary to know the behaviour of the source function as $\tau_\nu \to \infty$ in order to write the lower boundary condition. For this purpose, let us assume that the source function can be expanded as a power series around a given point, that is,

$$S_\nu(\tau_\nu) = \sum_{n=0}^{\infty} \frac{(t_\nu-\tau_\nu)^n}{n!}\left[\frac{d^n S_\nu(t_\nu)}{dt_\nu^n}\right]_{\tau_\nu}. \tag{2.54}$$

Substituting this expression into (2.44), it is easy to show that the outgoing intensities will be given by

$$I_\nu^+(\tau_\nu,\mu) = \sum_{n=0}^{\infty} \mu^n \left[\frac{d^n S_\nu(t_\nu)}{dt_\nu^n}\right]_{\tau_\nu} \tag{2.55}$$

For the incoming intensities, a more complicated expression is obtained, namely

$$I_\nu^-(\tau_\nu,\mu) = \sum_{n=0}^{\infty} \mu^n \left[\frac{d^n S_\nu(t_\nu)}{dt_\nu^n}\right]_{\tau_\nu} \left[1 - \frac{e^{-(\tau_\nu/|\mu|)}}{n!}\left\{(\tau_\nu/|\mu|)^n + n(\tau_\nu/|\mu|)^{n-1} + \cdots + n!\right\}\right]. \tag{2.56}$$

For $\tau_\nu \gg 1$, the expression in square brackets tends to one, hence over the full range $-1 \le \mu \le 1$ it holds that

$$\begin{aligned} I_\nu(\tau_\nu,\mu) = \sum_{n=0}^{\infty} \mu^n \left[\frac{d^n S_\nu(t_\nu)}{dt_\nu^n}\right]_{\tau_\nu} &= S_\nu(\tau_\nu) + \mu\left[\frac{dS_\nu(t_\nu)}{dt_\nu}\right] + \mu^2\left[\frac{d^2 S_\nu(t_\nu)}{dt_\nu^2}\right] \\ &+ \mu^3\left[\frac{d^3 S_\nu(t_\nu)}{dt_\nu^3}\right] + \cdots. \end{aligned} \tag{2.57}$$

Now we can calculate the mean intensity:

$$\begin{aligned} J_\nu(\tau_\nu) &= \frac{1}{2}\int_{-1}^{+1}\sum_{n=0}^{\infty} \mu^n \left[\frac{d^n S_\nu(t_\nu)}{dt_\nu^n}\right]_{\tau_\nu} d\mu \\ &= \frac{1}{2}\sum_{n=0}^{\infty}\left[\frac{d^n S_\nu(t_\nu)}{dt_\nu^n}\right]_{\tau_\nu}\int_{-1}^{+1}\mu^n d\mu = \sum_{k=0}^{\infty}\frac{1}{2k+1}\left[\frac{d^{2k} S_\nu(t_\nu)}{dt_\nu^{2k}}\right]_{\tau_\nu}. \end{aligned} \tag{2.58}$$

[9] For their definition and use in radiative transfer, see Kourganoff (1963).

Note that only even-order derivatives contribute to the even moment J_ν, like for the other even moments. For the odd moment H_ν, only the odd-order derivatives contribute and the same holds true for all the successive odd moments.

Deep in the atmosphere, just the leading term of the power series expansion is enough for a fairly accurate approximation of (2.37). Moreover, at the bottom, where the optical depth at all the frequencies is much larger than the corresponding photon mean free path, the values of the source function approach those of the Planck function. Therefore, over the full range $-1 \leq \mu \leq 1$, we can assume that

$$I_\nu^+(\tau_\nu, \mu) \approx B_\nu(\tau_\nu) + \mu \left[\frac{dB_\nu(t_\nu)}{dt_\nu}\right]_{\tau_\nu}, \tag{2.59}$$

$$J_\nu(\tau_\nu, \mu) \approx B_\nu(\tau_\nu), \tag{2.60}$$

$$H_\nu^+(\tau_\nu, \mu) \approx \frac{1}{3}\left[\frac{dB_\nu(t_\nu)}{dt_\nu}\right]_{\tau_\nu} \tag{2.61}$$

and

$$K_\nu(\tau_\nu, \mu) \approx \frac{1}{3} B_\nu(\tau_\nu). \tag{2.62}$$

These approximations, accurate up to the second order, are referred to as the *diffusion approximation*. The name comes from the fact that, in the case of the flux, its expression has the form of a diffusion process: the flux of the quantity transported is equal to the product of a diffusion coefficient times the spatial gradient of a physically related quantity. In the present instance, as shown in section 2-5 of Mihalas (1978), at great depth the flux is given by

$$H_\nu = \frac{1}{3}\frac{\partial B_\nu}{\partial \tau_\nu} = -\frac{1}{3}\frac{1}{\chi_\nu}\frac{\partial B_\nu}{\partial T}\frac{dT}{dz}, \tag{2.63}$$

that is, the flux of the radiant energy is proportional to the gradient of the temperature.

It follows from (2.60) and (2.62) that $K_\nu/J_\nu = 1/3$. The value of this ratio, known as the Eddington factor, is exactly 1/3 in TE. Out of TE, variable Eddington factors can be introduced, which are at the basis of efficient numerical methods to solving the RTE.[10]

2.6.3 *At the Surface of the Atmosphere: The Eddington–Barbier Approximation*

We will now pay attention to the surface of the atmosphere. Instead of the whole power expansion of the source function, let us assume the linear form

$$S_\nu(\tau_\nu) = S_0 + S_1\tau_\nu, \tag{2.64}$$

where the coefficients S_0 and S_1 may be frequency dependent. Then the emergent intensity at the surface of a semi-infinite atmosphere will be

$$I_\nu^+(0, \mu) = \int_0^\infty (S_0 + S_1\tau_\nu)\, e^{-\tau_\nu/\mu} d\tau_\nu = S_0 + S_1\ \mu = S_\nu(\tau_\nu = \mu). \tag{2.65}$$

This equation is known as the *Eddington–Barbier approximation* for the specific intensity: the value of the emergent intensity is equal to that of the source function at optical depth unity along the line of sight.

[10] See Section 3.7.2.

For most stars, we cannot observe the specific intensity but only the flux, because in general it is not possible to resolve spatially their surface. Therefore, we shall use the Eddington–Barbier approximation for the emergent flux, namely

$$H_\nu^+ = \frac{1}{4}\left(S_0 + \frac{2}{3}S_1\right) = \frac{1}{4}\, S_\nu(\tau_\nu = 2/3), \tag{2.66}$$

which states that the value of the emergent flux corresponds to that of the source function at optical depth 2/3.

The Eddington–Barbier approximation assumes that the source function is linear over the whole region of integration. However, because of the weighting of the source function through the atmosphere, as shown by (2.42) or (2.50), actually it can be a good approximation even if the source function is approximately linear only over a much reduced region. Note also that we can apply the Eddington–Barbier approximation to the intensity or flux emerging from any slab (an intermediate slab in the atmosphere, or a slab in the interstellar medium) whenever the source function is known and linear, together with the relevant boundary conditions. Sometimes in the literature it is said that the emerging radiation is characteristic of the value of the source function at $\tau_\nu = 1$, other times that it is characteristic of the source function value at $\tau_\nu = 2/3$. Both statements are right: in the first case, we are speaking of the specific intensity, in the second of the flux.

The Eddington–Barbier approximation can be used to obtain information on the source function at different heights in the atmosphere from measurements of the specific intensity. Figure 2.5 shows an observer looking at different points on the surface of a star. When looking directly at the centre of the stellar disk, the observed emergent radiation is propagating along a line of sight parallel to the outward radial direction through the point of emission. As said previously, this radiation has formed at $\tau_{\nu_1} \approx 1$, corresponding to a geometrical depth $\Delta S_{\nu_1} = 1/\chi_{\nu_1}$, where χ_{ν_1} is the opacity at frequency ν_1. When looking along line of sight that form an angle $\vartheta = arccos\ \mu$ with the outward radial direction, the observer sees photons that have travelled again a unit optical path, but have been generated now at heights closer to the surface. Therefore, looking along line of sight with $0 < \mu \leq 1$, we can obtain information on the variation of S_{ν_1}. If the assumption of LTE can be made, the distribution of the temperature within the atmosphere can be determined. Observations at a different frequency ν_2 will allow us to cover an extended range in height because of the different opacities χ_{ν_1} and χ_{ν_2}. Thus we can get information on all the layers, embraced between the most optically thin and the most optically thick frequencies. Combining the redundant information from many frequencies, it is possible to obtain the temperature structure of a spatially resolved atmosphere like that of the Sun (this procedure can be generalized to make it independent of the Eddington–Barbier approximation).

2.6.4 *The Centre-to-Limb Variation*

The Eddington–Barbier relationship easily explains a well-known fact of solar observations: the centre appears brighter than the limb. Taking into account (2.65), we can write

$$\frac{I_\nu(0,\mu)}{I_\nu(0,1)} = \frac{S_0 + S_1\mu}{S_0 + S_1} \equiv f(\theta). \tag{2.67}$$

Thus, as far as both coefficients are positive, we expect that the intensity reaching us from a direction with $\mu < 1$ will be always smaller than that from a direction with $\mu = 1$. Therefore, the stellar disk will show the so-called *limb darkening*. In more general terms,

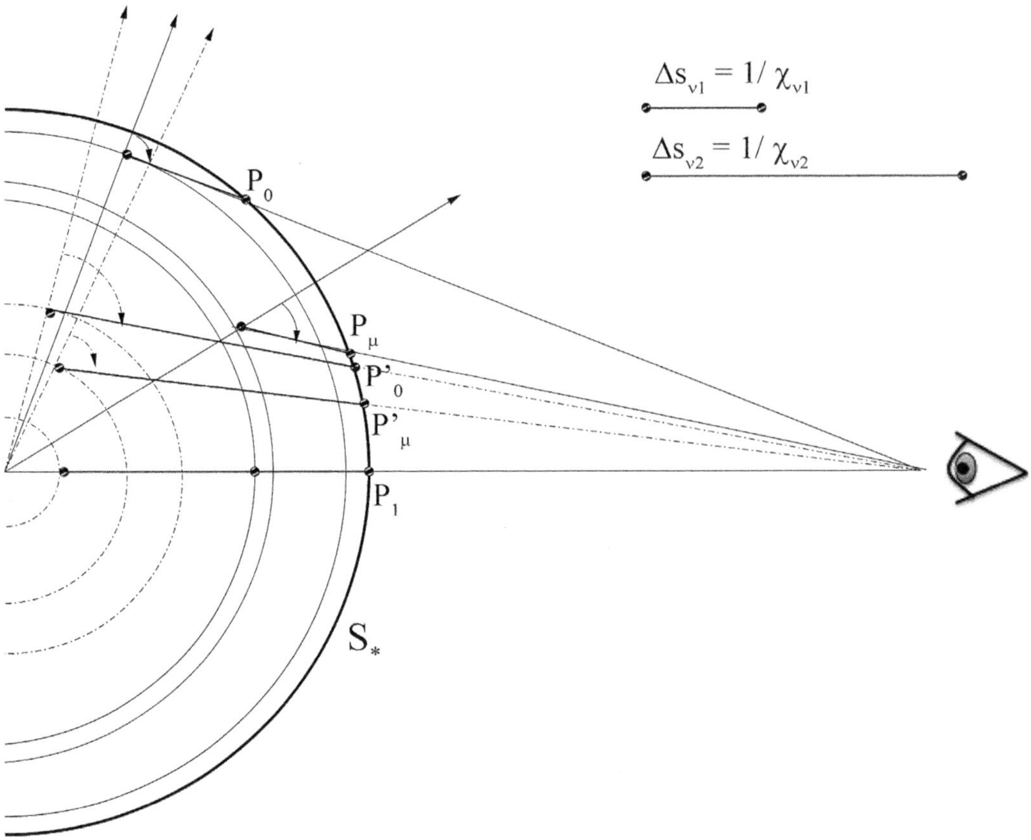

FIGURE 2.5. An observer looks at the stellar disk along different lines of sight at frequency ν_1 (solid lines) and ν_2 (dash-dotted lines).The monochromatic opacity χ_{ν_1} is larger than χ_{ν_2}, so that the geometrical path $\Delta s_{\nu_1} = 1/\chi_{\nu_1}$ is less than $\Delta s_{\nu_2} = 1/\chi_{\nu_2}$, both corresponding to unit optical depth. The different lines of sight for ν_1 through the points P_1, P_μ and P_0 (P_1, P'_μ and P'_0 for ν_2) on the stellar surface S_* correspond to specific intensities that form angles $\vartheta = 0$, $\vartheta = arccos\ \mu$ and $\vartheta = \pi/2$ with the respective outward radial directions.

we expect to observe a variation of the specific intensity between the centre and the limb of the solar disk, and speak of *centre-to-limb* variation.

A typical limb darkening law has the form

$$I_\nu(0, \mu) = I_\nu(0, 1)(1 - \epsilon + \epsilon\ cos\theta), \tag{2.68}$$

where ϵ varies with wavelength and the stellar temperature. For the Sun in the visible spectral range, ϵ is taken as 0.6.

2.7 Qualitative Account of Spectral Line Formation in Stellar Atmospheres

Now we have at hand all the elements for a qualitative discussion of the formation of spectral lines. Consider the particular instance sketched in Figure 2.6. The observer is looking at a stellar atmosphere, in which the temperature decreases outwardly. We have seen in Section 2.6.3 that in lack of spatial resolution only the emergent flux can be observed, and that the corresponding monochromatic radiation at frequency ν escapes

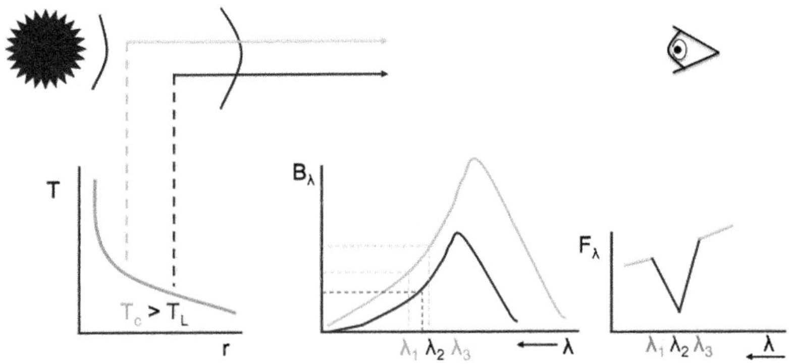

FIGURE 2.6. Sketch of the formation of spectral lines in a stellar atmosphere. In the three panels, grey lines refer to the continuum, black lines to the spectral line. Note that in the mid- and right panels, the λ-axis is inverted, from shorter wavelengths on the right to longer on the left.

from layers at an optical depth $\tau_\nu \approx 2/3$. By denoting with the superscript L the line and C the continuum, we can write $\tau^C = \chi^C r_C \sim 1$ and $\tau_\nu^L = \phi_\nu \chi^L r_L \sim 1$, where r_C and r_L are the corresponding geometrical depths, measured from the surface, and ϕ_ν the line profiles that will be zero outside a narrow range around the line central frequency. As it will be shown later in Section 2.9.1, the opacity χ^L of the spectral lines (bound–bound transitions) is larger by orders of magnitude than the opacity χ_ν^C of the continuum (bound–free transitions). As $\chi^L \gg \chi^C$, it follows that $r_L \ll r_C$ and therefore, when looking at continuum frequencies, the observer sees much deeper layers inside the atmosphere that, in the instance considered, have a higher temperature (see the leftmost panel of Figure 2.6). The Planck spectral distribution $B_\lambda(T)$ for the two values of the temperature corresponding to $T(\tau^C \sim 1)$ and $T(\tau^L \sim 1)$ is sketched in the middle panel. As the black body curves never cross, the continuum spectral distribution will be greater than that of the line at each wavelength. (For the sake of a straightforward illustration, we will assume that the Planck function is the source function both for the continuum and the spectral line.)

When scanning the spectral distribution recorded, from longer to shorter wavelengths (note that in the figure the λ-axis is inverted), the observer will sample wavelengths inside and outside the spectral range covered by the line. If λ_1 and λ_3 are outside of the latter, the radiation emerging at these wavelengths correspond to the Planck function for the continuum at these wavelengths. Going from λ_1 to λ_3 across the line spectral range, radiation will emerge from progressively outer layers, until the outermost layer is reached in correspondence of the maximum line opacity at its nominal wavelength λ_2, where the temperature is $T(\tau^L \sim 1)$. Herein on the line opacity will decrease and therefore inner layers will be sampled, until the border of the spectral line at λ_3, where the depth of formation of the continuum is reached again.

The behaviour of the flux with wavelength, sketched in the rightmost panel, is a mapping of the Planck function $B_\lambda[T(r)]$ into the flux spectral distribution, marked by the absorption trough due to the spectral line.

From the preceding qualitative sketch, it is easy to realize that we may have spectral lines in emission under particular conditions. For instance, if there is a temperature rise in the outer layers and the source function is coupled to temperature, the flux will be higher in the line than in the continuum. However, we may have emission lines also without a temperature rise in the outer layers. When observing a star, we are integrating the flux over the stellar disk, and the total energy flux escaping from the star at frequency

ν is proportional to $r^2\mathcal{F}_\nu$. Let us assume that the difference in opacity is so large that the stellar radius changes significantly when observing at the line nominal wavelengths. In this case, the increase in r^2 may more than compensate for the decrease of $\mathcal{F}_\nu$ and produce an emission line.

2.8 Local Thermodynamical Equilibrium

Under conditions of TE, the velocity distribution of the particles of the system, both atoms (or ions) and free electrons, is given by the Maxwell–Boltzmann (MB) equilibrium distribution, characterized by a unique value of the temperature. Although the evidence of a flux of radiation flowing through the border of the stars implies a temperature gradient in their outer layers, the space scale of the latter is usually much greater than the mean free path of the particles. Therefore, it is justifiable to make the simplifying hypothesis that in the neighbourhood of any point, the *external* state of the stellar material (i.e., that associated with the external degrees of freedom of the particles) must be in TE at a local temperature that determines point by point the MB distribution

$$f(v)\, dv = \frac{dN(v)}{N_{tot}} = \left(\frac{m}{2\pi k_B T}\right)^{3/2} 4\pi\, v^2\, e^{-mv^2/2k_B\, T}\, dv. \tag{2.69}$$

The distribution function $f(v)$ denotes the fraction of particles of a given species (electrons, ions, etc.) of mass m moving with velocity between v and $v + dv$. N_{tot} is the total number of particles (usually expressed per unit volume, i.e., the particle density number), k_B the Boltzmann constant and T the local temperature.

For a semiclassical model of stellar matter, in which the particles have well-defined position and moment as in classical mechanics, but are endowed with an internal structure of energy levels (either discrete or continuous), the available states corresponding to all the possible degrees of excitation and ionization may be considered the *internal* degrees of freedom of the system. Then, like for the external state, it will be matter of specifying the distribution of the internal energy among the quantum states available. In other words, to determine the distribution functions that express the number of particle per unit volume (atomic populations) of the atoms of a given species in the ith degree of ionization and, for a given ionization stage, in the jth excited level. The processes that determine the relevant distribution functions by populating and depopulating the energy levels involve the radiation field, which manifestly departs from its equilibrium Planck distribution. Only in the particular case that the inelastic collisions among atoms and free electrons overcome the corresponding radiative processes, the corresponding distribution functions take on their equilibrium form. The Saha ionization equation

$$\left(\frac{N_j}{N_{j+1}}\right) = \left(\frac{h^2}{2\pi m_e k_B}\right) N_e \frac{U_j}{U_{j+1}}\, T^{-3/2}\, e^{\chi_j/k_B T} = 2.07 \times 10^{-16}\, N_e \frac{U_j}{U_{j+1}}\, T^{-3/2}\, e^{\chi_j/k_B T} \tag{2.70}$$

yields the ratio between the populations of two consecutive ionization states, j and $j+1$, of an atom (or molecule). The Us are the partition functions of the corresponding ionization stage χ_j is the energy between the ground level of the jth state j and that of the ground level of the next ionization stage, N_e and m_e, are respectively the electron particle density and mass; and h is the Planck constant.

The Boltzmann excitation equation

$$\left(\frac{n_l}{n_u}\right) = \left(\frac{g_l}{g_u}\right) e^{\chi_{lu}/k_B T} \tag{2.71}$$

gives the ratio between the populations of two excitations levels of the same ionization state, where l represents the lower level (closer to the ground level) and u the upper one, smf the excitation energy χ_{lu} is the difference between the energies of the l and u levels.[11] g_l and g_u are the corresponding statistical weights.

In those regions of a stellar atmosphere where all the distribution functions, both for the external and the internal degrees of freedom, have their equilibrium values, matter can be considered to be locally in TE conditions. Such a less restrictive condition is called as LTE. This simplifying assumption is indeed a very strong one, because it implies to determine the previous distribution functions point by point locally, without reference to the behaviour of the physical system as a whole. Although such an approximation clearly lacks internal consistence, it is nevertheless an effective computational expedient.

Sometimes the Saha equation is combined with the Boltzmann equation to express the population of an excited level i of an ionization state j with respect to the population of the ground level of the next ionization state $j+1$. The LTE population $n^*_{i,j}$[12] will then be given by

$$n^*_{i,j} = n_{0,j+1} N_e\ 2.07 \times 10^{-16} \frac{g_{i,j}}{g_{0,j+1}}\ T^{-3/2}\ e^{\chi_{i,j}/k_B T} = n_{0,j+1}\ N_e\ \phi(T), \qquad (2.72)$$

where $\phi(T)$ is the so-called Saha factor. We use the *actual* population $n_{0,j+1}$ to compute the LTE population $n^*_{i,j}$. This is because the population of the low ionization state in TE is determined by the number of collisions between electrons and ions in the next ionization state, i.e., is proportional to the product $n_{0,j+1}N_e$.

Likewise, the Boltzmann equation is written sometimes as the ratio between the population of an excited level with respect to the total population of its ionization state, that is,

$$\left(\frac{n_{i,j}}{N_j}\right)^* = \frac{g_{i,j}}{U_j}\ e^{-\chi_{i,j}/k_B T}. \qquad (2.73)$$

In LTE, the preceding distribution functions are complemented by the Kirchhoff's law $\eta_\nu/\chi_\nu = B_\nu(T)$.

One cannot state a priori whether the LTE assumption holds valid. However, it is a reasonable approximation when the following conditions take place:

- If radiation is trapped, the radiation field will be in equilibrium with matter.
- If the amount of energy escaping by radiation is small compared with the radiative energy content, LTE can be considered. An intense emerging radiation field on the contrary favours departure from LTE.
- If the number of collisions is sufficiently large, the Maxwell–Boltzamnn, Saha, Boltzmann and Kirchoff equations hold valid.
- When densities are high, the number of collisions will also be high, so that high densities favour LTE.

2.9 Absorption and Emission Coefficients

To know the source function, we need to know the emission and absorption coefficients introduced from a macroscopic point of view in Section 2.3. However, we know that they

[11] Because excitation energies of individual levels may be given with respect either to the fundamental level or the continuum, care must be taken to put the right sign of the exponent. It must be such that the increase of the difference favours the population of the lower-lying level.

[12] As customary, the asterisk denotes LTE values.

are the result of absorption and emission of photons at a microscopic level. Following Einstein's phenomenological theory of radiative transitions (Einstein, 1917), we will consider three kinds of transitions (photon–atom interactions) that produce a change in the population of the atomic energy levels and the radiation field:

- *Spontaneous emission*: an atom in a higher energy level decays spontaneously to a lower one with emission of photon, because of the limited lifetime of the excited level.
- *Stimulated emission*: an atom in a higher energy level decays to a lower one due to the perturbation by a photon of a given frequency, with the emission of another photon of almost the same frequency as that of the perturber.
- *Absorption*: a photon is absorbed by an atom that undergoes a transition from a lower to a higher energy level.

The atomic states involved may be either bound or free, so that we may have bound–bound, bound–free and free–free transitions. Bound–bound transitions take place between two quantum levels with (relatively) well-defined energy, and the energy of the photon emitted is equal to the difference between the energies of the two levels. Thus only a narrow range of frequencies will be involved in the transition (either an absorption or an emission), whose signature will show as an alteration in the photon energy spectrum at these frequencies, i.e., a *spectral line.*

Bound–free transitions involve a quantum state consisting of an atom in a discrete energy level and a free electron, whose energy is not quantized. The energy of the photons (either emitted or absorbed) will be that required to ionize the atom from its initial quantum state, plus the kinetic energy delivered to the free electron. Thus there is a minimum energy threshold corresponding to the ejection of an electron with zero kinetic energy; above the threshold any value for the kinetic energy of the free electron is allowed. In this case, we have a continuous spectrum.

Free–free transitions involve a change in a system formed by an ion and a free electron. When the electron passes close to an ion, its velocity changes (in module and/or direction), which implies an acceleration. As it is known, an accelerated charge radiates or absorbs energy and, consequently, a photon is either emitted or absorbed with an energy equal to the change of the kinetic energy of the electron. As the kinetic energy of the free electron is not quantized, the frequency spectrum of the free–free transitions extends from zero to infinity, without any threshold.

2.9.1 *Line Coefficients*

In order to compute the absorption and emission coefficients, we will start with the case of a bound–bound transition between a lower-level l and an upper-level u ($E_l < E_u$). According to Einstein's theory, the probabilities per unit time that an atom absorbs or emits a photon with frequency in the range $\nu, \nu + d\nu$ into a solid angle $d\Omega$ are

$$dP_\nu^{abs} = B_{lu}\, I_\nu\, \phi_\nu\, d\nu \frac{d\Omega}{4\pi}, \tag{2.74}$$

$$dP_\nu^{st} = B_{ul}\, I_\nu\, \varphi_\nu\, d\nu \frac{d\Omega}{4\pi} \tag{2.75}$$

and

$$dP_\nu^{sp} = A_{ul}\, \psi_\nu\, d\nu \frac{d\Omega}{4\pi}. \tag{2.76}$$

Here B_{lu} and B_{ul} are the probabilities of transitions due to the specific intensity of the radiation field, A_{ul} that of a spontaneous transition $u \to l$ and $d\Omega/4\pi$ is the probability that the photon comes from (or is emitted into) a direction within the solid angle $d\Omega$. The atomic coefficients B_{lu}, B_{ul} and A_{ul} are the Einstein–Milne coefficients for absorption, stimulated emission and spontaneous emission, respectively.[13] The profiles ϕ_ν, φ_ν and ψ_ν, normalized to unity, account for the frequency distribution across the spectral line. In general, they are different; however, if complete frequency redistribution (CFR) is assumed, namely that the frequency and direction of an absorbed photon are uncorrelated with those of the reemitted photon, it holds that $\phi_\nu = \varphi_\nu = \psi_\nu$. Although CFR works well for most astrophysical problems, sometimes partial frequency redistribution (PFR) must be taken into account.

The number of absorptions and stimulated emissions within a volume dV during a time interval dt can be computed from (2.74) and (2.75), namely

$$N^{abs} = n_l \; dP_\nu^{abs} \; dV \; dt \tag{2.77}$$

and

$$N^{st} = n_u \; dP_\nu^{st} \; dV \; dt. \tag{2.78}$$

Because of the common dependence of absorption and stimulated emission on the radiation field, it is customary to include both of them into a single macroscopic absorption coefficient.

Thus the energy involved in each process will be equal to the number of transitions times the energy of each transition, so that the energy balance between absorptions and stimulated emissions is given by

$$dE_\nu = n_l \; h\nu \; dP_\nu^{abs} \; dV \; dt - n_u \; h\nu \; dP_\nu^{st} \; dV \; dt = \frac{h\nu}{4\pi} \; \phi_\nu \left(n_l B_{lu} - n_u B_{ul}\right) I_\nu \; d\nu \; d\Omega \; dV \; dt. \tag{2.79}$$

From the definition of the absorption coefficient, given by (2.26), we have

$$dE_\nu^{abs} = k_\nu \; I_\nu \; d\nu \; d\Omega \; dS \; ds \; dt, \tag{2.80}$$

where $dS\,ds = dV$ and we write k_ν instead of χ_ν, because we consider pure absorption only. Thus for a bound–bound transition $l \to u$, we have

$$k_\nu = \frac{h\nu}{4\pi}\phi_\nu \; \left(n_l B_{lu} - n_u B_{ul}\right), \tag{2.81}$$

and likewise for the emission coefficient

$$\epsilon_\nu = \frac{h\nu}{4\pi}\phi_\nu \; A_{ul}. \tag{2.82}$$

It can be shown that in TE, where the emission and absorption processes are individually compensated because of detailed balance, the following relations among the Einstein coefficients take place:

$$A_{ul} = \frac{2h\nu^3}{c^2}\frac{g_l}{g_u} \; B_{ul} \tag{2.83}$$

[13] Differently from that of the original ones, the dimension of the Einstein–Milne coefficients B_{lu} and B_{ul} is $M^{-1}T$ instead of $M^{-1}L$.

and

$$g_l \, B_{lu} = g_u \, B_{ul}. \tag{2.84}$$

However, the Einstein coefficients are atomic properties and therefore the preceding relations hold valid in general, not only under TE conditions. We can then define the cross section for the transition $l \to u$ as

$$\sigma_{lu}(\nu) = \frac{h\nu}{4\pi} \phi_\nu B_{lu}. \tag{2.85}$$

In order to calculate the transition probabilities, we need the Einstein's coefficients. We know from quantum mechanics that B_{lu} is related to $\langle u|\hat{H}|l\rangle$, where $|u\rangle$ and $|l\rangle$ are the wave functions of the levels u and l, and $\hat{H}$ is the Hamiltonian of the interaction. The largest contribution comes from the first term of the Hamiltonian, corresponding to the dipole approximation. Further terms (electric quadrupole, magnetic dipole) are much smaller, and are responsible for the so-called forbidden transitions, which may be important if the dipole term is zero and the collisional rate is very low. (If collisions are frequent, they will remove the atom from its actual energy level before a low probability radiative transition takes place.)

The quantum mechanical computation of the radiative transitions is lengthy and requires the knowledge of the relevant wave functions. The following semiclassical approach to radiative transitions is very suitable and widely in use. Lorentz (1900) put forward the simple model of the *classical oscillator* to account for the interaction between atoms and a radiation field: a radiating atom can be represented by a dipole wherein an electron of charge e, and mass m is bound to the nucleus by an elastic restoring force. Later, Ladenburg (1921) proposed a quantum interpretation of the anomalous dispersion consistent with Bohr's atomic model. He postulated the equivalence between the level populations of the absorbing atoms and the oscillators of the classical model.[14]

While in the foregoing semiclassical approximation the total (i.e., frequency integrated) cross section turns out to be $\sigma_{lu} = \pi e^2/mc$ (the same for all transitions), the quantum mechanical result is

$$\sigma_{lu} = \frac{\pi e^2}{mc} f_{lu} = \frac{h\nu}{4\pi} B_{lu}, \tag{2.86}$$

where the dimensionless factor f_{lu}, called after W. Pauli the *strength of the oscillator*, may be interpreted as a measure of the strength of the transition; in other words, as the probability of the transition. Equation (2.86) shows the link between the oscillator strength and the Einstein coefficient B_{lu}. The intensity of a spectral line is proportional to the oscillator strength of the corresponding transition. In the case that there are several levels involved in the transition, the strength of the oscillator is given by the Thomas–Reiche–Kuhn sum rule

$$f_{lu} = \frac{1}{g_l} \sum_{l',u'} g_{l'} f_{l',u'} \tag{2.87}$$

We can determine f_{lu} simply by comparing our calculations with laboratory measurements or even stellar spectra observations. Of course, in some cases it is also possible to carry out the corresponding quantum mechanical calculations.

[14] An extensive discussion of the Lorentz'oscillator, the anomalous optical dispersion and the Ladenburg relation can be found in Crivellari (2016a) and Crivellari (2016b).

The same calculation as the preceding provides the frequency profile ϕ_ν of the transition, which has the form of a Lorentzian function. Then (2.85) can be written as

$$\sigma_{lu}(\omega) = \frac{\pi e^2}{m_e c} f_{lu} \frac{\gamma}{\Delta\omega^2 + (\gamma/2)^2}, \tag{2.88}$$

where γ is the damping constant, $\Delta\omega \equiv (\omega - \omega_0), \omega = 2\pi\nu$, and ω_0 corresponds to the central line frequency. Thus, intrinsically, spectral lines are not infintely sharp because, according to Heisenberg's uncertainty principle, upper levels have a finite energy width. This is called the *natural broadening* of the line. The corresponding profile is Lorentzian and not Gaussian. However, the latter is frequently used in astrophysics for the spectral line profiles because of the effects due to additional line broadening mechanisms, to be considered in the next sections.

Therefore, for a bound–bound transition (i.e., a spectral line), we have eventually

$$k_\nu^L = \sigma_{lu}(\nu)\left(n_l - \frac{g_l}{g_u} n_u\right) \tag{2.89}$$

and

$$\epsilon_\nu^L = \sigma_{lu}(\nu) \frac{2h\nu^3}{c^2} \frac{g_l}{g_u} n_u. \tag{2.90}$$

Note that the structure of (2.89) and (2.90) has the form of a *cross section* × *atomic population.*

In LTE, we can apply the Boltzmann equation, so that

$$k_\nu^L = \sigma_{lu}(\nu)\, n_l^* \left(1 - e^{-h\nu_{lu}/k_B T}\right), \tag{2.91}$$

where the factor $\left(1 - e^{-h\nu_{lu}/k_B T}\right)$ accounts for the stimulated emission.

Model atmosphere codes necessarily require model atoms, for which atomic data such as energy levels, ionization energies and oscillator strengths are fundamental. Useful base data can be found, inter alia, in web sites such as that of the National Institute of Standards and Technology,[15] that mantained by Peter van Hoof[16] or the atomic and molecular line lists by R. L. Kurucz.[17]

2.9.2 *Continuum Coefficients*

Let us consider now bound–free and free–free transitions. In the first case, an atom in the bound state i of the kth ionization stage is photoionized by the absorption of a photon with the result of an atom in the ground state of the $(k+1)$th ionization stage and a free electron. The corresponding populations (number densities) are given by n_i, n_k and n_e. The inverse process of recombination consists of the encounter of the free electron and an ion, with the subsequent capture of the electron by the ion, the emission of a photon and the increase of the population n_i. It is, however, possible that the interaction of the electron and the ion does not produce a recombination, but just the change of the direction and kinetic energy of the free electron. In this case, because of the acceleration or deacceleration of an electric charge, a photon will be emitted or absorbed. Note that, while the number of photoionizations is proportional to the population n_i, that of the inverse

[15] www.nist.gov/pml/atomic-spectra-database
[16] www.pa.uky.edu/~peter/newpage/
[17] www.re3data.org/repository/r3d100011427

processes is proportional to the number of encounters between free electrons and ions, given by the product $n_e \times n_k$. Note also that $n_e \times n_k \propto n_i^*$ according to Saha's equation. Thus, likewise as in the computation for the bound–bound transitions,[18] we have

$$k_\nu^{bf} = \sigma_{ik}(\nu) \left(n_i - n_i^* \, e^{-h\nu/k_B T}\right) \qquad \nu \geq \nu_{ik} \tag{2.92}$$

$$\epsilon_\nu^{bf} = \sigma_{ik}(\nu) \frac{2h\nu^3}{c^2} \, n_i^* \, e^{-h\nu/k_B T} \qquad \nu \geq \nu_{ik} \tag{2.93}$$

$$k_\nu^{ff} = \sigma_{kk}(\nu) \, n_e \, n_k \, \left(1 - e^{-h\nu_{lu}/k_B T}\right) \qquad \forall\nu \tag{2.94}$$

$$\epsilon_\nu^{ff} = \sigma_{kk}(\nu) \, n_e \, n_k \frac{2h\nu^3}{c^2} \, n_i^* \, e^{-h\nu/k_B T}. \qquad \forall\nu \tag{2.95}$$

We shall stress again that there is a threshold frequency (energy) for the bound–free processes, whereas free–free processes may involve any frequency value.

The problem is the computation of the relevant cross sections. Hydrogen is the main component of stellar atmospheres, as was first shown by C. H. Payne in 1925, and its cross section was calculated by H. A. Kramers, who obtained the expression (which holds valid also for hydrogeneic ions)

$$\sigma_\nu^{bf} = 2.815 \times 10^{29} \frac{Z^4}{n^5 \nu^3} \, g_{bf} \qquad \nu \geq \nu_{nk}, \tag{2.96}$$

where Z is the ion charge, n is the principal quantum number of the involved level and g_{bf} the so-called Gaunt factor. This cross section is given in units of cm^2 per atom.

Figure 2.7 shows the form of the H bound–free cross section. We see that (a) for each principal quantum number n, there is a threshold wavelength, corresponding to the ionization energy from that level; (b) the maximum value of the cross section is reached

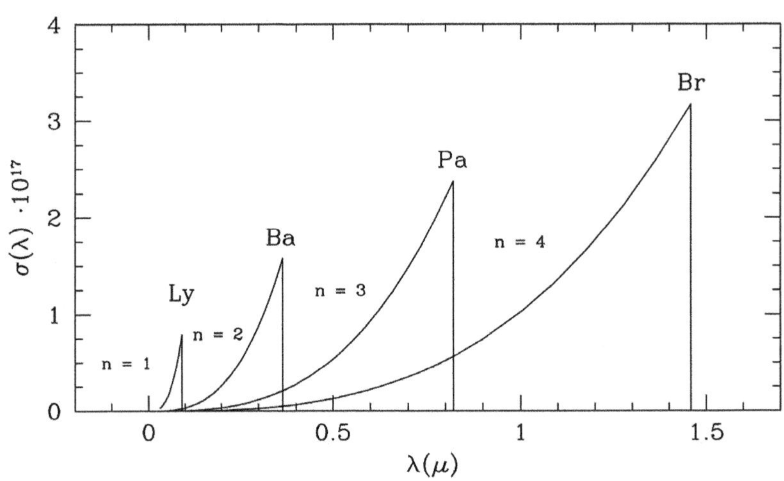

FIGURE 2.7. The bound–free absorption cross section of neutral hydrogen for the first four levels, identified by their principal quantum number n. Ordinate: the cross section per atom in cm^2; abscissa the wavelength in microns. The positions of the Lyman, Balmer, Paschen and Bracket discontinuities are shown.

[18] The generalization of the Einstein's relations to bound–free processes is due to E. A. Milne (Milne, 1924).

at the threshold energy, and it decreases with the third power for higher energies (this implies that the probability of the transition increases for decreasing electron velocities); (c) the cross-section value at threshold energy increases for increasing n (decreasing ionization); and (d) at a given energy, the cross section is larger for lower quantum numbers (if the photon energy is sufficient to ionize the atom from that level).

The Gaunt factor for bound–free H transitions is given by

$$g_{bf} = 1 - \frac{0.3456}{(\lambda R)^{1/3}} \left(\frac{\lambda R}{n^2} - \frac{1}{2} \right), \tag{2.97}$$

where R is the Rydberg constant.

The free–free cross section for hydrogenic atoms takes the form

$$\sigma_\nu^{ff} = 3.7 \times 10^8 \frac{Z}{T^{\frac{1}{2}} \nu^3} g_{ff} \quad cm^5 \qquad \forall \nu. \tag{2.98}$$

Note that the cross section is measured here in units of cm^5 per ion and per electron, as it depends on an ion–electron encounter. The constants have here the same meaning as before, and the Gaunt factor takes the form

$$g_{ff}(H) = 1 + \frac{0.3456}{(\lambda R)^{1/3}} \left(\frac{\lambda k_B T}{hc} + \frac{1}{2} \right). \tag{2.99}$$

Rupert Wildt discovered in 1939 that the negative hydrogen ion H^- is the principal opacity source for cool stars. The dissociation energy of H^- ($0.755\ eV$) is much smaller than the ionization energy of H ($13.60\ eV$). Nevertheless, it is the main opacity source in those atmospheres, in which a large number of both neutral H atoms and free electrons are present. The latter come from metals with a low ionization threshold. Such conditions are met in the atmospheres of solar-type stars. For hotter stars, hydrogen is almost completely ionized, while in the atmospheres of cooler stars free electrons are lacking because metals remain neutral.

The second most abundant element in the Universe is helium, and it plays an important role in hot atmospheres, where the large amounts of photons can excite and ionize their atoms. Ionized helium has a hydrogenic structure and therefore the relevant equations are similar to those for hydrogen but for a higher nuclear charge, which implies a deeper potential well and higher energies. Neutral helium, on the contrary, has a more complex atomic structure, and complexity increases with increasing atomic number. This results in a complex shape for the cross section as a function of the energy of the incident photon, due to the multiple resonances between initial and final atomic states. In Figure 2.8, we show the cross section for the 2p^4 ^{1}D level of O III as a function of wavelength, where many strong peaks and troughs appear as a consequence of the aforementioned resonances.

Free electrons are a strong opacity source in hot stars, due to the high ionization degree of the atoms. Their interaction with the radiation takes place through scattering processes (Thompson scattering). For photon energies lower than those of X-rays, their cross section is constant and given by

$$\sigma_e = 6.65 \times 10^{-25} cm^2. \tag{2.100}$$

Values for atomic energy levels, cross sections and oscillator strengths can be found in the reports of two projects resulting from a big international effort to make these fundamental values available: the Opacity Project and the Iron Project (see http://cdsweb.u-strasbg.fr/topbase/home.html) and also in the pages of the National Institute

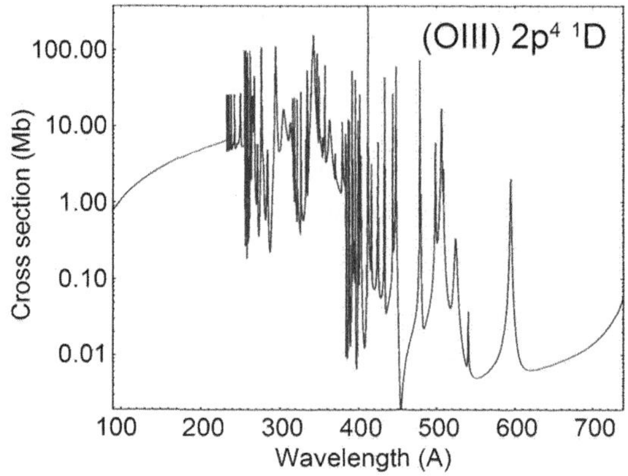

FIGURE 2.8. Radiative bound–free cross section of an excited level of O III (from Urbaneja, 2004).

of Standards and Technology (www.nist.gov/physlab/data/asd.cfm) or in the personal effort by Peter van Hoof (www.pa.uky.edu/~peter/newpage/).

Molecules play an increasing important role at temperatures typical of the atmospheres of cool stars. A large number of atomic and molecular data are also made accessible through the Virtual Atomic and Molecular Data Center, which groups a large number of databases (including some of those cited previously) through a common entry point (http://portal.vamdc.eu; Dubernet et al., 2016. Both atomic and molecular data provided by R. P. Kurucz (see Section 2.9.1) have been extensively used in stellar atmosphere modelling.

2.10 The Statistical Equilibrium Equations

We have seen how to calculate the cross sections for radiative transitions, but we still need to know the atomic (or molecular) populations in order to calculate the absorption and emission coefficients, hence the source function. We will do this under the assumption of statistical equilibrium for the energy levels population. We assume that the population of any level is constant with time. Thus, for any energy level (of atoms, molecules or free particles)

$$\frac{dn_i}{dt} + \nabla(n_i) = 0, \tag{2.101}$$

where n_i is the population density of level i within the volume element considered and $\nabla(n_i)$ is the net flux of the particles of the level i crossing the boundary surface of the volume element, which we will assume equal zero. Inside the volume, the energy levels are constantly populated and depopulated by different processes, so that the steady-state constraint will impose[19]

$$\sum_{j\neq i} n_i P_{ij} = \sum_{j\neq i} n_j P_{ji}. \tag{2.102}$$

[19] Note here the difference to detailed balance, where opposite processes balance each other. In statistical equilibrium, all processes populating a level balance with all processes depopulating it.

Here P_{ij} is the rate of processes depopulating level i through transitions from i to any other level j, whereas P_{ji} is the rate of processes populating it. If we discriminate between radiative (involving the radiation field) and collisional transitions, we can write

$$\sum_{j\neq i} n_i\,(R_{ij}+C_{ij}) = \sum_{j\neq i} n_j\,(R_{ji}+C_{ji}). \tag{2.103}$$

As customary, we will denote by the index k the continuum level of bound–free and free–bound transitions, while i and j are bound levels of the same ionization stage. Although continuum transitions (transitions that imply either ionization to the successive ionization state or recombination to the previous one) can involve any level of the upper ionization stage, usually only its ground level is considered. Thus, for a level i of a given ionization state,

$$\sum_{j\neq i} n_i\,(R_{ij}+C_{ij}) + n_i\,(R_{ik}+C_{ik}) = \sum_{j\neq i} n_j\,(R_{ji}+C_{ji}) + n_k\,(R_{ki}+C_{ki}). \tag{2.104}$$

There is one such equation for each energy level considered, and the resulting system takes on the form

$$\begin{array}{rrrrrrl}
n_1\sum\limits_{j\neq 1}^{N} P_{1j} - & n_2P_{21} - & \ldots - & n_iP_{i1} - & \ldots - & n_NP_{N1} = & 0\\
-n_1P_{12} + & n_2\sum\limits_{j\neq 2}^{N} P_{2j} - & \ldots - & n_iP_{i2} - & \ldots - & n_NP_{N2} = & 0\\
\ldots\ldots & \ldots\ldots & \ldots\ldots & \ldots\ldots & \ldots\ldots & \ldots\ldots & \ldots\\
-n_1P_{1i} - & n_2P_{2i} - & \ldots + & n_i\sum\limits_{j\neq i}^{N} P_{ij} - & \ldots - & n_NP_{Ni} = & 0\\
\ldots\ldots & \ldots\ldots & \ldots\ldots & \ldots\ldots & \ldots\ldots & \ldots\ldots & \ldots\\
-n_1P_{1N} - & n_2P_{2N} - & \ldots - & n_iP_{iN} - & \ldots - & n_N\sum\limits_{j\neq N}^{N-1} P_{Nj} = & 0
\end{array} \tag{2.105}$$

This is an homogeneous linear system of N algebraic equations, of which only $N-1$ are independent. Therefore, we must replace one of them by an independent closure relation, for example the conservation of the particle number,[20] namely

$$n_1+n_2+\ldots+n_i+\ldots+n_N = N_s = \mathcal{A}_s\, N_{\rm tot}, \tag{2.106}$$

where N_s is the number density of all the particles of the element s, $\mathcal{A}_s$ is its relative abundance and $N_{tot} \equiv \sum_s N_s$ is the total number density of the particles of all the elements considered. The system can be written in matrix form as

$$\boldsymbol{A}\cdot\mathbf{n} = \mathbf{b}, \tag{2.107}$$

whose solution is

$$\mathbf{n} = \boldsymbol{A}^{-1}\,\mathbf{b}. \tag{2.108}$$

[20] The conservation of the charge may be used as an alternative constraint.

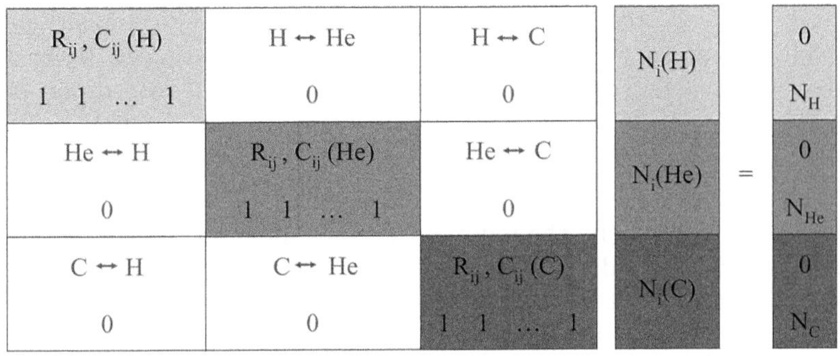

FIGURE 2.9. The structure of the system of statistical equilibrium equations, where only H, He and C are considered for simplicity's sake.

$\boldsymbol{A}$ is the coefficients matrix, containing the rates corresponding to all transitions between the levels considered, $\mathbf{n}$ is the vector of level populations and vector $\mathbf{b}$ contains the independent terms.

In Figure 2.9, we show the structure of the system equations. For each element included (here only H, He and C), the matrix $\boldsymbol{A}$ consists of full submatrices, whose elements are the rates of the transitions of the relevant elements, plus a last row with the coefficients of the particle conservation constraints. The vector $\mathbf{n}$ of the occupation numbers contains the populations (the unknowns) of all the energy levels of all the elements considered; the components of the vector $\mathbf{b}$ consist, for each element, of strings of zeros plus the total number density. In principle, the elements of the full matrix corresponding to cross-terms (for instance, transitions from an H to an He energy level) will be zero. Therefore, the equations of statistical equilibrium can be solved separately for each element. However, in case of molecules we will have cross-terms different from zero. For example, if we consider a CO molecule, there will be a submatrix that includes not only the coefficients of the transitions among all its energy levels: also those corresponding to molecule dissociations (CO $\rightarrow$ C+O) or molecules formation (C+O $\rightarrow$ CO) will be different from zero. This situation is more and more important as the temperature in the atmosphere decreases. Charge transfer is another process that can involve different elements. These processes involve the transfer of one electron from one ion of a given element to that of another element, producing an increase of the ionization of one element and a decrease of another one (for example, from O+ and Si++ to O++ and Si+). Of course, if we apply the statistical equilibrium equations to nuclear reactions there will be cross-terms representing the transformation of some elements into others. These cross-terms will make much larger the computational cost of the inversion of the coefficient matrix and consequently of the numerical solution of the equations.

2.10.1 *Radiative Rates*

To solve the equations, we must calculate the elements of matrix $\boldsymbol{A}$, i.e., the radiative and collisional transition rates. The former can be computed from the absorption and emission (both spontaneous and stimulated) probabilities, introduced in Section 2.9.1 in terms of the Einstein's coefficients. The total probability per atom of a transition $i \rightarrow j$ due to the absorption of a photon will be the integral over all the frequencies and directions of (2.74). Thus the bound–bound absorption radiative rate (per atom and unit time) will be given by

$$R_{ij} = \int_0^\infty B_{ij}\ \phi_\nu\ d\nu \frac{1}{4\pi} \oint I_\nu\ d\Omega = \int_0^\infty B_{ij}\ \phi_\nu\ J_\nu\ d\nu = B_{ij}\ \bar{J}_\nu. \tag{2.109}$$

Taking into account (2.85), we can write

$$R_{ij} = 4\pi \int_0^\infty \frac{\sigma_{ij}(\nu)}{h\nu} J_\nu d\nu. \tag{2.110}$$

Likewise, for a bound–free transition $i \to k$ we have

$$R_{ik} = 4\pi \int_0^\infty \frac{\sigma_{ik}(\nu)}{h\nu}\ J_\nu\ d\nu. \tag{2.111}$$

The results for the emission radiative rates, respectively bound–bound and free–bound, are

$$R_{ji} = 4\pi \left(\frac{n_i}{n_j}\right)^* \int_0^\infty \frac{\sigma_{ij}(\nu)}{h\nu}\ \left(\frac{2h\nu^3}{c^2} + J_\nu\right)\ e^{-h\nu/k_BT}\ d\nu \tag{2.112}$$

and

$$R_{ki} = 4\pi \left(\frac{n_i}{n_k}\right)^* \int_0^\infty \frac{\sigma_{ik}(\nu)}{h\nu}\ \left(\frac{2h\nu^3}{c^2} + J_\nu\right)\ e^{-h\ nu/k_BT}\ d\nu, \tag{2.113}$$

where the term J_ν in the parentheses represents the contribution of induced emission and $2h\nu^3/c^2$ that of spontaneous emission, and the asterisks indicate that the ratio is calculated with the Saha formula.

2.10.2 *Collisional Rates*

We also need to calculate the corresponding rates for the collisional transitions. Let us denote by $Q_{ij}(v)$ the excitation cross section for the collision between an atom originarily in its j level and an electron with velocity v. (Q_{ij} is usually measured in units of πa_0^2, where a_0 is the Bohr radius.) By integration over the electron velocity distribution function $f(v)$, we obtain the collisional excitation rate per atom and unit time, namely

$$C_{ij} = \int_0^\infty \pi a_0^2\ Q_{ij}\ v\ n_e\ f(v)\ dv = n_e\ q_{ij}(T), \tag{2.114}$$

where n_e is the electron number density. We need to know Q_{ij} (or q_{ij}) that can be obtained either from laborious calculations or by difficult laboratory measurements. In the case that we don't know them for the transitions of interest, we can use van Regermorter's formula

$$q_{ij}(T) = C_0\ T^{\frac{1}{2}}\ \left[14.5 f_{ij} \left(\frac{E_H}{E_{ij}}\right)^2\right]\ u_0\ e^{-u_0}\ \Gamma_e(u_0), \tag{2.115}$$

where $u_0 = E_{ij}/k_BT$, $\Gamma_e(u_0) = max\,[\bar{g}, 0.276e^{u_0} E_1(u_0)]$ and E_1 is the first exponential integral. This expression is valid for ions (Γ_e has a different form for neutral atoms) and allowed transitions.

Because of detailed balance, in TE upwards and downwards transitions compensate exactly each other, so that $n_i^*\ C_{ij} = n_j^*\ C_{ji}$ and hence

$$\frac{C_{ij}}{C_{ji}} = \left(\frac{n_j}{n_i}\right)^*. \tag{2.116}$$

However, C_{ij} and C_{ji} depend only on atomic properties and the electron velocity distribution function. Therefore, (2.116) holds valid also out of TE (as far as the electron velocity distribution is Maxwellian).

2.11 Non local Thermodynamic Equilibrium (NLTE)

LTE is a computational expedient that makes it possible to compute straightforwardly the absorption and emission coefficients via the Boltzmann and Saha laws. However, as discussed at the end of Section 2.8, LTE holds valid only under particular conditions that are not fulfilled in most cases. In order to compute the occupation numbers of the atomic populations out of LTE, we must resort to the statistical equilibrium equations (SEEs) introduced in Section 2.10. This introduces an essential difficulty, because the occupation numbers strongly depend on the radiation field, which in turn depends on the occupation numbers via the absorption and emission coefficients. This loop is sketched in Figure 2.10. Our aim is to compute the emergent flux or, when possible, the specific intensity of the radiation field at the stellar surface. To do that, we must solve the RT equation, which implies the need to know its source function, i.e., the absorption and emission coefficients that require the determination of the relevant occupation numbers via the SEEs. Among the coefficients of the latter, there are the radiative rates, which include integrals of the radiation field over the whole frequency range. This implies that the effect on the radiation field of a given transition may affect the rates of other transitions[21]. Moreover, atomic and molecular data are required through the whole problem, and their knowledge plays sometimes a critical role in the problem.

There are basically four ways to solve the coupling between the equations of statistical equilibrium and the radiative transfer equation:

- We can assume LTE, as in the first case of Section 2.5.1, so that the local source function is the Planck function $B_\nu(T)$, and the atomic populations are given by the Saha and Boltzmann equilibrium distributions.
- We can use the Λ-iteration (the succesive application of (2.53)) and start with an educated guess of the initial source function $S_\nu^{(0)}$ (e.g., the LTE source function) to compute the radiation field from the formal solution $J_\nu^{(0)} = \Lambda[S_\nu^{(0)}]$. After computing

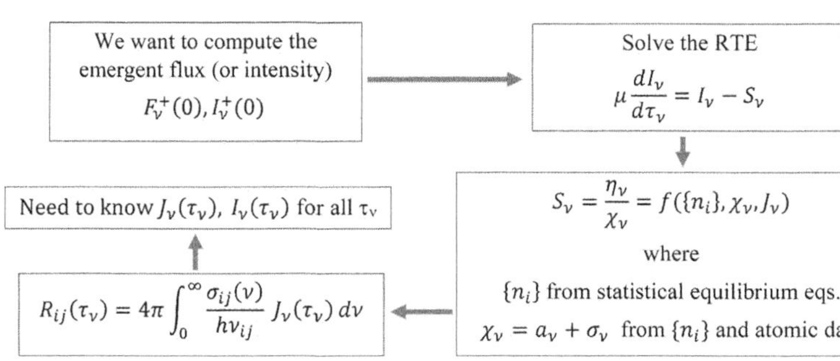

FIGURE 2.10. Scheme illustrating the coupling between the radiation field and the atomic populations.

[21] The integrals will be actually limited only to those frequencies for which the cross section of the transition is not zero. However, we prefer here to consider the full range to emphasize the interplay among different transitions.

the coefficients of the SEEs with these values, the updated source function $S_\nu^{(1)}$ is obtained, and the procedure iterated.

- It is well known, however, that the Λ-iteration converges too slowly, especially for optical thick transitions. In order to improve the convergence rate, accelerated lambda iteration (ALI) methods can be introduced. This issue is exhaustively discussed in Sections 3.7.4 through 3.7.10.
- A widely used alternative approach is complete linearization (CL), based on the Newton–Raphson method for the successively better approximation of the zeros of a real-valued function. If $\bar{x}$ is a solution of the equation $f(x) = 0$, the Taylor expansion of $f(x)$ around $\bar{x}$ allows us to write

$$f(\bar{x}) = \sum_{n=0}^{\infty} \left.\frac{d^n f}{dx^n}\right|_{x=x_0} (\bar{x} - x_0)^n = f(x_0) + \left.\frac{df}{dx}\right|_{x=x_0} + \cdots \approx f(x_0) + f'(x_0)\delta(x_0) = 0. \tag{2.117}$$

The solution of this equation is

$$\delta(x_0) = (\bar{x} - x_0) = -\frac{f'(x_0)}{f(x_0)}. \tag{2.118}$$

Thus, starting from an initial estimate x_0, we can calculate the difference between the latter and the correct solution $\bar{x}$, which will yield a new estimate $x_1 = x_0 + \delta(x_0)$.

In the case of the SEEs, starting from

$$\boldsymbol{A}_0\ \mathbf{n}_0 = \mathbf{b}_0, \tag{2.119}$$

we obtain

$$\left(\mathbf{n}_0 - \boldsymbol{A}_0^{-1}\mathbf{b}_0\right) + \sum_j \left.\frac{\partial\left(\mathbf{n} - \boldsymbol{A}^{-1}\mathbf{b}\right)}{\partial x_j}\right|_0 \delta(x_j) = \sum_j \left.\frac{\partial\left(\mathbf{n} - \boldsymbol{A}^{-1}\mathbf{b}\right)}{\partial x_j}\right|_0 \delta(x_j) = 0, \tag{2.120}$$

which is a system of equations whose unknowns are the variables x_j corresponding to the unknowns of the SEEs, namely the level occupation numbers and the value of the radiation field at any frequency of interest. The Newton–Raphson method is in general stable and has a high convergence rate, provided one starts with initial values near to the region of convergence (which in many cases requires a previous model atmosphere close to the target one). However, the solution of the SEEs implies a very large number of unknowns, a drawback that sets a limit to the number of atomic and molecular transitions that can be taken into account. Therefore, alternative techniques to reduce the dimension of the system equation must be envisaged. (See, for example, Mihalas, 1978, or Section 3.8.)

2.12 The Structure Equations

So far, we have treated the coupling between radiative transfer and the statistical equilibrium equations assuming a given temperature, pressure and density distribution inside the atmosphere. In the most general case, however, in order to determine such variables, we must compute a model atmosphere, and this requires three more equations. The first one will be the equation of state, given in our case by the perfect gas law. The other two will be the momentum and energy conservation equations, which we will reduce here to the hydrostatic and radiative equilibrium equations. Departure from hydrostatic equilibrium is considered in Section 5.2; at the end of this section, we will briefly discuss convection, the other form of energy transport that can occur in stellar atmospheres.

2.12.1 *Momentum Equation and Hydrostatic Equilibrium*

A mass element in the atmosphere is subject to forces that push it upwards and others that pull it inwards. If all the forces acting balance exactly, the element does not experience any acceleration and keeps at rest. Otherwise, it will be accelerated along the direction determined by the dominant force. We will assume in the following that no net horizontal force is present, so that the motion will be only along the radial direction.

Consider now that the outward forces are due to the gas and radiation pressure gradients and that the inward force is gravity. In terms of accelerations, we will have at any point r in the atmosphere

$$\frac{\partial v}{\partial t} + v\frac{\partial v}{\partial r} = -\frac{1}{\rho}\frac{dP}{dr} - \frac{Gm_r}{r^2} + g_{rad}, \tag{2.121}$$

where m_r is the mass contained inside the sphere of radius r, G is the gravitational constant and g_{rad} is the acceleration of the mass element due to radiation pressure, related to the second moment of the specific intensity. By proper manipulation, we obtain

$$\left(1 - \frac{a^2}{v^2}\right) v\frac{dv}{dr} = -\frac{da^2}{dr} + \frac{2a^2}{r} - \frac{Gm_r}{r^2} + g_{rad}. \tag{2.122}$$

The left member of the equation is determined by the velocity field (a is the sound velocity). The two first terms of the right member are due to the gas pressure, the third one to gravity and the fourth to radiation pressure. The term 2 a^2/r will always be small, and da^2/dr dominates. As the temperature decreases outwards, $-da^2/dr$ will be positive and have the same sign as g_{rad}, opposing gravity.

If the forces in the right member of (2.122) are in balance, the left member will be equal to zero, and we recover the equation of hydrostatic equilibrium

$$\frac{dP}{dr} = -\rho(r)\ (g - g_{rad}) = -\rho(r)\ g_{\text{eff}}. \tag{2.123}$$

If the column mass m (mass per unit area) is used as the independent variable, (2.123) takes on the simpler form

$$\frac{dP}{dm} = -g_{\text{eff}}. \tag{2.124}$$

The Pressure Scale Height

If we consider only the gas pressure in the simplifying instance of an isotermal atmosphere, we will have

$$dP_g = -g\ \rho\ dr, \tag{2.125}$$

hence

$$\frac{dP_g}{P_g} = -g\frac{\rho}{P_g}\ dr = -\frac{g\ \mu\ m_H}{k_B T}\ dr. \tag{2.126}$$

The left term in (2.126) comes from the equation of state $P_{gas} = (k_B T/m_H \mu)\rho$, where μ is the average mass of the particles in units of the hydrogen atom mass m_H.

By integrating at constant T, we obtain

$$\int_{P_0}^{P} \frac{dP_g}{P_g} = -\int_{r_0}^{r_1} \frac{g\ \mu\ m_H}{k_B T}\ dr, \tag{2.127}$$

from which it follows

$$\frac{P}{P_0} = e^{-(g\ \mu\ m_H/k_B T)(r_1 - r_0)} = e^{-\Delta r/H_p}, \tag{2.128}$$

hence

$$P = P_0\ e^{-\Delta r/H_p}. \tag{2.129}$$

The parameter H_p is the pressure scale height $H_p^{-1} = g\ \mu\ m_H/k_B T$, that is, the distance over which the pressure decreases by a factor $1/e$. For high g and low T, H_p is small, consequently the pressure will decrease rapidly over shorter distances. The result is a more 'compact' atmosphere, in which a large number of collisions will occour. Therefore, the internal conditions inside a compact stellar atmosphere will fairly approach LTE.

2.12.2 *Energy Equation*

Radiative Equilibrium

By integrating the radiative transfer equation

$$\mu \frac{I_\nu}{dr} = -\chi_\nu\,(I_\nu - S_\nu) \tag{2.130}$$

over angles and assuming that the absorption and emission coefficients, and hence the source function, do not depend on direction, we obtain

$$\frac{1}{2}\int_{-1}^{1} \mu \frac{dI_\nu}{dr}\ d\mu = \frac{dH_\nu}{dr} = -\frac{1}{2}\ \chi_\nu \int_{-1}^{1} (I_\nu - S_\nu)\ d\mu = -\chi_\nu\,(J_\nu - S_\nu)\,. \tag{2.131}$$

The successive integration over frequencies yields

$$\frac{d}{dr}\int_0^\infty H_\nu\ d\nu = -\int_0^\infty \chi_\nu\ \ (J_\nu - S_\nu)\ d\nu. \tag{2.132}$$

In a plane-parallel atmosphere, the flux integrated over all the frequencies, $H = \int_0^\infty H_\nu d\nu$, is constant with depth and therefore the condition of radiative equilibrium (RE)

$$\int_0^\infty (J_\nu - S_\nu)\ d\nu = 0 \tag{2.133}$$

is achieved.

The RE constraint is intrinsically linked with the temperature distribution inside the atmosphere via the source function. As discussed in the example (A) of Section 1.9.4, when the absorption and emission of radiation are due to both true absorption and scattering, the source function has the form

$$S_\nu = \frac{\kappa_\nu}{\kappa_\nu + \sigma_\nu}\ B_\nu + \frac{\sigma_\nu}{\kappa_\nu + \sigma_\nu}\ J_\nu, \tag{2.134}$$

where κ_ν is the absorption coefficient (bound–bound, bound–free and free–free processes) and σ_ν the coefficient that accounts for the scattering processes. By substitution into (2.132), it follows that

$$\int_0^\infty \kappa_\nu \; J_\nu \; d\nu = \int_0^\infty \kappa_\nu \; B_\nu \; d\nu. \tag{2.135}$$

This alternative formulation of the RE constraints shows two important physical facts: (i) the scattering terms have cancelled out, because scattering returns to the radiation field as much energy as it has removed; and (ii) the thermal emission $\int_0^\infty \kappa_\nu \; B_\nu \; d\nu$ is set at a value determined by the mean specific intensity, that is to say by the solution of the RT equation.

Therefore, (2.135) makes it possible to determine the temperature distribution $T(r)$. We start by assuming a given structure, defined by the values of $\kappa_\nu(r)$, $T(r)$ and hence $B_\nu[T(r)]$ at all the depth points in the atmosphere. Successively we solve the RT equation to obtain $J_\nu(r)$ and check whether the RE constraint is fulfilled. If it is so, the temperature distribution is consistent with radiative equilibrium; otherwise, we must update the input distribution T(r), using (2.135) as the corrector. Different methods can be used to correct the temperature. Among these are the inclusion of the RE equation of radiative equilibrium into the source functions of the RT equations and the statistical equilibrium equations, or the method of the thermal balance of electrons (Hubený and Mihalas, 2015; Kubát et al., 1999).

The Grey Atmosphere

Let us consider now the paradigm instance of a grey atmosphere (i.e., whose opacity is frequency independent) in LTE. We will make the further assumption that the Eddington approximation for the bolometric moments J and K, namely $J = 3K$, holds valid throughout the atmosphere. The well-known result is the grey temperature distribution

$$T^4(\bar{\tau}) = \frac{3}{4} \; T_{\rm eff}^4 \left(\bar{\tau} + \frac{2}{3} \right). \tag{2.136}$$

The Eddington approximation is correct at the bottom of the atmosphere, where the radiation field is almost isotropic. Thus the temperature structure given by (2.136) will be valid from the deepest layers outwards, until the depth where the radiation field departs markedly from isotropy. The exact solution of the grey problem was obtained by Hopf (1928), who derived the general form of the temperature distribution, that is,

$$T^4(\bar{\tau}) = \frac{3}{4} T_{\rm eff}^4 \left(\bar{\tau} + q(\bar{\tau}) \right), \tag{2.137}$$

where $q(\bar{\tau})$ is a slowly varying function of $\bar{\tau}$, called the Hopf function. It varies between $1/\sqrt{(3)} = 0.57$ at $\bar{\tau} = 0$ and 0.71 at $\bar{\tau} = \infty$. At the surface, we have $T(0) = 0.81 \; T_{\rm eff}$ ($0.84 T_{\rm eff}$ in the Eddington approximation) and $T(\bar{\tau} = 2/3) = T_{\rm eff}$.

In the top panel of Figure 2.11, we show the comparison of a model with $T_{\rm eff} = 8,000$ K calculated with the MARCS code (Gustavsson et al., 2008) in LTE and with plane-parallel geometry and the grey temperature structure. The agreement is in general good, particularly at the bottom of the atmosphere, where the conditions for the grey atmosphere are better fulfilled.

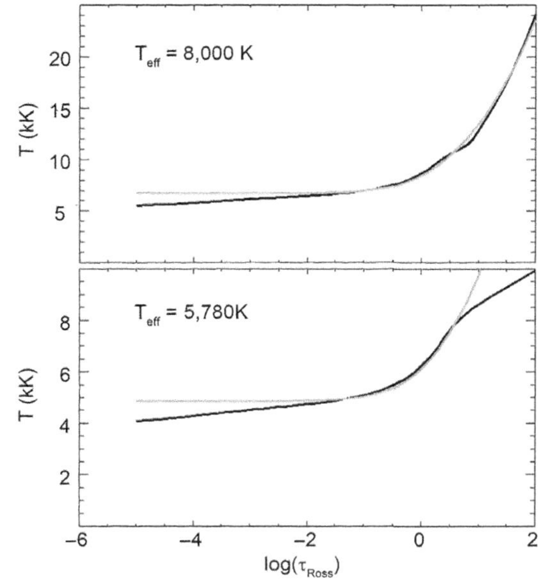

FIGURE 2.11. Top: temperature structure for a grey model atmosphere at $T_{\text{eff}} = 8000$ K (grey) and an MARCS model (black) at the same T_{eff}. We see the excellent agreement at the bottom of the atmosphere; bottom: same comparison for a solar model. The strong deviation at the bottom of the atmosphere illustrates the breakdown of radiative equilibrium.

Convection

In the bottom panel of Figure 2.11, the same comparion as in the top panel is shown, now for a solar mode with $T_{\text{eff}} = 5,780$ K. We note the strong departure of the grey atmosphere with respect to the MARCS model. The reason is that here radiative equilibrium breaks down: energy is transported not only by radiation, but also by convection. Radiative transfer carries on energy outwards, provided that the temperature gradient is large enough. The latter is determined by the bolometric luminosity L of the star, which imposes an upper limit. If this limit is exceeded, an instability sets up, which produces macroscopic motions inside the gas, known as *convection.* We will not discuss here the physics of convection, but only examine the conditions under which convection can develop in a stellar atmosphere.

Let us consider an element of gas, initially in equilibrium with its surroundings, that undergoes an upward displacement due to any small perturbation. When moving slowly towards the surface along a distance dr, the element remains in pressure equilibrium (*i.e.*, $P_{el} = P_{out}$) by expanding adiabatically. Because the pressure decreases outwards, the density ρ_{el} of the element will decrease, too. If at the new position $\rho_{el} > \rho_{out}$, gravity will pull the element downwards and equilibrium will be restored. Otherwise, the element will continue ascending until it is eventually destroyed, mixing with the environment and releasing its internal energy. By means of the perfect gas law, the equilibrium condition on density become a condition on temperature. As initially the two temperatures were equal, the change of temperature inside the element shall be larger than that of the outer temperature. We have assumed that the displacement is adiabatic and the surroundings are in radiative equilibrium, hence the condition for stability again convection will be $\mid \nabla T \mid_{ad} > \mid \nabla T \mid_{rad}$. By making use of the equation of state for the perfect gas, we can express the foregoing condition in the form first introduced by K. Schwartzschild, namely

$$\left| \frac{d\ lnT}{d\ lnP} \right|_{ad} > \left| \frac{d\ lnT}{d\ lnP} \right|_{rad}. \tag{2.138}$$

The Schwartzschild's criterion states that convective motions develop when the absolute value of the radiative gradient is larger than that of the adiabatic one. It shall be noted that the former is a spatial gradient of the temperature, while the latter represents the temperature variation within the element.

Recall that we consider a system consisting of matter and radiation, whose total pressure is the sum of the gas and the radiation pressure. Therefore, the adiabatic equations for the perfect gas must be generalized, and the adiabatic coefficient $\gamma = C_P/C_V$ is replaced by Chandrasekhar's adiabatic exponents Γ_1, Γ_2 and Γ_3 (Chandrasekhar, 1957, pages 55–59). We know that the radiative and the adiabatic gradients are given respectively by

$$\left(\frac{dT(r)}{dr} \right)_{rad} = \frac{3\kappa}{4acT^3} \frac{L(r)}{4\pi r^2} \tag{2.139}$$

and

$$\left(\frac{dT(r)}{dr} \right)_{ad} = \frac{\Gamma_2 - 1}{\Gamma_2} \frac{T(r)}{P(r)} dP(r)dr, \tag{2.140}$$

where κ is the opacity coefficient, $L(r)$ the stellar luminosity at point r, a the radiation constant and Γ_2 the second adiabatic exponent. If now we take into account (2.139) and (2.140) together with (2.126), it follows that

$$\nabla_{ad} \equiv \left(\frac{d\ lnT}{d\ lnP} \right)_{ad} = \frac{\Gamma_2 - 1}{\Gamma_2} = -\frac{H_p}{T} \left(\frac{dT}{dr} \right)_{ad}, \tag{2.141}$$

where H_p is the pressure scale height. For a monoatomic, nonrelativistic perfect gas, it holds that

$$\Gamma_2 = \frac{C_p}{C_v} = \frac{5}{3} = \gamma, \tag{2.142}$$

and all the adiabatic exponents are equal.

The preceding results show that convection is favoured wherever the opacity is large (k effect) or $\Gamma_2 \approx 1$ (Γ effect). Opacity becomes large when a notable fraction of hydrogen atoms are in their upper excitation levels, which happens at about the same conditions where ionization occurs with the consequent decrease of Γ_2, because of the departure from the condition of monoatomic gas. In those atmospheric regions where hydrogen (but also abundant elements such as helium or iron) changes their ionization degree, extensive convective zones develop. In the Sun, hydrogen recombines, going from the bottom to the upper layers (where it is mostly neutral) and convection appears. In hot stars, hydrogen remains fully ionized throughout the whole atmosphere, which is in radiative equilibrium. If conditions are favourable, abundant elements such as helium or iron may also originate convective zones.

So far we have considered an equation of state that links only the three variables P, T and ρ. If we take into account a change in the composition of the stellar material due to ionization, and characterize the composition by the average weight μ of the gas particles, the equation of state will be now in the form $\rho = \rho(P, T, \mu)$. Its differential expression is

$$d\ ln\ \rho = \alpha\ d\ ln p + \delta\ d\ lnT + \varphi\ d\ ln\mu. \tag{2.143}$$

The coefficients α, δ and φ are derived from the thermodynamics of adiabatic processes. The previous condition, given by (2.138), shall be replaced now by

$$\left|\frac{d\ lnT}{d\ lnP}\right|_{ad} + \frac{\varphi}{\delta}\left|\frac{d\ ln\mu}{d\ lnP}\right| > \left|\frac{d\ lnT}{d\ lnP}\right|_{rad}. \tag{2.144}$$

Equation (2.144) is the more general *Ledoux's criterion* (Ledoux, 1947) of stability.

To conclude this summary discussion, we will consider the case in which the convective elements do not stop suddenly, but continue their upward motion into regions where the criterion of stability is not fulfilled, until they eventually dilute into the environment, releasing their internal energy. This penetration of the convective elements into radiative layers is called convective overshooting, and usually parametrized as a fraction of the pressure scale height H_P, that is,

$$\alpha_{os} \equiv \frac{l_{os}}{H_P}, \tag{2.145}$$

where l_{os} is the path travelled inside radiative layers. The overshooting parameter α_{os} takes on usually values between 0.1 and 0.3 and is commonly fixed by fitting observed characteristics of stars and clusters (e.g., the position of the Terminal-Age Main Sequence), because it has a strong effect on stellar evolution.

2.13 Line Broadening

By solving the equations discussed in the previous sections, we can compute the emergent stellar flux and compare it with the observed spectrum. However, even if we use the right stellar parameters (effective temperature, gravity, abundances, etc.), the observed and the computed spectrum may be different. The reason is that spectral lines may be broadened by a variety of physical phenomena. This is illustrated in Figure 2.12, where the observed spectrum of HD 36591 is compared to the synthetic spectrum and calculated with the proper parameters but introducing a high rotational velocity that widens and blends the spectral lines.

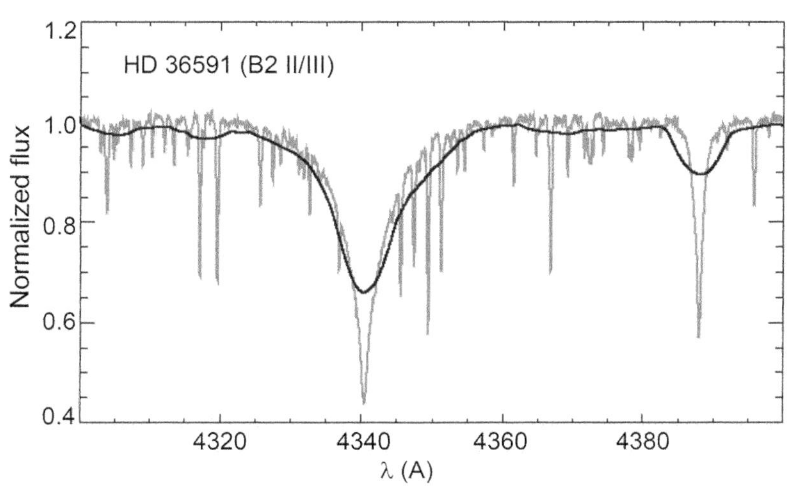

FIGURE 2.12. The observed (light grey) and synthetic (black) spectrum of HD 36591. The difference is due to the broadening of the spectral lines introduced by a projected rotational velocity of 300 km s^{-1}.

TABLE 2.1. The different broadening mechanisms according to their origin and view.

View	Origin:	Intrinsic	Extrinsic
Microscopic		Natural Collisional (pressure)	Thermal Microturbulence
Macroscopic			Rotation Macroturbulence

The broadening of the spectral lines may have its origin in the natural broadening of the atomic levels, consistent with Heisenberg's uncertainty principle (*intrinsic* origin), or in the Doppler effect due to the atomic motion (*extrinsic* origin).

In the first case, we know that

$$\Delta t \Delta E \geq \frac{h}{2\pi}, \tag{2.146}$$

where h is the Planck function, Δt the atomic level lifetime and ΔE the indetermination on energy. A bound–bound transition involving this atomic level will span a range in energy of at least ΔE (further indetermination may be introduced by the other level involved in the transition). If the lifetime of a level is reduced by some cause, the uncertainty on energy will increase with the consequent broadening of the spectral line.

In the second case, the broadening is due to the cumulative effect of an ensemble of atoms moving with a continuous velocity distribution, measured with respect to the observer's (or any other) reference frame. Atoms at rest will produce a spectral line centred at its nominal wavelength, while those moving with a nonzero velocity will originate a shifted line. The convolution of the atoms' individual absorption coefficients will result again in a broader spectral line.

Likewise, we can consider a microscopic picture when the broadening of the atomic energy levels affects the absorption coefficient itself, and a macroscopic picture when the atomic absorption coefficient is not modified. In the first case, the total energy absorbed in the spectral line is modified, whereas in the second it is not. Table 2.1 gives an overview of the cases that we will considered in the next sections.

2.13.1 *Microscopic Line Broadening*

Natural Broadening

We have already mentioned in Section 2.9.1 that, according to Heisenberg's principle of indetermination, excited atomic levels cannot be infinitely sharp because of their finite lifetime. Recalling (2.88), we can write the Lorentzian absorption profile due to natural broadening as

$$\alpha^{nat} = \frac{2\pi e^2}{mc}\frac{\gamma/2}{\Delta\omega^2 + (\gamma/2)^2} = \frac{e^2}{mc}\frac{\gamma/4\pi}{\Delta\nu^2 + (\gamma/4\pi)^2} = \frac{e^2\lambda^2}{mc^2}\frac{\gamma\lambda^2/4\pi c}{\Delta\lambda^2 + (\gamma\lambda^2/4\pi c)^2}. \tag{2.147}$$

The radiative damping constant γ for a transition from a lower-level l to an upper-level u, both excited, is $\gamma = \gamma_l + \gamma_u$, where γ_l and γ_u are the corresponding widths.[22]

[22] In a fundamental paper, Weisskopf and Wigner (1930) calculated the natural line width of spectral lines on the basis of Dirac's quantum theory of light. A summary of the subject can be found in Mihalas (1978, pages 277–278).

Collisional Broadening

The atoms in the stellar atmosphere will suffer encounters (collisions) with other atoms, ions, molecules and free electrons. These encounters will perturb the atomic levels, producing a shift in their energies that will depend on the nature of the interaction and the distance between the atom and the perturber. The combined effect of all these shifts will result in a broadening of the spectral lines. The total broadening will depend, in addition to the nature of the interaction and the inter particle distance, on the number of encounters. Therefore, this so-called collisional broadening (or pressure broadening) is a good indicator of the density (or pressure) in the stellar atmosphere, hence of the stellar gravity.

It is usually assumed that the energy shift of the atomic levels can be expressed as a power law like

$$\Delta\nu = C_n/R^n, \tag{2.148}$$

where n is an exponent that depends on the nature of the interaction, R is the distance between the perturbed atom and the perturbing particle and C_n is a constant that has to be calculated for each transition and type of interaction and that determines its strength.

The shape of the absorption profile is usually obtained by the impact approximation, where we assume that the duration of a collision is small compared with the time between collisions. We assume then that when an unperturbed atom emits a photon, it behaves as an oscillating dipole during a time interval Δt. The resulting photon can be represented as the product of a sinusoide and a box function. The resulting frequency spectrum (its Fourier transform, see e.g, chapter 2 of Gray (2008)) is a sinc function with an amplitude $sinc^2(\pi\Delta t(\nu-\nu_0))$ centered at the frequency of the unperturbed sinusoide ν_0 and with a width $\Delta\nu = 1/\Delta t$.

Now we assume that each perturbing particle produces a sudden phase shift in the sinusoide. The photon can be seen as the result of splitting the original wave in pieces, with a phase shift in each of them. The photon is now the sum of the partial sinusiodes. Its frequency spectrum is the sum of the individual Fourier transforms, each of them broader than the original one as $\Delta\nu_j = 1/\Delta t_j$. The resulting profile will be proportional to the $sinc^2$ function corresponding to each Δt_j, weighted by the probability distribution of the Δt_j and integrated over all possible Δt_j values, namely

$$\alpha_\nu \propto \int_0^\infty \Delta t^2 \left(\frac{sin\ \pi\ \Delta t(\nu-\nu_0)}{\pi\ \Delta t\ (\nu-\nu_0)}\right)^2 e^{-\Delta t/\Delta t_0}\frac{d\Delta t}{\Delta t_0}, \tag{2.149}$$

which gives

$$\alpha_\nu = const \times \frac{\gamma_n/4\pi}{(\nu-\nu_0)^2+(\gamma_n/4\pi)^2}, \tag{2.150}$$

which is a Lorentzian function with a damping parameter $\gamma_n = 2/\Delta t_0$ that depends on the nature of the interaction. The damping constant γ_n is related to the C_n constant in (2.148). Due to its Lorentzian profile, collisional broadening can be easily combined with natural broadening (with the same profile) through the convolution of their respective functions, whose damping parameters are γ_1 and γ_2. The result is another Lorentzian with total damping parameter $\gamma_t = \gamma_1 + \gamma_2$.

Table 2.2 gives a summary of the different types of collisional broadening that may occur in the stellar atmospheres, together with the index of the power law in (2.148). The Stark effect (the perturbation of the atomic energy levels by an external electric field) is important in atmospheres with a large number of free protons and electrons, i.e., in hot

TABLE 2.2. The different types of collisional broadening.

n	Type	Lines affected	Perturbers
2	Linear stark	H and hydrogenic atoms in hot stars	p^+, e^-
3	Resonance	All lines	Atoms of the same species
4	Quadratic stark	Nonhydrogenic atoms in hot stars	p^+, e^-
6	van der Waals	Atoms in cool stars	H

stellar atmospheres.[23] Resonance broadening is produced by encounters of atoms of the same species and will be important only for elements with a relatively high abundance (especially H). In most cases, however, impacts with H atoms are much more frequent (for instance, in cool atmospheres, where H is mostly neutral), and these interactions are governed by van der Waals forces. Approximated expressions for γ_3, γ_4 or γ_6 can be found for example in chapter 11 of Gray (2008) and chapter 8 of Hubený and Mihalas (2015), where more extensive discussions of the collisional broadening are presented.

The linear Stark broadening is particularly important for H atoms when there are free electrons and ions (it is also important for hydrogenic atoms, but decreases with the nuclear charge, so that in practice only HeII is also significantly affected). We have to consider both the slow protons and the fast electrons. The former can be treated in the nearest neighbour approximation (considering only the effect of the proton that is closest to the H atom and neglecting the rest, as their inlfuence will decrease with r^{-2}), although for accurate calculations we should consider the correct distribution of perturbers. Electrons are usually treated in the impact approximation. The accurate calculation of the Stark broadening profile results in complicated functions that require tabulation and thus cannot be easily treated together with other mechanisms that can be analytically represented (by a Lorentzian, as we have seen, or by a Gaussian, as we discuss in the next subsections).

Thermal Broadening

In a stellar atmosphere, atoms move with velocities (projected onto the observer's line of sight) due to thermal motion. This will produce a Doppler shift $\Delta\lambda$ in the wavelength (or frequency) of the absorbed or emitted photon. The distribution of $\Delta\lambda$ will be given by the velocity distribution, so that

$$\frac{dN}{N} = \frac{1}{\pi^{1/2}\Delta\lambda_D}\, e^{-(\Delta\lambda/\Delta\lambda_D)^2}\, d\lambda, \tag{2.151}$$

where dN/N is the fraction of atoms moving with velocities between v and $v + dv$ and $\Delta\lambda_D$ is the Doppler shift corresponding to the peak of the Maxwellian velocity distribution at temperature T for atoms of mass m, that is,

$$v_0 = \left(\frac{2k_B T}{m}\right) \tag{2.152}$$

and

$$\Delta\lambda_D = \frac{v_0}{c}\lambda_0 = 4.301 \times 10^{-7}\lambda_0(\frac{T}{\mu})^{1/2}, \tag{2.153}$$

[23] We will stress that the linear Stark broadening is stronger than the quadratic one. The terms *linear* and *quadratic* refer to the first and second order of approximation in the Hamiltonian considered in the perturbation theory.

where c is the speed of light and μ the atomic weight (in units of atomic mass) of the element considered.

The total energy absorbed by the line will be $(\lambda^2/c)(\pi e^2/mc)f$ (in λ units); in frequency units, it will be $(\pi e^2/mc)f$ as we have seen in (2.86). Thus the total absorption coefficient will be

$$\alpha_\lambda \, d\lambda = \frac{\pi^{1/2} e^2}{mc} f \frac{\lambda_0^2}{c} \frac{1}{\Delta\lambda_D} e^{-(\Delta\lambda/\Delta\lambda_0)^2} \, d\lambda \tag{2.154}$$

or, alternatively,

$$\alpha_\nu \, d\nu = \frac{\pi^{1/2} e^2}{mc} f \frac{1}{\Delta\nu_D} e^{-(\Delta\nu/\Delta\nu_0)^2} \, d\nu. \tag{2.155}$$

The profile of the line absorption coefficient, shaped by thermal broadening, is Gaussian.

Microturbulence

Let us assume that in the atmosphere there is an ensemble of cells moving in all directions with a Gaussian velocity distribution (as thermal motions), whose size is smaller than the photons' mean free path, as illustrated in Figure 2.13. Such a situation is known as *microturbulence.* The absorbing atoms will see the photons shifted according to the relative velocity between the emitting and the absorbing elements. As the velocity distribution of the latter is assumed to be Gaussian, the distribution of the relative velocities will be Gaussian, too. Therefore, the effect of microturbulence will be similar to that due to thermal broadening and the cumulative profile brought about by the sum of the two mechanisms will be akin to that of either (2.154) or (2.155), with velocity $v_{tot} = \sqrt{v_0^2 + \xi^2}$, where v_0 is the thermal velocity and ξ the microturbulent velocity.

We shall remark here that microturbulence has nothing to do with hydrodynamical turbulence. It is an ad hoc parameter introduced to fit the intensity of the observed spectral lines, particularly the strong ones. These lines can be saturated, that is, they cannot absorb more photons at the line centre or adjacent frequencies even if the number of absorbing atoms is increased. Microturbulence allows the atoms that cannot absorb at line centre to absorb at frequencies shifted from the core, increasing in such a way that the total amount of energy is absorbed by the line. In cool stars, microturbulence seems to be associated with convective motions and 3-D effects, whereas in hot stars its origin is still unclear, although it has been proposed to be associated to subsurface convective motions (Cantiello et al., 2009).

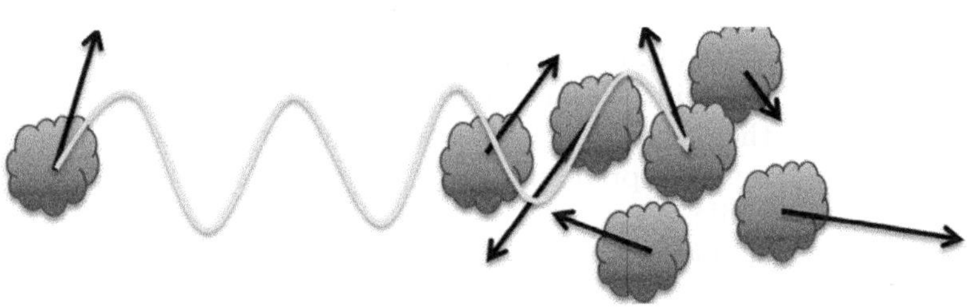

FIGURE 2.13. Schematic picture of the microturbulence assumption. Volume elements of sizes smaller than the photon mean free path move with a Gaussian velocity distribution (black arrows represent the velocity vector). The frequency of a photon (light grey) emitted by one of these elements is seen Doppler shifted when absorbed by another element. This shift adds to the shift due to the thermal motion of the atoms.

Hjerting and Voigt Functions

The global atomic coefficient will be the convolution of all the coefficients previously considered:

$$\alpha_{tot} = \alpha_{nat} * \alpha_{coll} * \alpha_{therm} * \alpha_{micro}. \tag{2.156}$$

Natural and collisional broadening jointly will yield a Lorentzian profile (but for the correct calculation of the linear Stark broadening), while thermal and microturbulent broadening will give a Gaussian profile. Then we will have

$$\alpha_\nu = \frac{\pi e^2}{mc} f \frac{\gamma/4\pi^2}{\Delta\nu^2 + (\gamma/4\pi)^2} * \frac{1}{\pi^{1/2}\Delta\nu_D} e^{-(\Delta\nu/\Delta\nu_D)^2}, \tag{2.157}$$

which can be written as

$$\alpha_\nu = \frac{\pi^{1/2}e^2}{mc}\frac{f}{\Delta\nu_D} H(u,a) \tag{2.158}$$

or

$$\alpha_\lambda = \frac{\pi^{1/2}e^2}{mc}\frac{\lambda_0^2 f}{\Delta\lambda_D} H(u,a), \tag{2.159}$$

where $H(u, a)$ is the Hjerting function, whose arguments are

$$u = \frac{\Delta\nu}{\Delta\nu_D} = \frac{\Delta\lambda}{\Delta\lambda_D} \tag{2.160}$$

and

$$a = \frac{\gamma}{4\pi\Delta\nu_D} = \frac{\gamma\lambda_0^2}{4\pi c\Delta\lambda_D}. \tag{2.161}$$

Note that u is related to the Gaussian part, whereas a contains the damping constant of the Lorentzian function.

The Hjerting function is

$$H(u,a) = \int_{-\infty}^{+\infty} \frac{\gamma/4\pi^2}{(\Delta\nu - \Delta\nu')^2 + (\gamma/4\pi^2)} e^{-(\Delta\nu'/\Delta\nu_D)^2} d\nu' = \frac{a}{\pi}\int_{-\infty}^{+\infty} \frac{e^{-u_1^2}}{(u-u')^2 + a^2} du'. \tag{2.162}$$

It can be expressed as a power expansion, namely

$$H(u,a) = H_0(u) + aH_1(u) + a^2H_2(u) + a^3H_3(u) + \cdots, \tag{2.163}$$

where the $H_j(u)$s are tabulated functions. $H_0(u)$ has a Gaussian form and is predominant at the profile centre, whereas in the far wings the even terms of the expansion tend to zero so that we can use the asymptotic approximation for the odd terms. In the wings, the Lorentzian part of the profile predominates, because a Lorentzian has more extended wings than a Gaussian profile.

Instead of the Hjerting function, it is customary to use the Voigt function

$$V(u,a) = \frac{1}{\pi^{1/2}\Delta\nu_D} H(u,a). \tag{2.164}$$

Thus the core of the line profile will be essentially Gaussian because of thermal or microturbulent broadening, while the wings will be predominantly Lorentzian, due to natural (when density is very low) or (very likely) to collisional broadening. In the case

Process	Natural	Collisional (or pressure)				Thermal	Microturbulence
		Resonance	van der Waals	Linear stark	Quadratic stark		
Shape	Lorentzian	Lorentzian	Lorentzian			Gaussian	Gaussian
Physics	Heisenberg uncertainty principle	Collisions with identical atoms	Collisions with other atoms	Ions and e- electric fields on hydrogenic atoms	Ions and e- electric fields on non-hydrogenic atoms	Atoms thermal velocity distribution	Heuristic: gas cells of smaller size than the photon mean free path.
Comment	Small broadening	Mainly collisions with H		Dominant in hot stars		Jointly with the Lorentzian, results a Hjerting or Voigt profile	

FIGURE 2.14. Summary of the spectral line broadening processes affecting the atomic absorption coefficient.

of Stark broadening, whose profile is neither Gaussian nor Lorentzian, we must perfom a numerical convolution with the Hjerting or Voigt function. A summary of the foregoing microscopic processes is given in Figure 2.14.

2.13.2 *Macroscopic Line Broadening*

Up to now we have seen processes that modify the atomic absorption profile, hence the total energy absorbed by the spectral line. We will now introduce two processes that alter the form of the line once it has been formed, without affecting the atomic coefficient nor modifying the total energy absorbed by the line.

Macroturbulence

Assume that in the atmosphere we have an ensemble of cells moving in all directions with a Gaussian velocity distribution (as the thermal motion) but now with a size *larger* than the photon mean free path, as illustrated in Figure 2.15. Now the emitting and absorbing atoms share the velocity of the cell and the only Doppler shifts to apply are those due to thermal motion. Each cell behaves as an independent atmosphere, and produces a spectrum similar to that of a static atmosphere (i.e., including all the broadening processes that we have seen), except that it will be shifted by the relative velocity between the cell element and the observer. However, as the cell elements have a Gaussian velocity distribution, each spectrum will be displaced by a different amount and the net result will be a broadened line. This is called *macroturbulence*. The total energy absorbed will be the same as in absence of macroturbulence. Note that, as in the case of microturbulence, macroturbulence is not necessarily related to hydrodynamic turbulence. It is introduced to explain the extra broadening observed in some lines. In cool stars, it is associated to the motion of large convective cells, whereas in hot stars its origin is still unclear (see, however, Simón-Díaz et al., 2010, Grassitelli et al., 2016, and Simón-Díaz et al., 2017 for recent lines of research).

As each cell behaves as an independent atmosphere, the emerging spectrum will be the convolution of the emergent intensity with the velocity distribution of the cells, namely

$$I_\nu = I_\nu^0 * \Theta(\Delta\lambda), \tag{2.165}$$

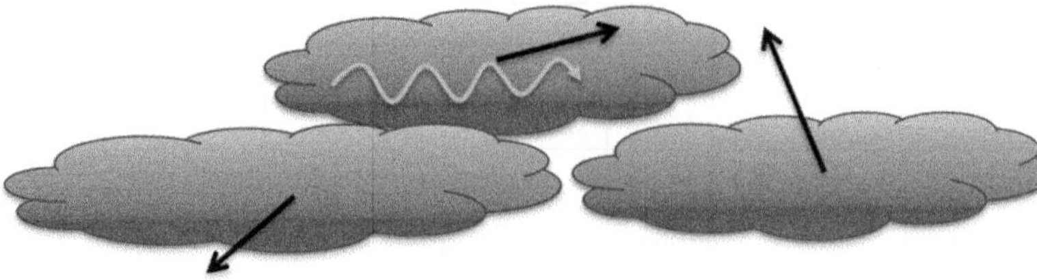

FIGURE 2.15. Illustration of the macroturbulence phenomenon. Cell elements of a size larger than the photon mean free path move with a Gaussian velocity distribution. The photon is emitted from one atom within one of those elements and is absorbed by another atom of the same cell. The relative velocity and corresponding Doppler shift will depend only on the relative velocity between the atoms due to their thermal motion. Black arrows represent the velocity vectors and the light grey sinusoide the photon.

so that the emergent flux will be

$$F_\nu = \oint I_\nu^0 * \Theta(\Delta\lambda)\ cos\theta\ d\Omega. \tag{2.166}$$

We have assumed in the preceding that the cells have a Gaussian velocity distribution; however, it is common to assume that actually it is matter of a radial-tangential distribution, because it fits better the observed profiles. That is,

$$\begin{aligned} \Delta(\lambda) &= A_R\ \Theta_R(\Delta\lambda) + A_T\Theta_T(\Delta\lambda) \\ &= \frac{A_R}{\pi^{1/2}\ \zeta_R\ cos\theta}\ e^{-(\Delta\lambda/\zeta_R cos\ \theta)^2} + \frac{A_T}{\pi^{1/2}\ \zeta_T\ cos\theta}\ e^{-(\Delta\lambda/\zeta_T cos\theta)^2}, \end{aligned} \tag{2.167}$$

where ζ_R and ζ_T are the macroturbulent velocities in the radial and tangential direction to the stellar surface, A_R and A_T are the areas covered by the corresponding cells (which may be taken as the weight we assign to each direction) and θ is the angle of the line of sight with the normal to the surface. It is usual to consider $A_R = A_T$ and $\zeta_R = \zeta_T$.

Rotation

When the star rotates, the emergent intensity $I_\nu(0,\mu)$[24] from each point of the stellar surface will appear Doppler shifted by an amount corresponding to the fraction of rotational velocity projected onto the line of sight. Points at the stellar limb will move radially towards or away from the observer, producing a maximum displacement (bluewards or redwards) of the emitted spectrum, affected also by limb darkening. Points at the centre of the stellar disk will move perpendicular the line of sight, producing no wavelength shift at all. The net result seen by the observer is a broadened spectral line, as illustrated in Figure 2.16. Finally, we have to consider that if the rotational axis of the star is inclined with respect to the line of sight by an angle i, the maximum velocity observed will be $v\ sin\ i$.

Similarly to the case of macroturbulence, the emergent flux will be given by the convolution of the original emergent spectrum with the corresponding broadening profile (here the rotation profile, $G(\lambda)$), namely

$$\mathcal{F}_\lambda = \int_{-\infty}^{+\infty} I_\nu(\lambda - \Delta\lambda)\ G(\Delta\lambda)\ d\Delta\lambda = I_\lambda\ *\ G(\lambda). \tag{2.168}$$

[24] Broadened by all the foregoing microscopic processes.

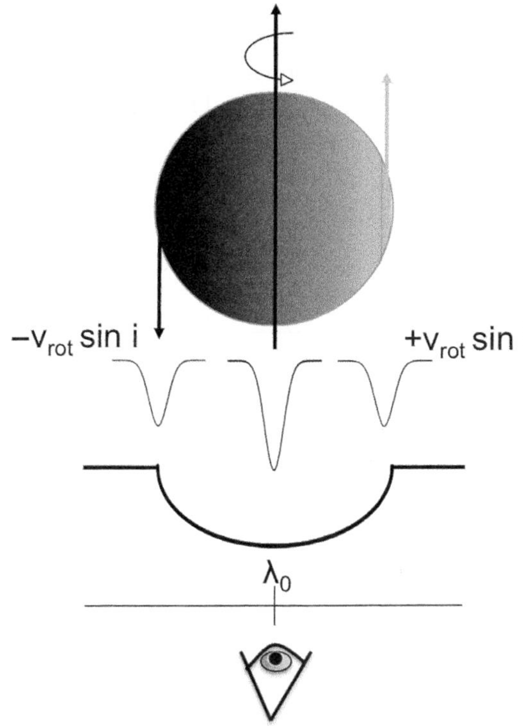

FIGURE 2.16. Illustration of the rotational broadening. Points approaching and receding from the observer will produce blue- and redshifted spectra, with a lower flux contribution due to the limb darkening effect. The maximum displacement will be given by the projection of the rotational velocity on the line of sight. The final effect as seen by the observer will be a broadened spectral line.

The derivation of the rotation profile can be found, e.g., in chapter 18 of Gray (2008). The result is

$$G(\Delta\lambda) = \frac{2(1-\epsilon)\left[1-(\Delta\lambda/\Delta\lambda_L)^2\right]^{1/2} + \frac{1}{2}\,\pi\epsilon\,\left[1-(\Delta\lambda/\Delta\lambda_L)^2\right]}{\pi\,\Delta\lambda_L(1-\epsilon/3)}, \tag{2.169}$$

where $\Delta\lambda_L = \lambda/c\ v\ sin\ i$ is the maximum Doppler shift produced by the projected rotational velocity, and ϵ is the coefficient of the limb darkening law, $I_c = I_c^0(1-\epsilon+\epsilon\ cos\ \theta)$.

The rotational profile is completely different from the macroturbulent one, and has a more parabolic form, so that a spectral line broadened by rotation joins with the local continuum in a steeper way than one broadened by macroturbulence. When considered together, rotation and macroturbulence have to be convolved.

2.13.3 *Summary of Broadening Processes*

Figure 2.17 gives an overview of all broadening processes that we have seen. Note that this is not a complete list, and other effects not considered here may be present, such as magnetic fields, pulsations, surface inhomogeneities, geometrical distortions or effects of stellar companions, that may contribute in different ways to broaden or modify the shape of spectral lines in stars. Nevertheless, the equations, methods and physical processes that we have seen in the present chapter allow us to reproduce the spectra of a large

	Natural	Collisional	Thermal	Rotation	Microturb.	Macroturb.
Origin	Uncertainty principle	Atomic interaction	Thermal velocity dispersion	Stellar rotation	Convection? (cool stars)	Convection (cool stars) Pulsation? (hot stars)
Shape	Lorentzian	Lorentzian (mostly)	Gaussian	own	Gaussian	Radial-tangential
Parameter	A_{ik}, τ	Gravity (pressure)	T	Rotational velocity	Abundances (needed for)	
Changes absorbed energy?	Yes	Yes	Yes	No	Yes	No
Objects	Atoms	Cool stars (van der Waals); hot stars (Stark)	Low-density hot plasmas	Stars	Stars	Stars

FIGURE 2.17. Summary of the spectral lines broadening processes seen in the present chapter. A_{ik} and τ are the Einstein coefficient for spontaneous emission and the lifetime of the atomic energy level.

number of stars, thus making possible the determination of their stellar parameters and physical properties.

REFERENCES

Cantiello, M., Langer, N., Brott, I., et al. 2009. *Astron. Astrophys*, **499**, 279

Chandrasekhar, S. 1957. *An Introduction to the Study of Stellar Structure.* Dover

Crivellari, L. 2016a. *Eur. J. Physics*, **37(5)**, 055404

Crivellari, L. 2016b. *Eur. J. Physics*, **37(5)**, 055405

Crivellari, L. 2018. *Serb. Astron. J.*, **196**, 1

Dubernet, M. L., Zwölf, C. M., Moreau, N., et al. 2016. *Journal of Physics B*, **49**, 7

Einstein, A. 1917. *Phys. Zeit.*, **18**, 121

Grassitelli, L., Fossati, L., Langer, N., et al. 2016. *Astron. Astrophys,* **593**, A14

Gray, D. 2008. *The Observation and Analysis of Stellar Photospheres*, 3rd edition. Cambridge University Press

Gustafsson, B., Edvardsson, B., Eriksson, K., et al. 2008. *Astron. Astrophys,* **486**, 951

Hopf, E. 1928. *Zeitschrift für Phys.*, **46**, 374–382

Kourganoff, V. 1963. *Basic Methods in Transfer Problems.* Dover

Kubát, J., Puls, J. and Pauldrach, A. W. A. 1999. *Astron. Astrophys,* **341**, 587

Hubený, I. and Mihalas, D. 2015. *Theory of Stellar Atmospheres (An Introduction to Astrophysical Non-Equilibrium Quantitative Spectroscopic Analysis).* Princeton University Press

Ladenburg, R. 1921. *Zeit. Phys.*, **4**, 451, English transl. in van der Waerden, B. L., 1967 *Sources of Quantum Mechanics.* Dover

Ledoux, P. 1947. *Astrophys. J.*, **105**, 305

Lorentz, H. 1900. *Rapports préesentés au Congrès International de Physique*, **3**, 1

Mihalas, D. 1978. *Stellar Atmospheres*, 2nd edition, W. H. Freeman and Co.

Milne, E. A. 1924. *Phil. Mag. Series 6, Vol. 47*, **277**, 209

Payne, C. H. 1925. *Stellar Atmospheres.* Harvard Observatory Monographs, Shapley, H., ed.

Simón-Díaz, S., Herrero, A., Uytterhoeven, K., et al. 2010. *Astrophys J.,* **720**, L174

Simón-Díaz, S., Godart, M., Castro, N., et al. 2017. *Astron. Astrophys,* **597**, A22

Urbaneja, M. A. 2004. *B Supergiants in the Milky Way and Nearby Galaxies: Models and Quantitative Spectroscopy.* PhD Thesis, Instituto de Astrofísica de Canarias, ISBN: 84-689-0021-4

Weisskopf, V. and Wigner, E. 1930. *Zeitschrift für Physik,* **63**, *54*

Wildt, R. 1939. *ApJ*, **89**, 295.

3. Numerical Methods in Radiative Transfer

OLGA ATANACKOVIĆ

Abstract

This chapter considers a selection of numerical methods developed since 1960s for solving radiative transfer (RT) problems in stellar atmospheres and in all other diluted media where nonlocal thermodynamic equilibrium (NLTE) effects are important. Special emphasis is put on the solution of the radiative transfer equation (RTE) when the source function is given, because its so-called formal solution constitutes a necessary step in any iterative procedure for the solution of more general RT problems. The application of different methods to the spectral line formation problem, which requires the self-consistent solution of the RTE for the line(s) radiation field and the statistical equilibrium (SE) equation(s) for the atomic-level populations involved, is discussed for both linear and nonlinear problems.

3.1 Introduction

RT offers a link between the observed macroscopic properties of distant heavenly bodies (i.e., the radiant flux they emit) and the microscopic interactions between photons and gas particles that determine their physical conditions. The transport of radiation through a medium is governed by the RTE that describes the changes in the specific intensity of the radiation field as a result of its complex interactions with the medium it propagates through. When the sources and sinks of photons, expressed in terms of the emission and absorption coefficients, are known, the RTE reduces to a liner first-order differential equation with known coefficients, easy to solve. Because this case, referred to as the *formal solution* of the RTE, is the backbone of any iterative technique developed for more complex problems, numerical methods for the formal solution are described in Section 3.4.

It is impossible to consider here all the methods developed to solve general RT problems, in which the sources and sinks of photons are not known because they depend on the radiation field itself. We can, however, learn a lot by restricting ourselves to the methods developed for the problem of RT in spectral lines, which are the main diagnostic tool for interpreting astronomical spectra. The efficient solution of the line formation problem is one of the most challenging tasks of astrophysics. In general, the nonlocal coupling of the radiation field with the state of the medium requires the simultaneous solution of the RTEs and the SE equations for the level populations of the particles constitutive of the medium, which include integrals of the specific intensity of the radiation field over all the frequencies and directions. Thus under the conditions of NLTE, the RT problem becomes an integro-differential one. Different forms of the dependence of the absorption and emission coefficients on the radiation field bring about different mathematical problems. When this dependence is linear (e.g., in the Two-Level Atom case), the NLTE problem can be solved by using either direct or iterative methods. On the contrary, when the coupling is nonlinear (e.g., in the Multilevel Atom problem), iterative procedures are required. A number of powerful techniques for the solution of linear and nonlinear line transfer problems are described in Sections 3.5 through 3.7 and 3.8, respectively.

Despite the important advances of the last decades, the search for new, fast and reliable numerical methods for solving RT problems still deserves dedicated attention.

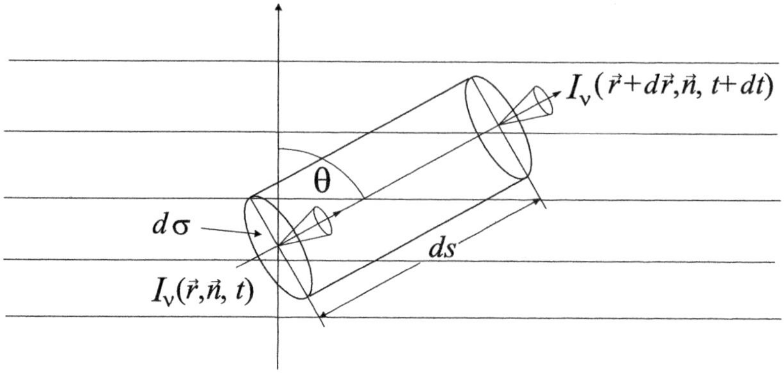

FIGURE 3.1. A radiation beam passing through the element of basis $d\sigma$ and height $ds = cdt$.

3.2 Radiative Transfer Equation (RTE)

We will derive and discuss briefly the general form of the RTE. Let us consider an element of material of length ds and cross section $d\sigma$ that absorbs and emits radiation (see Figure 3.1).

The beam of radiation incident normal to $d\sigma$ at position $\mathbf{r}$ and time t, propagating in direction $\mathbf{n}$ inside a solid angle $d\omega$ with frequency in the range $(\nu, \nu + d\nu)$ is described by its specific intensity $I(\mathbf{r}, t; \mathbf{n}, \nu)$.[1] The specific intensity in the same frequency interval and within the same solid angle emerging from the element at $\mathbf{r} + d\mathbf{r}$ at time $t + dt$ will be denoted by $I(\mathbf{r} + d\mathbf{r}, t + dt; \mathbf{n}, \nu)$. For a volume element of basis $d\sigma$ perpendicular to $\mathbf{n}$ and height ds, the difference between the amount of energy that leaves the element at position $\mathbf{r} + d\mathbf{r}$ at time $t + dt$ and the amount incident at position $\mathbf{r}$ at time t must equal the difference between the amount of energy emitted by the material and the amount absorbed, that is,

$$\begin{aligned}&[I(\mathbf{r} + d\mathbf{r}, t + dt; \mathbf{n}, \nu) - I(\mathbf{r}, t; \mathbf{n}, \nu)]d\sigma d\omega d\nu dt \\ &= [\eta(\mathbf{r}, t; \mathbf{n}, \nu) - \chi(\mathbf{r}, t; \mathbf{n}, \nu)I(\mathbf{r}, t; \mathbf{n}, \nu)]d\sigma ds d\omega d\nu dt. \qquad (3.1)\end{aligned}$$

Energy can be removed from the beam by both thermal absorption and scattering, so that the total macroscopic extinction coefficient $\chi(\mathbf{r}, t; \mathbf{n}, \nu)$ can be written as $\chi = \chi^{\text{th}} + \chi^{\text{s}}$. Likewise, the energy of the beam can be increased either by thermal emission or scattering of radiation from other directions. The total emission coefficient $\eta(\mathbf{r}, t; \mathbf{n}, \nu)$ will be given by $\eta = \eta^{\text{th}} + \eta^{\text{s}}$.[2]

The change in the specific intensity can be written as

$$dI = I(\mathbf{r} + d\mathbf{r}, t + dt; \mathbf{n}, \nu) - I(\mathbf{r}, t; \mathbf{n}, \nu) = \frac{dI}{ds} ds, \qquad (3.2)$$

where dI/ds is the total derivative of I with respect to the path length ds along the ray specified by the unit vector $\mathbf{n}$. Taking into account the variation of I with respect to all the variables, i.e., the space coordinates r_i $(i = 1, 3)$, the angles θ and φ that define the direction $\mathbf{n}$, the frequency ν and time t, (3.1) can be cast into the form

[1] The definition of specific intensity of the radiation field can be found in Section 1.5.1.
[2] The macroscopic radiative transfer coefficients χ and η are defined in Section 1.8.

$$\frac{dI}{ds} = \frac{\partial I}{\partial t}\frac{\partial t}{\partial s} + \sum_{i=1}^{3} \frac{\partial I}{\partial r_i}\frac{\partial r_i}{\partial s} + \frac{\partial I}{\partial \theta}\frac{\partial \theta}{\partial s} + \frac{\partial I}{\partial \varphi}\frac{\partial \varphi}{\partial s} + \frac{\partial I}{\partial \nu}\frac{\partial \nu}{\partial s}$$
$$= \eta(\mathbf{r}, t; \mathbf{n}, \nu) - \chi(\mathbf{r}, t; \mathbf{n}, \nu) I(\mathbf{r}, t; \mathbf{n}, \nu) \tag{3.3}$$

that holds valid in any inertial frame, both in static and moving media. Since s is a coordinate-independent path length, (3.3) applies in arbitrary coordinate systems (both Cartesian and curvilinear). As ds denotes an infinitesimal path length along the ray $\mathbf{n}$, it holds that $\partial t/\partial s = 1/c$.

If the coefficients χ and η are known, the RTE, whose most general form is given by (3.3), is a first-order partial differential equation (PDE), subject to boundary conditions. The method of characteristics makes it possible to transform a PDE into a family of ordinary differential equations (ODE), each one with its own initial condition. The ODEs can be integrated along the so-called characteristic curves (or just *characteristics*) to be identified here with the rays, as defined in Section 1.3.

In the most general case, radiative transfer is a seven-dimensional problem, as the specific intensity depends on three spatial coordinates, two angles, frequency and time. For a numerical solution, the discretization of all these variables is required. Even in the simplest cases, grids of an exceedingly large number of points become necessary, hence the need for opportune simplifications that we are going to examine in the following.

3.2.1 *Time Dependence*

Radiative transfer is, in general, a time-dependent process that needs to be treated simultaneously with (magneto)hydrodynamics. However, as the photon's flight time (the timescale of propagation over a mean free path length) is orders of magnitude shorter than any hydrodynamical characteristic timescale in a stellar atmosphere, the radiation field can be regarded as quasistationary (Hayek et al., 2010) and radiative transfer as a time-independent problem, hence $\partial/\partial t = 0$.

3.2.2 *Dynamics*

In a static medium, photons move on straight paths (rays). The opacity is isotropic, whereas the emissivity may depend on direction because of scattering. If the latter is isotropic, the emissivity will be isotropic, too.

For a moving medium, two different frames can be considered: the *laboratory* (either observer's or fixed) frame, in which the medium is seen to move, and the *comoving* (Lagrangian) frame (CMF), namely a frame moving with the medium. In the laboratory frame, photons are seen to move with constant direction and frequency, so that the differential operators in the left-hand side (LHS) of the RTE (3.3) take on a simpler form. However, due to a Doppler shift between frequencies measured in the observer's frame and the radiating atom's frame, i.e., because of the dependence of the frequency in the atom's frame on direction, the opacity and emissivity of the material are anisotropic. On the contrary, in the comoving frame, the source and sink terms in the right-hand side (RHS) of (3.3) are independent of motion, i.e., they are the same as in the static case, whereas the differential operators become quite awkward as they depend on the velocity field. The advantages and disadvantages of using either the observer's frame or the CMF are thoroughly discussed by Hubený and Mihalas (2015).

In this chapter, we will limit ourselves to the time-independent ($\partial/\partial t = 0$) and static ($\mathbf{v} = 0$) case. We are then left only with the spatial change of the specific intensity in the LHS of (3.3). The mathematical expression for $\partial/\partial s$ will depend on geometry. Since $(\mathbf{n})_i = \partial r_i/\partial s$, the sum in the LHS of (3.3) is the expansion of the scalar product $\nabla I \cdot \mathbf{n}$, which is by definition the *directional derivative* of I along $\mathbf{n}$. In curvilinear coordinates, due to the changes in the components of the propagation vector $\mathbf{n}$ along the coordinate basis vectors, $\partial/\partial s = \mathbf{n}\cdot\nabla + (d\mathbf{n}/ds)\cdot\nabla_{\mathbf{n}}$, where $\nabla_{\mathbf{n}}$ denotes differentiation with respect to the direction cosines of $\mathbf{n}$.[3] Hence, in the laboratory frame the time-independent RTE has the following form:

$$\Big[\mathbf{n}\cdot\nabla + (d\mathbf{n}/ds)\cdot\nabla_{\mathbf{n}}\Big] I_\nu(\mathbf{r},\mathbf{n}) = \eta_\nu(\mathbf{r},\mathbf{n}) - \chi_\nu(\mathbf{r},\mathbf{n}) I_\nu(\mathbf{r},\mathbf{n}). \tag{3.4}$$

3.2.3 *Geometry*

We will discuss here the simplifications made possible by the geometry and symmetry assumed.

At a first-order approximation, stellar atmospheres show a certain degree of symmetry. Giant and supergiant stars have extended outer layers comparable in size with their radii, and their atmospheres can be regarded as spherically symmetric. In the case of main sequence stars, however, the thickness of their atmosphere is very small compared to their radius so that curvature effects are negligible and we can assume that the atmosphere is stratified in plane-parallel layers. In the following, we will consider the RTE in both Cartesian and spherical coordinates, according to the geometry dictated by the symmetry assumed.[4]

(i) Cartesian Coordinates

In Cartesian coordinates, the propagation unit vector $\mathbf{n}$ is expressed by

$$\mathbf{n} = n_x\mathbf{e}_x + n_y\mathbf{e}_y + n_z\mathbf{e}_z = \sin\theta\cos\varphi\,\mathbf{e}_x + \sin\theta\sin\varphi\,\mathbf{e}_y + \cos\theta\,\mathbf{e}_z, \tag{3.5}$$

where θ is the polar angle formed by $\mathbf{n}$ and the z-axis, while φ is the azimuthal angle between the projection of $\mathbf{n}$ onto the (x, y) plane and the x-axis, measured counterclockwise from the latter. The unit vectors $\mathbf{e}_x$, $\mathbf{e}_y$ and $\mathbf{e}_z$ point to the direction of the three axes.

In the laboratory frame, the time-independent RTE in Cartesian coordinates has a relatively simple form. Under these assumptions, the RTE (3.4) reduces to

$$\begin{aligned}\Big[(1-\mu^2)^{1/2}\cos\varphi\frac{\partial}{\partial x} + (1-\mu^2)^{1/2}\sin\varphi\frac{\partial}{\partial y} + \mu\frac{\partial}{\partial z}\Big] I_\nu(x,y,z;\theta,\varphi) \\ = \eta_\nu(x,y,z;\theta,\varphi) - \chi_\nu(x,y,z;\theta,\varphi) I_\nu(x,y,z;\theta,\varphi),\end{aligned} \tag{3.6}$$

where $\mu \equiv \cos\theta, (-1 \le \mu \le 1)$. Equation (3.6) is a PDE in the three spatial coordinates (x, y, z) with three independent parameters: the two angles θ, φ and the frequency ν.

Let us now assume that the atmosphere is stratified in *horizontally homogeneous* plane-parallel layers, whose outward normal is oriented along the z-axis (Figure 3.2, left). Therefore, the problem becomes one-dimensional as all the quantities depend on the z-coordinate only. Because of the azimuthal symmetry, the specific intensity depends only on the zenithal (polar) angle $\theta = \mathbf{e}_z\cdot\mathbf{n}$. Hence the corresponding RTE takes the form

$$\mu\frac{dI_\nu(z,\mu)}{dz} = \eta_\nu(z,\mu) - \chi_\nu(z,\mu) I_\nu(z,\mu). \tag{3.7}$$

[3] The changes are due to the rotation of basis vectors with respect to the straight-line path determined by a vector $\mathbf{n}$. For details, see section 6.4 in Mihalas and Weibel Mihalas (1984).

[4] For more details on different geometries (including cylindrical geometry that is useful in the modelling of accretion disks and jets), see section 11.2 in Hubený and Mihalas (2015) and the paper by van Noort et al. (2002).

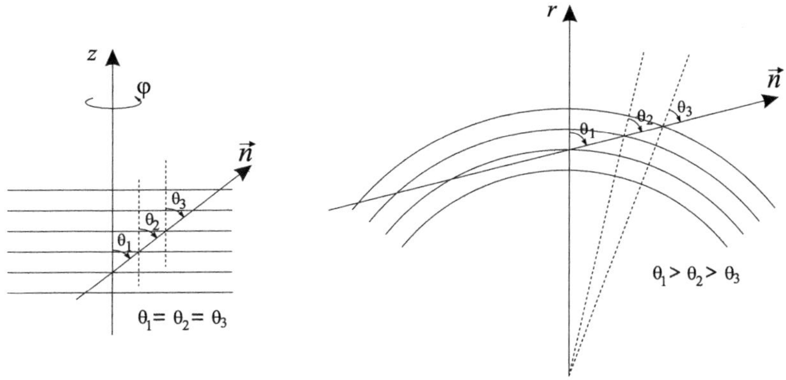

FIGURE 3.2. Plane-parallel vs. spherical geometry.

Each *specific* ordinary differential equation (3.7) is characterized by the pair of parameters (ν, μ).

(ii) Spherical Coordinates

For a spherical medium, in which the position of a point P is specified by the three spherical coordinates (r, Θ, Φ), the time-independent RTE in the laboratory frame is the PDE

$$\left[\cos\theta\frac{\partial}{\partial r} + \frac{\sin\theta\cos\varphi}{r}\frac{\partial}{\partial\Theta} + \frac{\sin\theta\sin\varphi}{r\sin\Theta}\frac{\partial}{\partial\Phi} - \frac{\sin\theta}{r}\frac{\partial}{\partial\theta} - \frac{\sin\theta\sin\varphi\cot\Theta}{r}\frac{\partial}{\partial\varphi}\right] \tag{3.8}$$
$$\times\ I_\nu(r,\Theta,\Phi;\theta,\varphi) = \eta_\nu(r,\Theta,\Phi;\theta,\varphi) - \chi_\nu(r,\Theta,\Phi;\theta,\varphi)I_\nu(r,\Theta,\Phi;\theta,\varphi),$$

where the angles θ and φ specify the direction of propagation $\mathbf{n}$.

In a spherically symmetric atmosphere, all the quantities depend on the r-coordinate only. If the atmosphere is stratified in homogeneous layers, they are also independent of φ. However, let us note that θ varies along the ray $\mathbf{n}$, i.e. $\frac{\partial\theta}{\partial s} \neq 0$ (see Figure 3.2, right). Hence only the first and fourth terms in the LHS of (3.8) are to be kept and the RTE becomes a PDE in the two variables, r and $\mu \equiv \cos\theta$, and is specified by the pair of parameters (ν, μ), namely

$$\left[\mu\frac{\partial}{\partial r} + \frac{1-\mu^2}{r}\frac{\partial}{\partial\mu}\right]I_\nu(r,\mu) = \eta_\nu(r,\mu) - \chi_\nu(r,\mu)I_\nu(r,\mu). \tag{3.9}$$

It must be stressed that μ plays here a double role, both as a variable and a parameter that characterizes any specific RTE.

3.3 Coupling of the RTEs and the Boundary Conditions

With the introduction of the monochromatic optical depth, defined by the differential relation

$$d\tau_\nu(z) = -\chi_\nu(z)\,dz, \tag{3.10}$$

and the source function, defined as the ratio of the total emissivity to the extinction coefficient, which are isotropic in most cases, that is,

$$S_\nu(\tau_\nu) \equiv \frac{\eta_\nu(\tau_\nu)}{\chi_\nu(\tau_\nu)}, \tag{3.11}$$

the RTE for a time-independent, static atmosphere stratified in homogeneous plane-parallel layers becomes [5]

$$\mu \frac{dI_\nu(\tau_\nu, \mu)}{d\tau_\nu} = I_\nu(\tau_\nu, \mu) - S_\nu(\tau_\nu). \tag{3.12}$$

The total emissivity $\eta_\nu = \eta_\nu^{\text{th}} + \eta_\nu^{\text{s}}$ is the sum of the thermal emission plus the contribution due to scattering. Under the assumption of local thermodynamic equilibrium (LTE) for matter, the thermal contribution is given at any point by the Kirchhoff–Planck relation $\eta_\nu^* = \chi_\nu^* B_\nu(T)$, where the asterisk denotes the LTE values that define the local thermodynamic state of matter. On the other hand, the anisotropy revealed by the presence of a flux shows the departure of the radiation field from its equilibrium distribution given by the Planck's law. As the values of the specific intensity are determined by the RTE, which accounts for the transport of photons, the scattering term includes the solution itself of the RTE. Consequently, the radiation field is decoupled from the local thermodynamic conditions of the medium and the physical state of the whole system at any point depends on those at all the other points via radiative transfer.

If the coefficients η_ν and χ_ν are known, hence the source function, the solution of a set of *independent* (uncoupled) RTEs (3.12), constitutes an *initial value problem.* However, as stated previously, in the general case all the specific RTEs are entangled in the source functions through the scattering integral and form a system of coupled integro-differential equations. Since the emissivity at one point depends, via scattering, on radiation coming from all the directions (along all the rays) passing through that point, the radiative transfer problem becomes a *boundary value problem* to be solved for all the rays at once. Thus, the solution of the system of equations (3.12) has to be carried out subject to certain *boundary conditions.*

Two cases of astrophysical importance are (i) a slab (in plane-parallel geometry) or a shell (in spherical geometry) of finite thickness; and (ii) a semi-infinite stellar atmosphere, i.e., a medium with an open upper boundary and of such a high opacity in the innermost layers that its lower boundary layer can be assumed to be at infinite optical depth.

(i) For a slab of total optical thickness T_ν ($\tau_\nu = 0$ at the side closer to the observer), the solution can be obtained if the intensities of radiation field incident on both faces of the slab are given. That is,

$$I_\nu^-(0, \mu) = I_\nu(\tau_\nu = 0, \mu < 0) \tag{3.13}$$

$$I_\nu^+(T_\nu, \mu) = I_\nu(\tau_\nu = T_\nu, \mu > 0). \tag{3.14}$$

(ii) For a semi-infinite atmosphere of a single star, we usually assume that there is no incident radiation at the surface

$$I_\nu^-(\tau_\nu = 0, \mu) = 0. \tag{3.15}$$

The lower boundary condition (at $\tau_\nu \gg 1$) is expressed by the boundedness requirement that

$$\lim_{\tau_\nu \to \infty} I_\nu(\tau_\nu, \mu) e^{-\tau_\nu/\mu} = 0. \tag{3.16}$$

[5] We will assume plane-parallel geometry throughout the text. However, let us note that the use of the RTE along the ray is often much more advantageous as it is not restricted to any particular symmetry.

In the numerical solutions, the so-called diffusion approximation is usually applied as the lower boundary condition, i.e., the intensities are expressed in terms of the local source function and its first-order derivative[6]:

$$I_\nu(\tau_\nu, \mu) = S_\nu(\tau_\nu) + \mu \frac{dS_\nu}{d\tau_\nu}. \tag{3.17}$$

3.4 Formal Solution of the RTE

The formal solution of the RTE under given boundary conditions is by definition its solution when the absorption and emission coefficients (hence the source function) are given. It not only yields the radiation field in a medium of prescribed properties, but also it constitutes a necessary step in any iterative procedure for the self-consistent computation of the radiation field and the state of the medium. In the following, we are going to consider the formal solution of the RTE (3.12) both in differential and integral form for a stellar atmosphere, whose thermal structure is assumed to be known.

3.4.1 *Second-Order Differential RTE and Three-Point Difference Equation Scheme*

The two-point boundary nature of the RT problem, namely the existence of two separate sets of boundary conditions, suggests to introduce two families of specific intensities propagating along any given direction (ray) in the two opposite ways $\pm\mu$. We will denote by $I^-_{\nu\mu} \equiv I_\nu(\mu < 0)$ the incoming and by $I^+_{\nu\mu} \equiv I_\nu(\mu > 0)$ the outgoing intensities. Then from (3.12) for each direction μ (now $0 < \mu \leq 1$) we obtain two *independent* RT equations, namely

$$+\mu \frac{dI^+_{\nu\mu}}{d\tau_\nu} = I^+_{\nu\mu} - S_\nu \tag{3.18}$$

and

$$-\mu \frac{dI^-_{\nu\mu}}{d\tau_\nu} = I^-_{\nu\mu} - S_\nu. \tag{3.19}$$

Following Feautrier (1964b), we can introduce the new variables

$$u_{\nu\mu} = \frac{1}{2}(I^+_{\nu\mu} + I^-_{\nu\mu}) \tag{3.20}$$

and

$$v_{\nu\mu} = \frac{1}{2}(I^+_{\nu\mu} - I^-_{\nu\mu}), \tag{3.21}$$

which are the symmetric and antisymmetric averages of the incoming and outgoing specific intensity, respectively. The former is the specific mean intensity along the ray, the latter is the specific flux.

By adding and subtracting (3.18) and (3.19) and taking into account (3.20) and (3.21), we obtain a system of two first-order differential equations for $u_{\nu\mu}$ and $v_{\nu\mu}$, that is,

$$\mu \frac{dv_{\nu\mu}}{d\tau_\nu} = u_{\nu\mu} - S_\nu, \tag{3.22}$$

[6] In the zeroth approximation, at large optical depths we have strict TE: $I^o_\nu = S_\nu$. In the first-order approximation, we can put $\mu dI^o_\nu/d\tau_\nu = I^1_\nu - S_\nu$ and, consequently, write $I^1_\nu = S_\nu + \mu dS_\nu/d\tau_\nu$. In many cases, however, higher-order derivatives must be employed to achieve a solution with the accuracy required.

and

$$\mu \frac{du_{\nu\mu}}{d\tau_\nu} = v_{\nu\mu}. \tag{3.23}$$

By substitution of (3.23) into (3.22), a second-order differential equation for $u_{\nu\mu}$ is obtained for each frequency ν and direction μ, namely

$$\mu^2 \frac{d^2 u_{\nu\mu}}{d\tau_\nu^2} = u_{\nu\mu} - S_\nu. \tag{3.24}$$

Equation (3.24) must be solved under the boundary conditions, at $\tau_\nu = 0$ and at $\tau_\nu = \tau_{\max}$, that follow directly from (3.20), (3.21) and (3.23) and can be cast always into the form

$$\left(\frac{du_{\nu\mu}}{d\tau_\nu}\right)_{\substack{\tau_\nu=0 \\ \tau_\nu=\tau_{\max}}} = a + b\, u_{\nu\mu}. \tag{3.25}$$

The numerical solution of (3.24) implies the discretization of all the relevant variables. A proper grid of depth points $\{\tau_l\}$, $l = 1, \ldots, N$ must be chosen,[7] together with a grid of frequency points $\{\nu_i\}, (i = 1, NF)$ and a set of directions (angles) $\{\mu_j\}, (j = 1, ND)$. Angles and frequencies are usually grouped into a single set of values denoted by the subscript k such that $k = j + (i-1)ND$. Thus, (3.24) can be written as the second-order equation

$$\mu_k^2 \left(\frac{d^2 u}{d\tau^2}\right)_{l,k} = u_{l,k} - S_{l,k}, \tag{3.26}$$

for any kth unknown meanlike intensity $u_{l,k}$ at each depth point l. If the source function is known, (3.26) represents a set of $NK = NF \times ND$ second-order ordinary differential equations at each depth point τ_l that must be solved for each pair (ν_i, μ_j).

Each differential equation (3.26) can be transformed into a difference equation by means of a simple three-point scheme[8] so that for each angle-frequency point k and at each depth point $l = 2, N-1$ we get the difference equation[9]

$$\frac{\mu_k^2}{\Delta\tau_{l-1/2,k}\Delta\tau_{l,k}} u_{l-1,k} - \frac{\mu_k^2}{\Delta\tau_{l,k}} \left(\frac{1}{\Delta\tau_{l-1/2,k}} + \frac{1}{\Delta\tau_{l+1/2,k}}\right) u_{l,k} + \frac{\mu_k^2}{\Delta\tau_{l+1/2,k}\Delta\tau_{l,k}} u_{l+1,k} = u_{l,k} - S_{l,k}, \tag{3.27}$$

where $\Delta\tau_{l-1/2,k} \equiv \tau_{l,k} - \tau_{l-1,k}$, $\Delta\tau_{l+1/2,k} \equiv \tau_{l+1,k} - \tau_{l,k}$ and $\Delta\tau_{l,k} \equiv (\Delta\tau_{l-1/2,k} + \Delta\tau_{l+1/2,k})/2$.

The difference equation (3.27) can be rewritten in the form

$$A_l\, u_{l-1} - B_l\, u_l + C_l\, u_{l+1} = L_l, \tag{3.28}$$

where the subscript k is omitted for simplicity's sake. The coefficients A_l, B_l and C_l are scalars that depend on μ and the monochromatic optical depth intervals only, and

[7] Because the opacity varies by orders of magnitude along the line profile, care must be taken to choose a proper equally spaced grid of depth points in $\log\tau$.

[8] See, e.g., Feautrier (1964a) or Mihalas (1978).

[9] Equation (3.27) is second-order accurate. A more accurate difference equation representation can be given by Hermite interpolation formulae as in Auer (1976), which leave, however, the tridiagonal structure unaltered.

$L_l = -S_l$ is given. In the corresponding difference equations for the boundary conditions (3.25)[10], it holds that $A_1 = 0$ (at the surface) and $C_N = 0$ (at the bottom). The resulting linear system of algebraic equations has a tridiagonal form and can be solved by means of a Gaussian forward elimination/back substitution (FEBS) scheme. Starting from the difference equation for the surface boundary condition, one can derive for all the other layers $(l, l+1)$, $l = 2, N-1$, the recursive linear relation

$$u_l = q_l + D_l\, u_{l+1} \tag{3.29}$$

using (3.28) and (3.29) derived for the previous layer $(l-1, l)$. The scalar coefficients q_l and D_l are computed during the forward elimination (FE) at all depth points $l = 1, N-1$ and stored for later use in the back substitution (BS) process. At the bottom, it holds that $C_N = 0$, hence $D_N = 0$ and $u_N = q_N$. Once u_N is obtained, all the other u_l are computed using (3.29).

The computational time scales as $N \times NF \times ND$. This rapid and stable method, based on the difference equation representation, was introduced by Feautrier (1964a) and is still widely employed.

3.4.2 *Second-Order Differential RTE and Two-Point Algorithm*

The formal solution described in the previous subsection is based on a three-point FEBS scheme. Let us consider now in more detail another simple FEBS scheme that uses a two-point algorithm to solve layer by layer the RTE in the second-order differential form. As we will see in Section 3.6.2, this scheme is at the basis of the local implicit algorithm that makes it possible to avoid cumbersome matrix inversions in the solution of the general NLTE line formation problem.

Instead of treating RT through the whole atmosphere with the boundary conditions specified on the two limiting surfaces, we can consider a series of one-layer two-point boundary problems.

We start from the first layer $(1, 2)$, where the upper boundary condition (3.25) can be written as

$$\left(\frac{du}{d\tau}\right)_1 = a_1 + b_1\, u_1. \tag{3.30}$$

According to (3.15), (3.20), (3.21) and (3.23), it holds that $a_1 = 0$ and $b_1 = 1/\mu$.

The relation between the meanlike intensities u at the two boundary points $l-1$ and l of each layer $(l-1, l)$, $l = 2, \ldots, N$ is given by the second-order Taylor's expansion:

$$u_l = u_{l-1} + \Delta\tau \left(\frac{du}{d\tau}\right)_{l-1} + \frac{1}{2}\Delta\tau^2 \left(\frac{d^2u}{d\tau^2}\right)_{l-1}. \tag{3.31}$$

Substituting in (3.31) the corresponding upper boundary condition

$$\left(\frac{du}{d\tau}\right)_{l-1} = a_{l-1} + b_{l-1}\, u_{l-1}, \tag{3.32}$$

as well as the expression for the second derivative given by (3.26) for $l-1$, we obtain for any pair of successive depth points $l-1$ and l the recursive relation

$$u_{l-1} = q_{l-1} + D_{l-1}\, u_l, \tag{3.33}$$

akin to (3.29).

[10] They can be derived by the use of the Taylor's expansion as is shown in Section 3.4.2.

The upper boundary condition for the next layer $(l, l+1)$ is derived from the Taylor's expansion of $(du/d\tau)_l$, that is,

$$\left(\frac{du}{d\tau}\right)_l = \left(\frac{du}{d\tau}\right)_{l-1} + \Delta\tau \left(\frac{d^2u}{d\tau^2}\right)_{l-1} + \frac{1}{2}\,\Delta\tau^2 \left(\frac{d^3u}{d\tau^3}\right)_{l-1}. \tag{3.34}$$

Substituting into (3.34) the third derivative, approximated by

$$\left(\frac{d^3u}{d\tau^3}\right)_{l-1} = \frac{1}{\Delta\tau}\left[\left(\frac{d^2u}{d\tau^2}\right)_l - \left(\frac{d^2u}{d\tau^2}\right)_{l-1}\right], \tag{3.35}$$

as well as the previous results for the first and second derivatives at $l-1$, we easily obtain the coefficients a_l and b_l of the boundary condition

$$\left(\frac{du}{d\tau}\right)_l = a_l + b_l\, u_l. \tag{3.36}$$

The coefficients q_{l-1} and D_{l-1} as well as a_l and b_l must be computed at each depth point $l = 2, \ldots, N$ in the FE process and stored for further use in the BS.

At the bottom, the coefficients a_N and b_N are known from the lower boundary condition and therefore the linear relation between u_N and $(du/d\tau)_N$ is known. The relation between u_N and $(d^2u/d\tau^2)_N$ is given by (3.26) for $l = N$. The Taylor's expansion for u_{N-1}, namely

$$u_{N-1} = u_N - \Delta\tau \left(\frac{du}{d\tau}\right)_N + \frac{\Delta\tau^2}{2}\left(\frac{d^2u}{d\tau^2}\right)_N, \tag{3.37}$$

where $\Delta\tau = \tau_N - \tau_{N-1}$, can now be combined with (3.33) for $l = N$, whose coefficients q_{N-1} and D_{N-1} have been computed and stored in the FE step, to compute straightforwardly the value of u_N. The recursive relation (3.33) yields then the values of u_{l-1} for $l = N, \ldots, 2$.

3.4.3 *Integral Form of the RTE*

Let us consider now the formal solution of the RTE in integral form. Because the source function is assumed known, (3.12) is a linear first-order differential equation with constant coefficients. Multiplying it by $e^{-\tau_\nu/\mu}$ and integrating between τ_1 and τ_2, we obtain

$$I_\nu(\tau_1, \mu) = I_\nu(\tau_2, \mu)e^{-(\tau_2-\tau_1)/\mu} + \int_{\tau_1}^{\tau_2} S_\nu(t_\nu)e^{-(t_\nu-\tau_1)/\mu}dt_\nu/\mu. \tag{3.38}$$

For simplicity's sake, we assume that the source function is isotropic.

Using the corresponding boundary conditions in (3.38), i.e., (3.15) for $\tau_2 = 0$ and (3.16) for $\tau_2 = \infty$, the following general expressions for the incoming $I_\nu^-(\tau_\nu, \mu)$ $(-1 \le \mu < 0)$ intensities and the outgoing $I_\nu^+(\tau_\nu, \mu)$ $(0 < \mu \le 1)$ intensities at any inner point $\tau_1 = \tau_\nu$ of a semi-infinite plane-parallel atmosphere for a given direction μ are obtained, that is,

$$I_\nu^-(\tau_\nu, \mu) = \int_0^{\tau_\nu} S_\nu(t_\nu)e^{-(\tau_\nu-t_\nu)/(-\mu)}\frac{dt_\nu}{(-\mu)} \tag{3.39}$$

and

$$I_\nu^+(\tau_\nu, \mu) = \int_{\tau_\nu}^{\infty} S_\nu(t_\nu)e^{-(t_\nu-\tau_\nu)/\mu}\frac{dt_\nu}{\mu}. \tag{3.40}$$

The integration of (3.39) and (3.40) over angles yields the μ-moments of the specific intensity. The zeroth moment of the specific intensity, i.e., the mean intensity, is given by

$$J_\nu(\tau_\nu) = \frac{1}{2}\int_{-1}^{1} I_\nu(\tau_\nu,\mu)d\mu = \frac{1}{2}\int_0^\infty S_\nu(t_\nu)E_1|t_\nu - \tau_\nu|dt_\nu$$
$$\equiv \Lambda_\tau\{S_\nu(t_\nu)\}, \tag{3.41}$$

where $E_1(t)$ is the first exponential integral function. The Λ-operator

$$\Lambda_\tau\{f(t)\} \equiv \frac{1}{2}\int_0^\infty f(t)E_1|t-\tau|dt \tag{3.42}$$

is a linear operator acting here on the source function, a continuous function of position. In the practice, the integral must be replaced by a finite sum. The result is a matrix that represents the original operator at a certain level of approximation.

The first and the second moments of the specific intensity, i.e., the Eddington flux H_ν and the K_ν integral, respectively, are defined as

$$H_\nu(\tau_\nu) = \frac{1}{2}\int_{-1}^{1} I_\nu(\tau_\nu,\mu)\mu d\mu \tag{3.43}$$

and

$$K_\nu(\tau_\nu) = \frac{1}{2}\int_{-1}^{1} I_\nu(\tau_\nu,\mu)\mu^2 d\mu. \tag{3.44}$$

The solution of the RTE given by (3.39) and (3.40) for the specific intensities or by (3.41) for the moment J_ν is *formal*, because the source function actually depends upon the radiation field itself, in other words upon the solution of the RTE. Equations akin to (3.41) can be written for the moments H_ν and K_ν.

For $\tau_\nu = 0$, equation (3.40) gives the emergent intensity from the surface of a semi-infinite atmosphere:

$$I_\nu^+(0,\mu) = \int_0^\infty S_\nu(t_\nu)e^{-t_\nu/\mu}\frac{dt_\nu}{\mu}. \tag{3.45}$$

Likewise, the emergent monochromatic radiative flux[11] $F_\nu(\tau_\nu = 0) = 4H_\nu(\tau_\nu = 0)$ will be given by

$$F_\nu^+(0) = 2\int_0^1 I_\nu^+(0,\mu)\mu d\mu = 2\int_0^\infty S_\nu(t_\nu)E_2(t_\nu)dt_\nu, \tag{3.46}$$

where $E_2(t_\nu)$ is the second exponential integral.

Equations (3.45) and (3.46) are of the uttermost importance: they not only yield quantities that can be directly compared with the observations, but also express formally the link between the radiation emitted by a star and the physical conditions in its atmosphere, which determine the source function.

The formal solution (3.38) for the radiation field along an outgoing ($\mu > 0$) ray (optical depth decreases in the direction of propagation) from a previous (*upwind*) point U to the local point L can be cast into the generic form

$$I_L = I_U e^{-(\tau_U - \tau_L)} + \int_{\tau_L}^{\tau_U} S(t)e^{-(t-\tau_L)}dt. \tag{3.47}$$

[11] See Section 2.2.

Here, $\tau \equiv \tau/|\mu|$ is the optical depth along a given ray (line of sight), and the frequency subscript is suppressed for notational simplicity. The equation for the incoming ($\mu < 0$) radiation field is analogous. If the point U is on the boundary, this equation gives the solution along a given ray passing through the whole medium at an angle cosine of μ. The actual computation of the integral in (3.47) requires the knowledge of the values of the source function on a finite number of discrete points $\{l\}$ between U and L. Instead of a single interpolating function (global fit), it is more convenient and more stable to use piecewise continuous interpolants (local fits) over multiple subintervals. One can break the integral in (3.47) into a sum over discrete intervals between the successive grid points (Mihalas et al., 1978). Separate expansions are used in the subintervals and continuity is preserved by requiring that the expansions match the values at the endpoint nodes. The source function $S(\tau)$ will be then interpolated along ray segments (*short characteristics*, or SCs) and the formal solution will be carried out depth by depth across the atmosphere. Thus the formal solution (3.47), applied to intervals delimited by the successive grid points τ_l and τ_{l+1} for the outgoing intensities ($\mu > 0$), is

$$I^+(\tau_l) = I^+(\tau_{l+1})e^{-\Delta\tau_l} + \int_{\tau_l}^{\tau_{l+1}} S(t)e^{-(t-\tau_l)}dt. \tag{3.48}$$

The formal solution for the incoming intensities ($\mu < 0$) between τ_{l-1} and τ_l reads as

$$I^-(\tau_l) = I^-(\tau_{l-1})e^{-\Delta\tau_{l-1}} + \int_{\tau_{l-1}}^{\tau_l} S(t)e^{-(\tau_l-t)}dt, \tag{3.49}$$

where $\Delta\tau_{l-1} = (\tau_l - \tau_{l-1})/|\mu|$. The choice of the interpolant (usually piecewise polynomial) will be dictated by an educated guess of the behaviour of the source function.

Olson and Kunasz (1987) developed an algorithm for the formal solution based on the SCs solution of the first-order differential RT equations and a three-point parabolic approximation of the source function, which is widely employed in the so-called Accelerated Lambda Iteration (ALI) methods to be described in Section 3.7. The integrals in the formal solution are computed on interval between pairs of successive grid points and expressed in terms of values of the source function at three points: $l-1$, l and $l+1$. Thus the integral in (3.48) is given as

$$\int_{\tau_l}^{\tau_{l+1}} S(t)e^{-(t-\tau_l)}dt = \alpha_l S_{l-1} + \beta_l S_l + \gamma_l S_{l+1}. \tag{3.50}$$

The coefficients α_l, β_l and γ_l depend on the approximation (polynomial representation of the source function) used. In the paper by Olson and Kunasz (1987), they are given for linear and parabolic fits.

Alternatively, there is an efficient *two-point* algorithm that follows naturally from the integration by parts of the integrals in the formal solution. Thus when applied to the integral in (3.48), it yields

$$\int_{\tau_l}^{\tau_{l+1}} S(t)e^{-(t-\tau_l)}dt = q_l S_{l+1} + p_l S_l + r_l S'_l. \tag{3.51}$$

The integration by parts is stopped according to the degree of the polynomial used to approximate $S(\tau)$ between two successive depth points. This formal solver is used by the Forth-and-Back Implicit Lambda Iteration (FBILI) method described in Section 3.7.10, and the corresponding coefficients are given in the paper by Atanacković-Vukmanović et al. (1997).

3.5 Linear RT Problems

As remarked in Section 3.3, radiative transfer is a *nonlocal* and in the general case an integro-differential two-point boundary problem. Namely, the source function of each specific RTE contains in most cases a scattering term that includes (explicitly or implicitly) the full set of specific intensities, i.e., the solution of the RTE itself. Moreover, the nonlocal coupling between the radiation field and the state of the medium through the source function is in general non linear, and the problem needs to be solved iteratively.[12] However, there are lot of important astrophysical instances where the dependence of the source function on the radiation field is linear, hence explicit. Then the system of coupled integro-differential equations can be solved either in a single step (direct solution) or by iterations. Although a direct solution of linear problems is, in principle, always possible, in many cases it may be incompatible with the standards of accuracy, stability and reliability required by the numerical computation because of the huge dimension of the problem. Therefore, iterative methods must be considered.

In this section, we give an overview of selected numerical methods developed since the 1960s to tackle the line formation problem. Although they aim to solve more general nonlinear case, these methods are always checked first on the linear instance of Two-Level Atom[13] line transfer, which serves as a benchmark problem because its exact solution is known.

For a Two-Level Atom line with overlapping continuum and with complete frequency redistribution (CRD) in the line profile, the source function in a constant property medium is given by

$$S_\nu(\tau) = \beta_\nu + \alpha_\nu J_\varphi(\tau), \tag{3.52}$$

i.e., it consists of a local term β_ν that takes into account the true creation of photons and a nonlocal, scattering term

$$J_\varphi(\tau) = \int_0^\infty \varphi_\nu J_\nu(\tau)\, d\nu = \int_0^\infty \varphi_\nu \, d\nu \, \frac{1}{2}\int_{-1}^{1} I_{\nu\mu}(\tau) d\mu \tag{3.53}$$

that couples all the specific RTEs (3.12) as it includes all their solutions. It is *the same* for each specific source function (3.52). The line profile φ_ν in (3.53) is defined here as the ratio between the monochromatic line opacity χ_ν^L and $\chi^L \equiv \int_0^\infty \chi_\nu^L d\nu$.[14] Accordingly, it is convenient to write the monochromatic line opacity as $\chi_\nu^L = \chi^L \varphi_\nu$, i.e., the product of a frequency independent 'mean line opacity' and a distribution function that accounts for the spectral broadening of the line. In the case of no background continuum, the total source function (3.52) reduces to the frequency independent line source function

$$S(\tau) = \varepsilon B + (1-\varepsilon) J_\varphi(\tau), \tag{3.54}$$

where the photon destruction probability ε[15] and the Planck function B (assumed to be constant over the frequencies of the narrow peaked line profile) are considered as given.

[12] The strategies required for the solution of nonlinear RT problems will be discussed in Section 3.8.

[13] The Two-Level Atom model constitutes a paradigm for the line formation problem and is widely studied in the astrophysical literature. An exhaustive exposition can be found in section 11-2 of Mihalas (1978). In this book, the topic is addressed in Section 1.9.4.

[14] The integral denoted by χ^L (finite because of the natural shape of χ_ν^L) is directly related to the atomic Einstein–Milne radiative transfer coefficients through the relation originally derived by Füchtbauer (1920).

[15] The branching parameter ε is defined in Section 1.9.4.

3.6 Direct Methods

Two alternative equivalent approaches to the *direct* solution of the linear line formation problem are possible, which lead to either differential or integral methods. The first case implies to make a guess on the behaviour of the specific intensities, which not only are the unknowns of the problem, but above all vary dramatically frequency by frequency. In the second case, hypotheses are made on the mathematical form of the source function, which in the instance under consideration includes a thermal contribution and a frequency independent scattering term J_φ, where the mean specific intensity is modulated by the given line profile φ_ν. In the following, we are going to compare the Feautrier's method with the Implicit Integral Method. Both are representative examples of the two aforementioned approaches.

3.6.1 *Feautrier's Method and Some of Its Variants*

Feautrier's method is the paradigm of differential methods. It uses the second-order differential form of the RTE (3.24), in which the source function (3.52) is not known as it includes the scattering integral (3.53) that couples all the meanlike intensities. After the discretization of all the variables, the integral is to be replaced by a quadrature sum, and the source function (3.52) then takes on the discretized form

$$S_{l,k} = \beta_{l,k} + \alpha_{l,k} \sum_{k'}^{NK} u_{l,k'}\, \varphi_{l,k'}\, w_{k'}, \tag{3.55}$$

where $w_{k'} = w_i \cdot w_j$, and w_i and w_j are the quadrature weights for frequencies and directions, respectively; $\alpha_{l,k}$ and $\beta_{l,k}$ are known at each depth and angle-frequency point. The introduction of (3.55) into the second-order equation (3.26) or into the corresponding difference equation (3.27) results in a system of linear difference equations, coupled through their source functions, which can be written in matrix form as

$$\boldsymbol{A}_l\, \mathbf{u}_{l-1} - \boldsymbol{B}_l\, \mathbf{u}_l + \boldsymbol{C}_l\, \mathbf{u}_{l+1} = \mathbf{L}_l. \tag{3.56}$$

The coefficients $\boldsymbol{A}_l$, $\boldsymbol{B}_l$ and $\boldsymbol{C}_l$ are matrices of dimension $NK \times NK$. The diagonal matrices $\boldsymbol{A}_l$ and $\boldsymbol{C}_l$ as well as the diagonal of the full matrix $\boldsymbol{B}_l$ contain the finite-difference representation of the differential operator, whereas the off-diagonal elements of $\boldsymbol{B}_l$ come from the nonlocal, scattering term in (3.55). The vector $\mathbf{L}_l$ of length NK contains only the known (thermal) source function terms.

The equations (3.56) for $l = 2, N-1$, supplemented by the corresponding boundary conditions, form a *block* tridiagonal system of equations that can be solved by means of an efficient Gaussian FEBS scheme. The recursive linear relations between the vectors $\mathbf{u}_l$ and $\mathbf{u}_{l+1}$ at each depth point l have the form

$$\mathbf{u}_l = \mathbf{q}_l + \boldsymbol{D}_l\, \mathbf{u}_{l+1}, \tag{3.57}$$

with the matrix coefficients

$$\begin{aligned} \boldsymbol{D}_l &= -(\boldsymbol{B}_l + \boldsymbol{A}_l \boldsymbol{D}_{l-1})^{-1} \boldsymbol{C}_l \\ \mathbf{q}_l &= (\boldsymbol{B}_l + \boldsymbol{A}_l \boldsymbol{D}_{l-1})^{-1} (\mathbf{L}_l - \boldsymbol{A}_l \mathbf{q}_{l-1}) \end{aligned} \tag{3.58}$$

that must be computed during the FE at all depth points $l = 1, N-1$ and stored for later use in the BS.[16] Their computation requires at each depth the inversion of a very large

[16] Note that in the formal solution of the RTE via Feautrier's method in Section 3.4.1, the second-order equations (3.26) and the corresponding difference equations (3.27) are independent,

$(NK \times NK)$ matrix, a risky operation from the numerical standpoint. At the bottom, it holds that $\boldsymbol{C}_N = 0$, hence $\boldsymbol{D}_N = 0$ and $\mathbf{u}_N = \mathbf{q}_N$. Once $\mathbf{u}_N$ obtained, all other $\mathbf{u}_l$ are computed using (3.57).

Feautrier's method has proven to be stable and has been largely employed in the computation of stellar atmosphere models. It suffers, however, from the severe drawback brought about by the need to invert huge matrices, with the consequent large computational cost (the computational time scales as $N \times NK^3$). Feautrier's method solves the RT equations depth by depth, grouping together all the angle-frequency information at each depth point, so that it can be used to solve the linear line formation problem not only in the case of complete but also in the most general instance of partial frequency redistribution. It is optimum for the instance of coherent scattering, where $NF = 1$, and when the number ND of angles required is small. However, in general, the number of frequency points is very large and the aformentioned drawbacks become severe. The problem can be overcome by the following alternative approaches.

An efficient direct solution of the problem in the case of complete redistribution has been proposed by Rybicki (1971). Since the scattering integral is independent of both angle and frequency, instead of considering the angle-frequency variation of the radiation field at a given depth, one can consider the variation with depth of the intensity at any given angle-frequency point. Rybicki's method is more economical than Feautrier's, because the computational time scales as NK instead of NK^3. For details, see Mihalas (1978) and the original paper by Rybicki (1971).

An alternative way of solving the problem at a lower computational cost is to use within a Feautrier's scheme the monochromatic variable (depth-dependent) Eddington factors (VEFs), to be defined in Section 3.7.2. Since only the mean intensities $J_\nu = \int_0^1 u_{\nu\mu} d\mu$ appear in the SE equations that determine the atomic level populations (hence the macroscopic RT coefficients) in the NLTE line formation problem, the angular information is redundant and can be eliminated by using the VEFs. The dimension of the problem is thus considerably reduced, as well as the computational time that scales here as $N \times NF^3$.

Finally, the memory storage and computing time can be drastically reduced by introducing frequency-independent iteration factors defined for a line as a whole, as will be shown in Section 3.7.3. Thanks to them, an exact and fast-convergent solution is achieved because the corresponding algorithm does not employ matrices.

3.6.2 *Implicit Integral Method*

Global integral methods arise from the consideration that the system of the RTEs resulting from the coupling of equations (3.24) is brought about by the scatteringlike integral J_φ. Therefore, instead of a solution for the vector $\mathbf{u}_l$, one may look for an equation for the scalar $J_\varphi(\tau)$, independent of both angles and frequencies.

At the heart of integral methods is the assumption that $J_\varphi(\tau_L)$ at any depth point L can be expressed *implicitly* by interpolating the so far unknown values of J_φ over the discrete set of depth points $\{\tau_l\}$ (see, e.g., Avrett and Loeser, 1969). The interpolation formula can be generalized as

$$J_\varphi(\tau_L) = \sum_{l=1}^{N} \sigma_l(\tau) J_\varphi(\tau_l), \tag{3.59}$$

because their source functions are *individual* given data, and the coefficients D_l and q_l of (3.29) are *scalars*. Here, due to the coupling of the RTEs, the coefficients $\boldsymbol{D}_l$ and $\mathbf{q}_l$ of (3.57) contain matrices. The computation of the former implies just ordinary divisions; on the contrary, matrix inversions are required for the latter. For more details on Feautrier's method, see section 6-3 of Mihalas (1978).

where the coefficients $\sigma_l(\tau)$ are usually given low-order polynomials defined on subintervals of the optical depth range.

By substituting in (3.59) the integral formal solution of the RTE, (3.39) and (3.40), with the source function given by (3.52), it follows that

$$J_\varphi(\tau_L) = \Gamma(\tau_L) + \sum_{l=1}^{N} \boldsymbol{M}(\tau_L, \tau_l) J_\varphi(\tau_l), \tag{3.60}$$

where the function $\Gamma(\tau_L)$ and the matrix $\boldsymbol{M}(\tau_L, \tau_l)$ are computed analytically from the given data. Equation (3.60) represents a linear system of N equations for the unknowns $J_\varphi(\tau_L)$s.

Integral methods, however, require again a cumbersome matrix formalism. To circumvent this drawback, Simonneau and Crivellari (1993) developed an efficient algorithm, the Implicit Integral Method (IIM), originally devised for the solution of the non-LTE line formation problem.

The IIM combines the previous integral representation of the radiation field with an FEBS scheme so that the RTE can be solved *implicitly* layer by layer. The global problem is thus reduced to a series of one-layer two-point boundary problems.

To set up the computational procedure, the medium is sliced into a series of layers (τ_l, τ_{l+1}) and the formal integration of the RTE is performed within each of them. The relevant boundary conditions are the incoming intensities $I^-_{\nu\mu}(\tau_l)$ incident onto the top and the outgoing intensities $I^+_{\nu\mu}(\tau_{l+1})$ incident onto the bottom. When starting the FE from the upper boundary surface, only one half of the data, i.e., the set of initial conditions $\{I^-_{\mu\nu}(\tau_1 = 0)\}$, is known. Therefore, it is not possible to evaluate *explicitly* the values of $J_\varphi(\tau_1)$ because the values of the outgoing intensities $I^+_{\mu\nu}(\tau_1)$ are so far unknown. This prevents the explicit solution for the first layer and consequently to go on with the FE because we cannot set the upper boundary condition for the next layer.

However, thanks to the linear coupling of the specific intensities in the scattering term, it turns out to be possible to derive for each layer (τ_L, τ_{L+1}) the coefficients of a linear relation between $J_\varphi(\tau_L)$ and a set of proper variables that account *implicitly* for the effects of the specific intensities incident at τ_{L+1}. Differently from Feautrier's method, where the linear relations (3.57) link the unknown vectors $\mathbf{u}_l$s by means of matrices, here the coefficients are scalars because they link scalar quantities.

At each step of the FE, the formal solution of the discrete set of RTE for the ensemble of the $\{I^\pm_{\nu_i,\mu_j}\}$, where (3.52) and (3.59) are taken into account, yields the fundamental relation

$$J_\varphi(\tau_L) = \sum_{i=1}^{NF} \sum_{j=1}^{ND} A_{\nu_i,\mu_j}(\tau_L) I^+_{\nu_i,\mu_j}(\tau_{L+1}) + C(\tau_L) + B(\tau_L) J_\varphi(\tau_{L+1}). \tag{3.61}$$

The scalar coefficients A, B and C for each pair of discrete frequency and direction are straightforwardly computed and stored for further use in the BS.

Moreover, we must derive another linear relation that gives each $I^-_{\nu_I,\mu_J}(\tau_L)$ in terms of the set of the unknown outgoing intensities $\{I^+_{\nu_i,\mu_j}(\tau_L)\}$, that is,

$$I^-_{\nu_I,\mu_J}(\tau_L) = \sum_{i=1}^{NF} \sum_{j=1}^{ND} R_{\nu_I,\mu_J,\nu_i,\mu_j}(\tau_L) I^+_{\nu_i,\mu_j}(\tau_L) + \alpha_{\nu_I,\mu_J} + \beta_{\nu_I,\mu_J} J_\varphi(\tau_L). \tag{3.62}$$

The set of equations (3.62) is the implicit form of the upper initial conditions for the next layer. The 'transmission' factor α_{ν_I,μ_J} accounts for the incident intensity reduced by exponential attenuation and the 'reflection' matrix $R_{\nu_I,\mu_J,\nu_i,\mu_j}$ for the emission and diffusion of photons inside the layer. They are scratch variables, used only to transmit information to the next layer.

Starting from the upper boundary condition for the first layer (τ_1, τ_2) that can be cast into the form $\alpha_{\nu_I,\mu_J}(\tau_1 = 0) = I^-_{\nu_I,\mu_J}(\tau_1 = 0)$, $\beta_{\nu_I,\mu_J}(\tau_1 = 0) = 0$ and $R_{\nu_I,\mu_J,\nu_i,\mu_j}$ $(\tau_1 = 0) = 0$, we compute the coefficients of (3.61) and (3.62) depth by depth until the bottom, where the outgoing specific intensities $I^+_{\nu_i,\mu_j}(\tau_N)$ are known. Using (3.62) at the last depth point ($l = N$) together with the definition of $J_\varphi(\tau)$, namely

$$J_\varphi(\tau_l) = \frac{1}{2}\sum_{I=1}^{NF} w_I \sum_{J=1}^{ND} w_J \varphi_{\nu_I,\mu_J}(\tau_l)\left[I^+_{\nu_I,\mu_J}(\tau_l) + I^-_{\nu_I,\mu_J}(\tau_l)\right], \tag{3.63}$$

it is possible to eliminate $I^-_{\nu_I,\mu_J}(\tau_N)$ and compute $J_\varphi(\tau_N)$. Then, using (3.61) and (3.62) in the BS process up to $l = 1$, we can achieve the full solution of the problem.[17]

The IIM takes into account in a natural way the two-stream representation of the radiation field for a two-point boundary problem. Because the algorithm mimics the physical process, its results are stable and its accuracy is very high, even for values of the non-LTE parameter ε in the source function as low as 10^{-8}. Among all the integral methods, it is the only one that avoids the inversion of the Λ-operator thanks to the implicit use of the (unknown) variables $\{I^\pm_{\nu\mu}(\tau_l)\}$. The Λ-operator is set up and used only *locally*, which makes the computational time scales as N instead of N^3. The necessary condition for applying the IIM is that the same scattering integral J_φ couples all the RT equations, for all frequencies and directions. It has been successfully employed for the LTE modelling of stellar atmosphere (Crivellari and Simonneau, 1994) and monochromatic RT in spherical geometry (Gros et al., 1997). The method can be also applied within an iterative procedure to solve more general RT problems such as line transfer with partial frequency redistribution (Crivellari and Simonneau, 1995) and the multilevel atom line formation problem (Crivellari et al., 2002) by introducing proper auxiliary functions (*iteration profiles*).

3.7 Iterative Methods

Radiative transfer is a 'chicken or egg' problem: to compute $I_{\nu\mu}$, we need to know S_ν, and to compute S_ν, we need to know $I_{\nu\mu}$. For its solution, iterative methods can be envisaged in which each iteration is split into two steps. In the first one, the specific intensity of the radiation field is obtained from the solution of the RTE, where the source function $S_\nu(\tau)$ is either known from the previous iteration or set equal to its equilibrium value in the first iteration. In addition to the specific intensities, in certain cases it may be necessary to compute some μ-moments of the latter or approximate operators to be used in the second step of iteration to update $S_\nu(\tau)$. The so-called Λ-iteration is the most straightforward algorithm to solve simultaneously the RT equation (3.12) together with (3.52) for the source function, but its convergence is infinitely slow when applied in most astrophysical instances. However, by introducing proper alterations in the basic scheme, it is possible to achieve an *extremely high* convergence rate. Before taking into consideration how to speed up its convergence rate, let us have a look at the basic Λ-iteration.

[17] Full details are given in the original paper by Simonneau and Crivellari (1993). An improved version of the method that employs more accurate interpolating strategy is given by Simonneau et al. (2012).

3.7.1 *Lambda Iteration*

For the sake of an illustrative exposition, we will consider the application of the Λ-iteration to the Two-Level Atom line formation with CRD and with no overlapping continuum. The physics of the problem is completely described by two coupled equations: the formal solution of the RTE (3.12) given by

$$J_\varphi(\tau) = \Lambda[S(t)], \tag{3.64}$$

where $\Lambda = \int \Lambda_\nu \varphi_\nu d\nu$ with Λ_ν defined by (3.42), and the frequency-independent line source function (3.54), directly derived from the single SE equation for the populations of the two levels, namely

$$S(\tau) = \varepsilon B(\tau) + (1-\varepsilon) J_\varphi(\tau), \tag{3.65}$$

where the possible dependence of ε on τ is omitted.

By substitution of (3.64) into (3.65), the problem can be stated in the form of the single integral equation

$$S(\tau) = (1-\varepsilon)\Lambda[S(t)] + \varepsilon B(\tau), \tag{3.66}$$

whose direct solution is

$$S(\tau) = [1-(1-\varepsilon)\Lambda]^{-1}\ [\varepsilon B(t)]\,. \tag{3.67}$$

The inversion of the Λ operator, which is computationally very expensive, can be avoided if the problem equations (3.64) and (3.65) are solved sequentially one by one. This procedure is customarily called Λ-iteration.[18] Starting from an initial guess $S^{(o)}(\tau)$ of the source function, the improved source functions are computed using the simple iteration scheme

$$S^{(n+1)}(\tau) = (1-\varepsilon)\ \Lambda\left[S^{(n)}(t)\right] + \varepsilon B(\tau). \tag{3.68}$$

The limit $\lim_{n\to\infty} S^{(n)}(\tau) \equiv S^{(\infty)}(\tau)$ stays for the exact solution.

The major advantage of the Λ-iteration is that it does not require the inversion of an operator (matrix). Moreover, the application of the Λ-operator to the known current source function $S^{(n)}(\tau)$ in (3.68) is just the formal solution of the RTE that can be performed by using any of the methods described in Section 3.4, either differential or integral. The previous construction of the Λ-operator is not necessary.

Equation (3.66) is a linear integral equation of the second kind, whose kernel for this specific problem is $K_1(|\tau - t|) = (1/2)\int \varphi_\nu^2 E_1[\varphi_\nu(\tau - t)]\, d\nu$. According to the general theory,[19] its solution is given by the related Neumann's series, provided it converges uniformly. Each nth term of the series includes the corresponding iterated kernel $K^{(n)}$ that can be written by canonical factorization as the product of n exponential integrals $(1/2)\, E_1 \mid \tau - t \mid$.

The repeated application of the Λ-operator to the source function is equivalent to take successively into consideration the 1st, 2nd, ..., nth term of the Neumann's series. According to the statistical interpretation of radiative transfer (see Section 1.10), each of these terms is linked to the probability that a line photon originated at optical depth τ can

[18] The mathematical grounds of the Λ-iteration date back to Liouville and Neumann. It was introduced in the practice of RT by Hopf (1928).

[19] See, e.g., chapter III of Courant and Hilbert (1937).

escape through the outer boundary surface after undergoing $1, 2, \ldots, n$ scatteringlike[20] events. Each factor $(1/2)\ E_1 \mid \tau - t \mid$ represents the probability distribution function, integrated over all the outward directions, which accounts for the probability that a photon after travelling undisturbed from τ to t will be scattered in the successive interval $(t, t + dt)$. As shown in Figure 1.2, this probability falls rapidly as $\mid \tau - t \mid \geq 1$.

Due to the nonlocal nature of RT, the line photons absorbed at τ in a scatteringlike process carry on information on the physical conditions of the neighbouring points at optical depth t. For the foregoing considerations, at each step of the Λ-iteration information is transmitted only from points within a volume of radius $\Delta\tau \sim 1$, so that the current solution is corrected at each depth point only over an optical path of order of unit. Therefore, in general, the convergence of the Λ-iteration is extremely slow for those media of large optical thickness in which scattering predominates.

Olson et al. (1986) examined the convergence properties of the Λ-iteration in terms of the maximum eigenvalue λ_{max} of the amplification matrix $(1 - \varepsilon)\Lambda$.[21] It can be shown that $\lambda_{max} = (1 - \varepsilon)(1 - T^{-1})$, where T is the total optical thickness. Mathematical consideration confirms that if ε is small and T large, λ_{max} is so close to unity that the convergence is extremely slow. This is illustrated, e.g., by figures 12.1 through 12.3 of Hubený and Mihalas (2015) or in figure 4 of the paper by Auer in Kalkofen (1984). A rather severe consequence of this fact is that in the solution of actual problems, we cannot decide when and even if the final (exact) solution is reached.

3.7.2 *Variable Eddington Factors (VEFs) and Their Application in Feautrier's Method*

We have pointed out in Section 3.6.1 that a large number of frequency and angle quadrature points is required for an accurate description of the radiation field. This implies a prohibitive amount of memory storage and computing time as the matrix operations required in the difference equations approach scale as the cube of the number of quadrature points. In order to circumvent this disadvantage, Auer and Mihalas (1970) introduced into Feautrier's method the so-called VEFs, which can be computed by a straightforward iterative scheme even in the instances of noncoherent scattering or non-LTE model atmosphere problems. The idea is to replace the RTE (3.12) by the system of its first two μ-moments, that is,

$$\frac{dH_\nu(\tau_\nu)}{d\tau_\nu} = J_\nu(\tau_\nu) - S_\nu(\tau_\nu) \tag{3.69}$$

and

$$\frac{dK_\nu(\tau_\nu)}{d\tau_\nu} = H_\nu(\tau_\nu). \tag{3.70}$$

The system of two preceding equations in the three unknowns J_ν, H_ν and K_ν can be closed by introducing the frequency-dependent VEFs, defined as

$$F_\nu(\tau_\nu) = \frac{K_\nu(\tau_\nu)}{J_\nu(\tau_\nu)} = \frac{1/2 \int I_\nu(\tau_\nu, \mu)\ \mu^2\ d\mu}{1/2 \int I_\nu(\tau_\nu, \mu)\ d\mu}. \tag{3.71}$$

Within an iterative procedure, the VEFs are computed from the current values of the μ-moments of the specific intensity, and therefore they yield an approximate closure.

[20] By scatteringlike process, we mean absorption followed by emission. (See Section 1.9.4.)

[21] The smaller the value of λ_{max} (which must be less than unity for the convergence of the procedure), the faster the convergence.

However, as they are the ratios of two homologous quantities, their current values are quite insensitive of the initial guess; in other words, they are *quasi-invariant* quantities. They converge in a few iterations to the exact values that make the closure relation exact.[22]

With the preceding definition, we are led to the second-order differential equation

$$\frac{d^2(F_\nu J_\nu)}{d\tau_\nu^2} = J_\nu(\tau_\nu) - S_\nu(\tau_\nu), \tag{3.72}$$

akin to (3.24). It can be transformed into a system of difference equations together with its boundary conditions and solved with Feautrier's method.

The simple iteration procedure is as follows:

(i) Starting with a known source function (either $S_\nu = B_\nu$ as a first guess or S_ν known from the previous iteration) we solve formally the RTE for the specific intensities $I_\nu(\tau_\nu, \mu)$ or the Feautrier variables $u_{\nu\mu}$. The time required for this formal solution scales as $N \times NF \times ND$.

(ii) For each depth point and each frequency, we compute the moments J_ν, H_ν and K_ν of the specific intensity and the Eddington factor F_ν given by (3.71).

(iii) The system of difference equations derived from (3.72) is solved by applying the FEBS scheme described in Section 3.6.1. Once obtained the new J_ν, we can update S_ν and repeat the procedure until convergence.

Let us note that, thanks to the VEFs, the dimension of the problem is considerably reduced as the tridiagonal grand matrix of Feautrier's method now contains matrices of dimensions $NF \times NF$ only. The time required for the Feautrier's solution with VEFs scales as $N \times NF^3$. The total computing time of the method scales as $N_{iter} \times (N \times NF^3 + N \times NF \times ND)$, which is much less than the total computing time of Feautrier's method, i.e., $N \times NF^3 \times ND^3$. Due to good quasi-invariant properties of VEFs, the number of iterations N_{iter} needed to get the 'exact' solution is usually very small.

Because the source function is isotropic (i.e., angle-independent), it is possible to remove the redundant angular information by using the VEFs. Moreover, in those line formation problems in which the source function is given by (3.52), the scattering integral is also frequency independent. This fact makes it possible to reduce further the redundant information, as will be shown in the next subsection.

3.7.3 *Iteration Factors Method*

The Iteration Factors Method (IFM) generalizes the idea of the VEFs, originally introduced in the monochromatic RT, to the iteration factors (IFs) to be used in the frequency-dependent line formation problems. For simplicity's sake, we shall present the method for the well-known instance of a line with complete redistribution and no background continuum.[23]

If we adopt the standard optical depth scale τ, defined by the differential relation $d\tau = -\chi^L dz$, it follows straightforwardly that $d\tau_\nu = \varphi_\nu d\tau$. Under the preceding assumptions, the RTE (3.12) becomes

$$\mu \frac{dI_\nu(\tau, \mu)}{d\tau} = \varphi_\nu \, [I_\nu(\tau, \mu) - S(\tau)] \, , \tag{3.73}$$

[22] The idea to iterate on the ratio of two unknown quantities (intensity moments), and not on the quantities themselves, appeared for the first time in the paper by Feautrier (1964c).

[23] Full details are given in Atanacković-Vukmanović and Simonneau (1994).

where the line source function $S(\tau)$ is given by (3.54). Equations (3.73) and (3.54), which contain all the physics of the problem, must be solved simultaneously. A procedure similar to that described in the previous subsection can be employed for their solution.

Starting from an initial estimate of the source function, the specific intensities $I_{\nu\mu}(\tau)$ for a chosen mesh of frequency, angle and optical depth can be obtained from the formal solution of (3.73). They are then used to compute the corresponding intensity moments and the iteration factors needed to close the system of the RTE moments and SE equation. Since now the two variables ν and μ enter the problem, in order to obtain J_φ we must integrate the RTE over both angles and frequencies. For each frequency ν, the first- and second-order μ-moments of (3.73) are given here by

$$\frac{dH_\nu}{d\tau} = \varphi_\nu(J_\nu - S) \tag{3.74}$$

and

$$\frac{dK_\nu}{d\tau} = \varphi_\nu H_\nu. \tag{3.75}$$

Multiplying (3.74) by φ_ν^2 and (3.75) by φ_ν, and by successive integration over frequencies, we obtain the system of the two differential equations

$$\frac{dH_{\varphi^2}}{d\tau} = J_{\varphi^3} - \varphi^3 S \tag{3.76}$$

and

$$\frac{dK_\varphi}{d\tau} = H_{\varphi^2}, \tag{3.77}$$

where we denote by

$$Q_{\varphi^n} = \int \varphi_\nu^n Q_\nu d\nu \quad \text{and} \quad \varphi^n = \int \varphi_\nu^n d\nu \tag{3.78}$$

the intensity and profile moments.

The system of equations (3.76) and (3.77) can be reduced to the single second-order differential equation

$$\frac{d^2 K_\varphi}{d\tau^2} = J_{\varphi^3} - \varphi^3 S \tag{3.79}$$

in the three unknowns K_φ, J_{φ^3} and J_φ (see (3.54)). Equation (3.79) describes the line radiative transfer and includes all the information about the coupling between the line photons and the Two-Level Atom populations. Two more equations, however, are necessary for the solution. As in the case of the monochromatic VEFs, the generalized (frequency-independent) Eddington factor

$$F(\tau) \equiv \frac{K_\varphi(\tau)}{J_\varphi(\tau)} \tag{3.80}$$

is introduced here to account for the anisotropy of the radiation field. The second closure relation between J_{φ^3} and J_φ is given in a natural way by the iteration factor

$$f_J(\tau) \equiv \frac{J_{\varphi^3}(\tau)}{J_\varphi(\tau)}. \tag{3.81}$$

As J_{φ^3} involves the profile φ_ν^3 that is more narrow than φ_ν and covers mainly the line core, the factor f_J represents the ratio of the photons in the core (J_{φ^3}) to the photons (J_φ) in the whole line. Using (3.80) and (3.81) in (3.79), together with (3.54), we eventually obtain one second-order differential equation for $K_\varphi(\tau)$ only, that is,

$$\frac{d^2 K_\varphi}{d\tau^2} = \frac{f_J - \varphi^3(1-\varepsilon)}{F} K_\varphi - \varphi^3 \varepsilon B. \tag{3.82}$$

The factors $F(\tau)$ and $f_J(\tau)$, as well as the coefficients $\gamma_l \equiv H_{\varphi^2}(\tau_l)/K_\varphi(\tau_l)$ of the two boundary conditions

$$\left(\frac{dK_\varphi}{d\tau}\right)_{\tau_l} = \gamma_l K_\varphi(\tau_l) \tag{3.83}$$

for $l = 1$ (surface) and $l = N$ (bottom), are computed from the formal solution of the RTE in the first part of each iteration. Then the RT moment equation (3.82), with the boundary conditions given by (3.83), can be solved for $K_\varphi(\tau)$ by means of either the three-point difference equation scheme of Section 3.4.1. or the two-point algorithm of Section 3.4.2. Because no matrix operation is needed, both the memory storage and the computational time required are very small. Once we obtain the new $K_\varphi(\tau)$, we can compute the mean intensity $J_\varphi(\tau)$ by applying (3.80) and the updated value of the source function by using (3.54).

An extremely fast convergence to the exact solution is achieved thanks to the solution of just one second-order differential equation, whose coefficients are the angle and frequency independent iteration factors F and f_J that, defined as the ratios of the relevant intensity moments, are good quasi-invariants of the problem.

The Iteration Factors Method was originally introduced for the computation of LTE stellar atmosphere models in radiative equilibrium (Simonneau and Crivellari, 1988), and in the more general case where convective transport is taken into account (Crivellari and Simonneau, 1991). Fieldus et al. (1990) generalized the method to include spherically extended line blanketed model atmospheres. The method has been developed for the Two-Level Atom line transfer problem by Simonneau and Atanacković-Vukmanović (1991) and Atanacković-Vukmanović (1991), and described in detail in the paper by Atanacković-Vukmanović and Simonneau (1994). It has been extended to the solution of the multilevel line transfer problem by Kuzmanovska-Barandovska and Atanacković (2010).

3.7.4 *ALI Methods*

ALI is the acronym for approximated (or accelerated) lambda iteration. As pointed out by Rybicki (1991), "it is not easy to define ALI precisely. ALI is not one monolithic idea, but comprises ideas from many sources, including: Cannon's method, the core saturation method, Scharmer's method, OAB[24] operators and accelerated convergence methods. In this respect ALI more resembles a 'tool kit' than a single 'method'." As we shall see in the following, the specific 'tools' can be combined within a particular ALI application. For a more detailed review on ALI methods, see the paper by Hubený (2003), the books by Kalkofen (1987) and Hubený and Mihalas (2015) and the references therein.

3.7.5 *Core Saturation Method*

This method is at the origin of the ALI's idea. In order to eliminate the cause of the slow convergence of the ordinary Λ-iteration, that is, the poor conditioning of the equations

[24] Olson et al., 1986.

due to a large number of scatterings, Rybicki (1972) proposed to eliminate scatterings in the line core, where they do not contribute significantly to the transfer process. He assumed the equality of the monochromatic mean intensity of the radiation field and the local source function at the optically thick line core frequencies (core saturation assumption), namely

$$J_\nu(\tau_\nu) = S_\nu(\tau_\nu). \tag{3.84}$$

By using (3.84) in the SE equations, only the photons in the line wings are treated explicitly. This improves the numerical conditioning of the SE equations, which retain the basic form of the original ones, but with modified coefficients. The solution of such *preconditioned* equations by means of the Λ-iteration has shown a substantial increase in the convergence rate. The main disadvantage of this method is the need of an adjustable parameter to specify the core saturation region, which has to be found empirically. If it is chosen far from its optimum value, a very slow convergence or even divergence may occur.

3.7.6 *Cannon's Method: Operator Perturbation Methods*

Cannon (1973) was the first to use approximate lambda operators (ALOs) and a perturbation 'operator splitting' technique in radiative transfer computations, thus providing a mathematical framework for the class of methods known as ALI. He replaced the exact[25] Λ-operator by

$$\Lambda \equiv \Lambda^* + (\Lambda - \Lambda^*), \tag{3.85}$$

where Λ^* is a simplified operator and $(\Lambda - \Lambda^*)$ a (small) error term computed by a perturbation technique. Because of that, ALI methods are sometimes also called operator perturbation methods. Cannon's approximate operator is a Λ operator evaluated with fewer angular-frequency quadrature points.

By introducing the approximate operator Λ^* defined by (3.85), the RTE (3.64) can be written as

$$J^{(n+1)} = \Lambda^*[S^{(n+1)}] + (\Lambda - \Lambda^*)[S^{(n)}]. \tag{3.86}$$

Likewise, the expression for the Two-Level Atom line source function (3.66) can be put into the form of the recursive relation

$$S^{(n+1)} = (1-\varepsilon)\Lambda^*[S^{(n+1)}] + (1-\varepsilon)(\Lambda - \Lambda^*)[S^{(n)}] + \varepsilon B. \tag{3.87}$$

Here the approximate operator Λ^* acts on the new (yet unknown) source function, whereas $\Lambda - \Lambda^*$ acts on the current value of the source function and may be easily obtained by the formal solution of the transfer equation. The solution of (3.87)

$$S^{(n+1)} = [1 - (1-\varepsilon)\Lambda^*]^{-1}[(1-\varepsilon)(\Lambda - \Lambda^*)[S^{(n)}] + \varepsilon B] \tag{3.88}$$

now contains an approximate Λ^* matrix that is easier to invert than the full Λ matrix.[26]

If $\Lambda^* = \Lambda$, one recovers the direct solution (3.67) via expensive matrix inversion, whereas if $\Lambda^* = 0$ one has the classical Lambda iteration (3.68). Of course, it is easier to invert Λ^* than the exact (full) Λ operator, with the consequent dramatic reduction of the number of operations per iteration. At the same time, Λ^* must be as close as possible

[25] The so-called exact lambda operator is itself an approximation based on some discretization.

[26] From the mathematical point of view, the ALI scheme is an application of the idea of an iterative solution of a large linear system by preconditioning.

to the exact Λ operator in order to keep the physical information and thus to reduce the number of iterations. The choice of Λ^* is arbitrary to a certain degree.

3.7.7 *Scharmer's Method*

Cannon's idea to replace the exact Λ-operator with a simplified one, together with the iterative computation of the error term $(\Lambda - \Lambda^*)$, was applied to stellar atmospheres for the first time by Scharmer (1981). Combining Rybicki's core saturation idea and the Eddington–Barbier relation with the perturbation technique, Scharmer derived two approximate Λ operators for the case of Two-Level Atom line transfer. Scharmer's ALI method, based on his nonlocal global operator, is at the heart of the well-known multilevel code MULTI (Carlsson, 1986).

3.7.8 *Diagonal Operator (Jacobi Method)*

On the basis of a numerical analysis of the problem under consideration, Olson et al. (1986) concluded that the diagonal part of the exact (full) Λ matrix represents an optimum approximate operator. Using a 'short characteristic' solution of the transfer problem, Olson and Kunasz (1987) derived a fast method to compute the exact diagonal of the Λ matrix. An important advantage of the so-called OAB operator[27] is that no free parameter is required to control the convergence of the process. Moreover, the matrix inversion in (3.88) is replaced by a simple scalar division in the case that the operator is diagonal, i.e.

$$S^{(n+1)} = \frac{\varepsilon B + (1-\varepsilon)(\Lambda - \Lambda^*)[S^{(n)}]}{1-(1-\varepsilon)\Lambda^*}. \tag{3.89}$$

Thanks to that the computing time per iteration is the smallest possible. Such a choice of the approximate operator corresponds to the iterative scheme known as the Jacobi method.[28]

Let us note that with a diagonal ALO, a matrix formalism is not necessary. Instead of $J = \Lambda S = (\Lambda - \Lambda^*)[S] + \Lambda^*[S]$, one can write at each depth point l an implicit linear relation

$$J_l = a_l + b_l S_l, \tag{3.90}$$

where the coefficients a_l and b_l (which correspond to $(\Lambda - \Lambda^*)S^{\text{old}}$ and Λ^*, respectively) are computed during the formal solution of the RTE and successively used in (3.54) to update the source function by means of the relation

$$S_l = \frac{\varepsilon B + (1-\varepsilon)\, a_l}{1-(1-\varepsilon)\, b_l}, \tag{3.91}$$

akin to (3.89).

A slight modification of the ordinary Λ iteration scheme provided by the Jacobi method reduces the number of iterations by a few orders of magnitude. For some more demanding problems (strong lines dominated by scattering), even the Jacobi method is not fast

[27] The OAB operator is used as the local-operator option of the multilevel transfer code MULTI (Carlsson, 1986). It has been adapted for spherical geometry by Puls and Herrero (1988).

[28] The Jacobi method (1845) is the simplest iterative method for solving the system of linear equations $\boldsymbol{A}\ \mathbf{x} = \mathbf{b}$. The matrix $\boldsymbol{A}$ can be decomposed into a diagonal component $\boldsymbol{D}$ and a nondiagonal remainder $\boldsymbol{R}$. The solution is then obtained iteratively via $\mathbf{x}^{(i+1)} = \boldsymbol{D}^{-1}(\mathbf{b} - \boldsymbol{R}\mathbf{x}^{(i)})$.

enough. Additional acceleration can be achieved by employing some purely numerical techniques such as that developed by Ng (1974) (the so-called Ng acceleration). Olson and Kunasz (1987) showed that a faster convergence (i.e., with many fewer iterations) can be obtained by using as an ALO the tridiagonal (or pentadiagonal) part of the exact Λ operator. However, the use of these nonlocal operators is computationally more expensive due to the unavoidable matrix inversions that considerably increase CPU time per iteration.

3.7.9 *Gauss–Seidel methods*

More efficient numerical methods for RT problems based on Gauss–Seidel iteration[29] were proposed and developed in two different ways in the papers by Trujillo Bueno and Fabiani Bendicho (1995) and by Atanacković-Vukmanović (1991) and Atanacković-Vukmanović et al. (1997).

While the Jacobi method uses the old (from the previous iteration) values of the source function to compute the coefficients (a and b) of the implicit linear relation (3.90) at all depth points and then to update the source function according to (3.91) at the end of each iteration step, in the Gauss–Seidel scheme one updates the value of the source function *as soon as* the coefficients a and b are available (known) at some point. Only the ingoing intensities (i.e., the corresponding coefficients a^- and b^- in Atanacković-Vukmanović et al., 1997) are computed with the old source function during the forward process. With the given boundary condition at the bottom, one can compute the coefficients of (3.90) at $l = N$ and use them in (3.91) to update S_N and, hence, the outgoing intensity at that point. Using the formal (SC) solution for the outgoing intensity at the next upper depth point, the procedure is repeated and the source function is corrected during the backward process together with the outgoing intensity sweeping up to the surface.

Trujillo Bueno and Fabiani Bendicho (1995) developed Gauss–Seidel method using three-point algorithm to solve the coherent scattering and Two-Level Atom line transfer with CRD. Their pure Gauss–Seidel (G–S) method (pure as the corrections are made during the backward process) is faster by a factor 2 than the Jacobi method, and by a factor 4 when the G–S iterations are continuously performed in both the incoming and outgoing passes (symmetric G–S method). They employed the successive overrelaxation (SOR) method to additionally increase the convergence rate. With the optimal value of the parameter ω (which lies between 1 and 2), the G–S method with SOR is about an order of magnitude faster than the Jacobi method.

3.7.10 *FBILI Method*

In this subsection, we will describe the Forth-and-Back Implicit Lambda Iteration (FBILI), an extremely simple and efficient method introduced by Atanacković-Vukmanović et al. (1997). It is intrinsically so fast that there is no need for additional mathematical acceleration.

[29] The Gauss–Seidel method for solving the system of linear equations $\boldsymbol{A}\,\mathbf{x} = \mathbf{b}$ has an improved convergence rate with respect to the Jacobi method. It uses the new (updated) values of vector $\mathbf{x}$ as soon as they are available, instead of waiting for the subsequent iteration as in the case of the Jacobi method. The G–S scheme can be written as

$$(\boldsymbol{L}+\boldsymbol{D})\mathbf{x}^{(i+1)} + \boldsymbol{U}\mathbf{x}^{(i)} = \mathbf{b} \quad \text{i.e.} \quad \mathbf{x}^{(i+1)} = (\boldsymbol{L}+\boldsymbol{D})^{-1}(\mathbf{b} - \boldsymbol{U}\mathbf{x}^{(i)}),$$

where $\boldsymbol{D}$, $\boldsymbol{L}$ and $\boldsymbol{U}$ are the diagonal, strictly lower-triangular and strictly upper-triangular parts of matrix $\boldsymbol{A}$, respectively.

Let us recall the main reason for the slow convergence of the Lambda iteration. From the previous iteration it carries on more information than necessary because the old source function $S^o(\tau)$ is used to compute the whole mean intensity of the radiation field. In Section 3.7.8, we have seen that the Jacobi method, which treats the source function implicitly and uses the old source function to compute only the nonlocal part of the mean intensity, i.e., the coefficient a_l of the *implicit* linear relation (3.90), has a much higher convergence rate. The FBILI achieves a dramatic acceleration of the convergence by using the old source function for the computation of the nonlocal part of the incoming mean intensity (coefficient a_l^-) only.

Let us consider the FBILI procedure in more detail. We start with the integral form of the RTE for the incoming intensities given by (3.49), i.e.,

$$I_l^- = I_{l-1}^- e^{-\Delta} + \int_{\tau_{l-1}}^{\tau_l} S(t) e^{-(\tau_l - t)} dt, \tag{3.92}$$

and perform an integration by parts under the assumption of a piecewise parabolic behaviour of the source function between the successive depth points $l-1$ and l. The integral thus expressed in terms of the source function and its derivatives at the points $l-1$ and l, after eliminating S'_{l-1} by means of[30]

$$S'_{l-1} = \frac{2}{\Delta\tau}(S_l - S_{l-1}) - S'_l, \tag{3.93}$$

allows us to rewrite (3.92) as

$$I_l^- = I_{l-1}^- e^{-\Delta} + q_l^- S_{l-1} + p_l^- S_l + r_l^- S'_l. \tag{3.94}$$

The coefficients p_l^-, q_l^- and r_l^- depend only on the known optical distance $\Delta \equiv \Delta\tau\varphi_\nu/\mu$, where $\Delta\tau = \tau_l - \tau_{l-1}$.

Equation (3.94) is an *implicit* linear relation between the incoming intensities $I_{\nu\mu}^-$ at point l and the yet unknown values of the *local* source function S_l and its derivative S'_l. The first two terms in the RHS of (3.94) represent the nonlocal part of the incoming intensities, which will be computed with the old values of the source function at previous depth points. Namely, proceeding from the given upper boundary condition ($I_1^- = 0$), the values of all the other I_{l-1}^- are obtained by recursive application of (3.94) with the old values of S and S' at all $\tau < \tau_l$.

By integration of (3.94) over frequencies and directions, we obtain the implicit linear relation

$$J_l^- = \tilde{a}_l^- + \tilde{b}_l^- S_l + \tilde{c}_l^- S'_l, \tag{3.95}$$

where the contribution of the nonlocal terms $I_{l-1}^- e^{-\Delta}$ and $q_l^- S_{l-1}$ is grouped in the coefficient $\tilde{a}_l^-$. The coefficients $\tilde{a}_l^-$, $\tilde{b}_l^-$ and $\tilde{c}_l^-$ are to be computed in the FE process and stored for further use in the BS. Moreover, we can rewrite (3.95) as

$$J_l^- = \left(\frac{\tilde{a}_l^-}{S_l^{(o)}} + \tilde{b}_l^-\right) S_l + \tilde{c}_l^- S'_l = b_l^- S_l + c_l^- S'_l \tag{3.96}$$

by introducing the ratio $\tilde{a}_l^-/S_l^{(o)}$ as a quasi-invariant iteration factor.[31] The iterative computation of this factor is fundamental to achieve an extremely high convergence rate.

30 The relation is derived by using the first-order Taylor's expansion of S'_l.

31 For more details, see section 5.1 of Atanacković-Vukmanović et al. (1997).

In the BS, we use for each layer (τ_l, τ_{l+1}) the formal solution (3.48) of the RTE for the outgoing intensities, namely

$$\begin{aligned} I_l^+ &= I_{l+1}^+ e^{-\Delta} + \int_{\tau_l}^{\tau_{l+1}} S(t) e^{-(t-\tau_l)} dt \\ &= I_{l+1}^+ e^{-\Delta} + q_l^+ S_{l+1} + p_l^+ S_l + r_l^+ S'_{l+1}, \end{aligned} \tag{3.97}$$

where we assume again a piecewise parabolic behaviour of S within each layer (τ_l, τ_{l+1}) and use the relation

$$S'_l = \frac{2}{\Delta\tau}(S_{l+1} - S_l) - S'_{l+1}, \tag{3.98}$$

akin to (3.93).

In the BS, we start from the bottom, where the outgoing intensities I_N^+ and hence J_N^+ are known. If we make the hypothesis that $I_N^+ = S_N$ and $S'_N = 0$, which is a valid assumption for a semi-infinite medium, the coefficients of the implicit relation $J_N^+ = a_N^+ + b_N^+ S_N$ are known; it holds that $a_N^+ = 0$ and $b_N^+ = 1$. The coefficients a_N^- and b_N^- are derived from (3.96) for $l = N$, where b_N^- and c_N^- have been computed and stored in the forward process. Hence we get the coefficients a_N and b_N of (3.90), which substituted in (3.91) for $l = N$ give the updated value of the source function $S(\tau_N)$. With $S(\tau_N)$ and the given values of $I_{\nu\mu}^+(\tau_N)$ and $S'(\tau_N)$, one can straightforwardly compute the coefficients of the linear relation (3.97) at the next upper depth point $N-1$.

Once we obtain I_{l+1}^+, S_{l+1} and S'_{l+1}, we proceed as follows for each next upper depth point l. By using (3.98), we can eliminate the derivative S'_l in (3.96) in order to get J_l^- as a linear function of S_l only. Then, integrating (3.97) over all the frequencies and directions and taking into account that all the terms except S_l are known, a similar expression for J_l^+ can be easily derived. Finally, at depth point τ_l we obtain the coefficients a_l and b_l of the linear relation (3.90), which substituted in (3.91) allow us to derive the new value of S_l. The derivative S'_l and the outgoing intensities I_l^+ are computed using (3.98) and (3.97), respectively. Thus the computation of the new source function together with the outgoing intensities is performed during the BS layer by layer to the surface. The process is iterated until the convergence is achieved.

The extra computational effort of the iterative computation of the coefficients of the implicit relations rather than that of the intensities themselves is negligible and results in the extremely fast convergence of the FBILI method. The iteration factor introduced in the coefficient b_l^- plays an extremely important role as it quickly attains its exact value leading to the exact and rapid solution of the whole procedure. Let us note also that the use of two-point short characteristics instead of the usual three-point ones in the formal solution of the RTE enables simpler and more efficient corrections to the current solution during the BS step. Figure 3.3 shows that, even under extreme non-LTE conditions such as $\varepsilon = 10^{-12}$, only 14 iterations are needed to reach the exact solution. Moreover, the correct thermalization depth ($L_T \approx 1/\varepsilon$) is attained already in the first iteration. Let us recall that the classical Λ iteration needs about $1/\varepsilon \approx 10^{12}$ iterations to reach the exact solution of this problem.

In Figure 3.4, the convergence rate of the FBILI method is compared with that of several ALI schemes (Atanacković-Vukmanović, 2007).

The FBILI method was first developed for the Two-Level Atom and Multilevel Atom line transfer problem with CRD by Atanacković-Vukmanović (1991) and Atanacković-Vukmanović et al. (1997). The Two-Level Atom line transfer problem was generalized

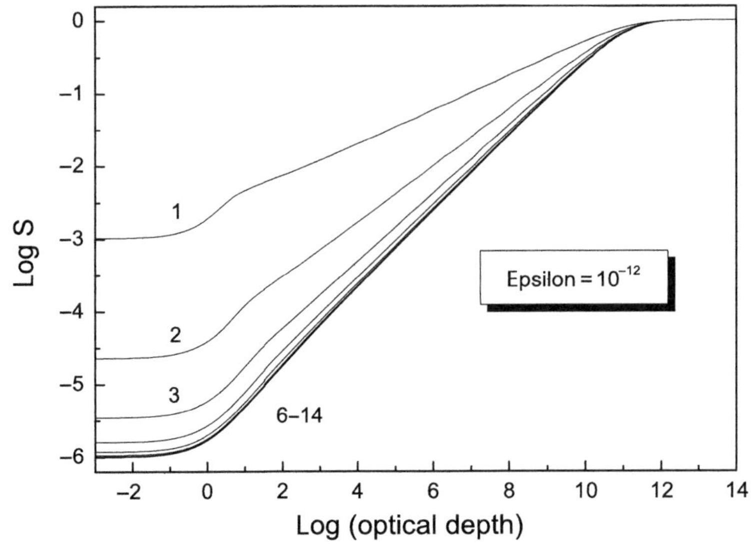

FIGURE 3.3. The Two-Level Atom line source function for a purely Doppler broadened profile φ_ν, $\varepsilon = 10^{-12}$ and $B = 1$ as function of optical depth through iterations.

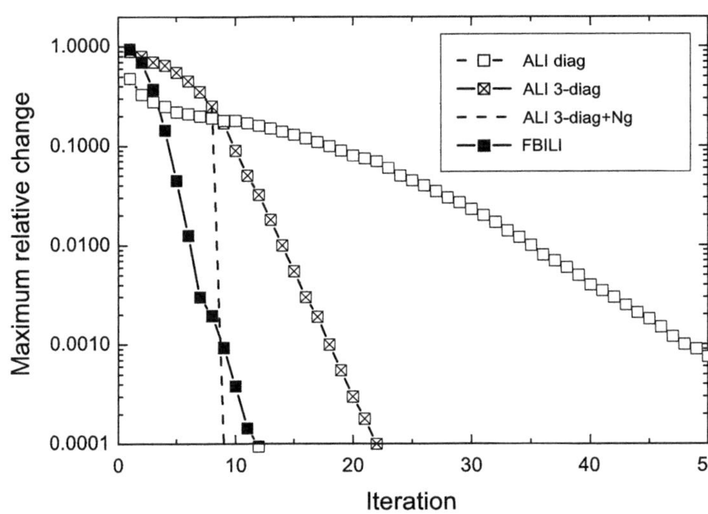

FIGURE 3.4. Maximum relative change of the source function as a function of the iteration number for various ALI schemes and the FBILI method. Two-Level Atom line transfer with no continuum is considered, with $\varepsilon = 10^{-4}$ and $B = 1$ and four points per decade.

to the case of partial frequency redistribution (PRD) by Atanacković-Vukmanović et al. (1997). The method was applied to spherically symmetric monochromatic transfer by Atanacković-Vukmanović (2003), to the Two-Level Atom line formation in 2-D by Milić and Atanacković (2014) and the line transfer in moving media by Pirković and Atanacković (2014). Its application to the Multilevel Atom line transfer is more explicit in Kuzmanovska et al. (2017). Due to its high accuracy and considerable savings of computational time and memory storage (that grow only linearly with the dimension of the problem), this method looks to be a far-reaching tool to deal with more complex problems when the RT has to be coupled with other physical phenomena.

3.8 Nonlinear RT Problems

In this section, we will briefly review the basic mathematical strategies used to solve those RT problems in which the coupling between the radiation field and the state of the medium is strongly *nonlinear*. In particular, we will discuss the paradigm case of the multilevel line formation problem. For the proper analysis of astronomical spectra, it is necessary to consider realistic atom models with many discrete energy levels and the line transitions among them. The main difficulty arises from the nonlocal and nonlinear coupling of the atomic level populations and the radiation field intensities in the corresponding line transitions. The simultaneous solution of the RT and the SE equations is required and can be realized only by means of an iterative procedure. The various methods that can be employed differ in the way the problem is linearized at each iteration step.

3.8.1 *Multilevel Line Formation Problem*

Let us consider only the bound–bound (line) transitions among the NL levels of an atom or ion in a static and plane-parallel atmosphere. The temperature and electron density as well as the continuum opacity χ^C and the emissivity η^C at the nominal line frequency are prescribed as a function of depth. We also assume that the coefficients χ^C and η^C are independent of frequency over the line width. For each line transition, the RTE is

$$\mu \frac{dI_{\nu\mu}}{d\tau_\nu} = I_{\nu\mu} - S_\nu, \tag{3.99}$$

where the monochromatic optical depth τ_ν is defined by (3.10).

The *total* source function S_ν is by definition

$$S_\nu = \frac{\eta^C + \eta^L_\nu}{\chi^C + \chi^L_\nu}. \tag{3.100}$$

By substituting the continuum and the line source function, defined respectively by $S^C \equiv \eta^C/\chi^C$ and $S^L_\nu \equiv \eta^L_\nu/\chi^L_\nu$, assuming complete frequency redistribution (see Section 2.9.1) and using $\chi^L_\nu = \chi^L \varphi_\nu$, we can recast (3.100) into the form

$$S_\nu = \frac{\beta}{\beta + \varphi_\nu} S^C + \frac{\varphi_\nu}{\beta + \varphi_\nu} S^L, \tag{3.101}$$

where $\beta \equiv \chi^C/\chi^L$.

As shown in Section 1.9.4, the frequency independent line source function for each transition ij between the bound levels i and j $(i < j)$ is given by

$$S^L_{ij} = \frac{n_j A_{ji}}{n_i B_{ij} - n_j B_{ji}} = \frac{2h\nu^3_{ij}}{c^2} \frac{1}{n_i g_j / n_j g_i - 1}. \tag{3.102}$$

Here, n_i and n_j denote the number density of the level i and j, g_i and g_j are the corresponding statistical weights; and A_{ji}, B_{ji} and B_{ij} are Einstein's radiative coefficients.

It is clear from (3.99), (3.101) and (3.102) that in a multilevel atom the specific intensities at the frequencies of the different spectral lines are coupled through the corresponding source functions that include the populations of the atomic levels. The values of the latter are determined by the solution of the SE (or rate) equations that describe, for each level, the balance between the processes that populate and depopulate the level. At each depth point, the individual SE equation for the level j reads

$$\sum_{i<j}[(n_i C_{ij} - n_j C_{ji}) + (n_i B_{ij} - n_j B_{ji})\bar{J}_{ij} - n_j A_{ji}]$$
$$+ \sum_{m>j}[(n_m C_{mj} - n_j C_{jm}) - (n_j B_{jm} - n_m B_{mj})\bar{J}_{mj} + n_m A_{mj}] = 0. \qquad (3.103)$$

The probability per unit time that an atom undergoes a transition between the levels i and j $(j > i)$ is given in terms of the Einstein's radiative coefficients A_{ji}, B_{ji} and B_{ij} as well as the inelastic collisional rates C_{ji} and C_{ij}. The atomic level populations depend on the radiation field via the radiative rates that include the scatteringlike integrals (mean intensities averaged over the line profiles) $\bar{J}_{ij}$s. For each line transition ij, they are defined as

$$\bar{J}_{ij}(\tau) = \frac{1}{2}\int_0^\infty \varphi_\nu(\tau)d\nu \int_{-1}^{1} I_{\nu\mu}(\tau)d\mu, \qquad (3.104)$$

and are computed on a standard optical depth scale τ.

For an NL-level atom model, the $NL - 1$ linearly independent SE equations (3.103) are closed by the particle conservation constraint, i.e.,

$$\sum_{j=1}^{NL} n_j = n_{\text{total}}, \qquad (3.105)$$

where n_{total} is the total number density of atoms of the chemical element under consideration.

The SE equations can be written formally as

$$\boldsymbol{R}\,\mathbf{n} = \mathbf{b}, \qquad (3.106)$$

where $\boldsymbol{R}$ is the rate matrix, $\mathbf{n}$ is a vector whose components are the values of the atomic-level populations and $\mathbf{b}$ is a vector with only one nonzero component.[32]

If the $\bar{J}_{ij}$s were given, the SE equations would be linear. But the $\bar{J}_{ij}$s, determined by the solution of the RTE, depend in turn on the values of the level populations. Therefore, the SE equations are intrinsically entangled with the RTE and the global problem results *nonlinear.* Because of this strong coupling, the simultaneous solution of the system of joint equations requires an iterative procedure. Linearization and preconditioning, the two most widely used approaches to achieve linearity within an iterative procedure, will be briefly described in the next two subsections.

3.8.2 *Linearization*

The exact solution of the system of strongly nonlinear, coupled RT and SE equations cannot be obtained of course in a closed form. Auer (1973) was the first to propose a fully self-consistent solution of the multilevel non-LTE line transfer problem by applying the complete linearization (CL) method, previously introduced by Auer and Mihalas (1969) for stellar atmosphere modelling.

For the numerical solution of the multilevel line formation problem in a given model atmosphere, it is necessary, after discretization of the independent variables (frequencies,

[32] For more details, see section 5-4 of Mihalas (1978).

angles and depths), to replace the second-order differential RT equations ((3.72) if the variable Eddington factors are used) with difference equations[33] and the scattering integrals by quadrature sums. Thus we obtain a system of coupled, nonlinear algebraic equations, that is, the system of the second-order RT difference equations for the mean intensities of radiation field J_i $(i = 1, NF)$ at NF chosen frequency points (in all lines and continua of the atom under study) and the SE equations for the populations n_j $(j = 1, NL)$ of NL selected atomic levels. This system can be formally written as

$$\boldsymbol{F}(\boldsymbol{\Psi}) = 0, \tag{3.107}$$

where $\boldsymbol{\Psi} = (\boldsymbol{\Psi}_1, \ldots, \boldsymbol{\Psi}_N)$ is the global state vector, consisting of the vectors

$$\boldsymbol{\Psi}_l = (J_1, \ldots, J_{NF}, n_1, \ldots, n_{NL}), \tag{3.108}$$

functions of all the $NF + NL$ unknown quantities at each depth point $l = 1, N$. Because of the nonlocal nature of the problem, each vector at point l is coupled with those at all the other depth points.

In order to obtain the required solution $\{\boldsymbol{\Psi}_l\}$, we start with some initial estimate $\boldsymbol{\Psi}_l^{(0)}$ and at each nth step of the unavoidable iterative procedure we solve the system of linearized equations for the corrections $\boldsymbol{\delta\Psi}_l$ to the current solution $\boldsymbol{\Psi}_l^{(n)}$. Assuming that $\boldsymbol{\delta\Psi}_l$ is small compared to $\boldsymbol{\Psi}_l^{(n)}$, we solve the system of linearized equations

$$F_l(\boldsymbol{\Psi}_l^{(n)} + \boldsymbol{\delta\Psi}_l) = F_l(\boldsymbol{\Psi}_l^{(n)}) + \sum_{k=1}^{K} \frac{\partial F_l}{\partial \Psi_{l,k}} \, \delta\Psi_{l,k} = 0. \tag{3.109}$$

The linearization of the RTEs in difference equation form brings about terms in δJ_ν, $\delta\chi_\nu$ and $\delta\eta_\nu$ (since $S_\nu = \eta_\nu/\chi_\nu$). The changes in the absorption and emission coefficients, $\delta\chi_\nu$ and $\delta\eta_\nu$, can be eliminated by means of the linearized SE equations to yield a system of coupled linear equations for the corrections δJ_ν only. The resulting system of linearized finite-difference RT equations has the block-tridiagonal matrix form

$$\boldsymbol{A}_l \; \boldsymbol{\delta}\mathrm{J}_{l-1} - \boldsymbol{B}_l \; \boldsymbol{\delta}\mathrm{J}_l + \boldsymbol{C}_l \; \boldsymbol{\delta}\mathrm{J}_{l+1} = \mathrm{L}_l, \tag{3.110}$$

which can be solved for the corrections $\boldsymbol{\delta}$Js by means of a recursive standard Gaussian elimination scheme. In (3.110) $\boldsymbol{A}_l$, $\boldsymbol{B}_l$ and $\boldsymbol{C}_l$ are matrices of dimension $NF \times NF$ at each depth l. Thus, thanks to the explicit elimination of the SE equations, the order of the basic matrices is reduced from $NF + NL$ to NF only. As the number of operations required to obtain the solution of this system scales as NF^3, the total computing time scales as $NF^3 \times N \times N_{iter}$. The solutions δJ_νs of (3.110) satisfy both the linearized RT and the linearized rate equations. The new (updated) values of J_ν are then used to derive new populations, which are in turn used in the formal solution of the RTE to obtain the new J_νs, and so on. The complete linearization method is exhaustively described in section 12-3 of Mihalas (1978).

The convergence rate of this CL method is extremely high. However, due to the huge size of the matrices that must be inverted, it is very expensive in terms of CPU time and memory storage. Therefore, a compromise must be sought between the request for a large discrete mesh in order to warrant an accurate numerical solution and the need

[33] It is much easier to linearize the RTE in the difference equation form than in the integral form.

to keep the dimension of the matrices as small as possible. The possibility of avoiding a matrix representation at all is an alternative option to be considered. These aims can be achieved by using the ideas suggested by the various methods that will be considered in the next subsection.

3.8.3 *Preconditioning*

Because of the overwhelming computational cost of global methods such as the CL, less expensive, simpler and more efficient methods were sought to solve more realistic and hence more complex problems. These methods employ in general a sequential iterative procedure akin to that proposed by Simonneau and Crivellari (1994). Such a strategy is based on the idea that a global problem can be split into a set of interconnected problems, self-consistent from the physical point of view. Each individual problem constitutes a block of a block diagram, representative of a specific physical instance that is described by the relevant equations. Each block is an 'atomic' problem, in the sense that it contains all the physical information necessary for its solution. The global solution requires therefore a sequential structure such that the output of each block, i.e., the solution of the corresponding system of equations, will be the input for the successive block. The procedure is iterated until the fulfilment of the constraints (usually conservation equations) that characterize the problem.

In the specific case of multilevel line formation, two blocks can be separated: the first one accounts for the transport of radiant energy, described by the RT equations; the second one for the radiative and collisional transitions among the atomic levels, described by the SE equations. Each block is solved sequentially by taking as known the solution (or part of the solution) of the other one. The coupling of the two blocks is performed at the end of the iterative procedure. The convergence rate of this algorithm, like in any akin procedure, depends on the amount of information exchanged between the two blocks.

One of the first attempts to facilitate the solution of the multilevel line transfer problem was the Equivalent-Two-Level-Atom (ETLA) approach developed by Avrett (1968) and described in Mihalas (1978) and Avrett and Loeser (1987). ETLA simplifies the use of the SE equations so that only one transition in the model atom at a time is combined with radiative transfer; the coupling of all the levels is achieved by iteration over all the transitions.

Soon after the introduction of the CL method, Cannon (1973) proposed the operator (splitting) perturbation technique that, later reformulated by Scharmer (1981), is well known in the field of RT as ALI.[34] From the mathematical point of view, ALI is an application of *preconditioning* to the iterative solution of a linear system of equations.[35] The implicit use of approximate operators in the non-LTE solution became the method of choice for solving problems of increasing complexity. These methods, however, still required the linearization of either both transfer and rate equations (Scharmer and Carlsson, 1985), or the rate equations only, when Scharmer's simpler ALO was generalized to the multilevel non-LTE line formation by Werner and Husfeld (1985).

[34] A survey of operator perturbation methods can be found in Kalkofen (1987).

[35] Preconditioning is the application of a transformation, called the *preconditioner*, that conditions a given problem into a form that is more suitable for numerical solving methods. It reduces the condition number of the problem, which in numerical analysis measures the sensitivity of the function to small changes in the input argument. A preconditioner $\boldsymbol{P}$ of a matrix $\boldsymbol{A}$ is a matrix such that $\boldsymbol{P}^{-1}\boldsymbol{A}$ has a smaller condition number than $\boldsymbol{A}$. The Jacobi preconditioner is one of the simplest, chosen as the diagonal of the matrix.

The approach, which eliminates linearization by employing preconditioning of the RT equations and introducing an ALO directly into the SE equations to make them linear from the outset, is due to Rybicki and Hummer (1991). The diagonal part of the exact Λ operator is used as an approximate operator Λ^* in the implicit relations between the integrated mean intensity $\bar{J}_{ij}$ and the (yet unknown) line source functions S_{ij}^L for each line transition, i.e.,

$$\bar{J}_{ij} = (\Lambda - \Lambda^*) S_{ij}^{L\ (o)} + \Lambda^* S_{ij}^L = a_{ij} + b_{ij} S_{ij}^L. \tag{3.111}$$

Because $S_{ij}^{L\ (o)}$ is the old value of the line source function, known from the previous iteration, the coefficients a_{ij} and b_{ij} are known. By substituting (3.111) into the SE equations (3.103), together with (3.102), one obtains the preconditioned SE equations

$$\sum_{i<j} [(n_i C_{ij} - n_j C_{ji}) + (n_i B_{ij} - n_j B_{ji}) a_{ij} - n_j A_{ji}(1 - b_{ij})]$$
$$+ \sum_{m>j} [(n_m C_{mj} - n_j C_{jm}) - (n_j B_{jm} - n_m B_{mj}) a_{jm} + n_m A_{mj}(1 - b_{jm})] = 0 \tag{3.112}$$

that are linear in the level populations. The form of the SE equations remains the same as before but with altered coefficients. The solution of (3.112) leads to the updated values of the level populations and thus of the source functions. The convergence is several orders faster than that of the classical Λ iteration. This Jacobi iteration scheme, customarily called Multilevel Accelerated Lambda Iteration (MALI), is probably the most widely used ALI method for the multilevel RT problems in the literature.

The MALI method was successfully applied to the solution of various multilevel RT problems, (e.g., in multidimensional multilevel line RT by Auer et al., 1994), in isolated solar atmospheric structures by Heinzel (1995), in multilevel RT with partial frequency redistribution by Uitenbroek (2001). The use of local operators requires less CPU time per iteration but more iterations than that of the global ones. Because of that, the MALI method is usually accelerated furthermore by means of some mathematical devices, of which the Ng acceleration[36] is the most frequently used. In order to provide optimal convergence with the Ng method, some preliminary experimentation is required to find out the optimal values of two necessary parameters: the iteration number at which the acceleration is switched on and the number of previous iterates.

One can say that, in general, the preconditioning makes the coupling linear thanks to certain approximations. A way to transform the original nonlinear equations into linear ones is to take as known some quantities computed in the previous iteration. All the iterative methods developed to solve the multilevel line transfer problems assume that all the line absorption coefficients in the formal solution of the RTE are known from the previous iteration. However, one may use the old level populations not only in the formal solution of the RTE, but also in the line-opacity-like terms of the SE equations. This assumption is employed in the generalization of the MALI to nonlocal ALOs, in the Implicit Integral Method developed for the multilevel atom line transfer by Crivellari et al. (2002) and in the generalization of the Iteration Factors Method to the multilevel case by Kuzmanovska-Barandovska and Atanacković (2010). In the two latter papers, this approximation makes it possible that the level populations and, consequently, the relevant line source functions can be expressed as a linear

[36] In the Ng acceleration method (Ng, 1974) the up-to-date solution is estimated on the basis of several consecutive iterations.

function of the full set of radiation field mean intensities in all line transitions. These two methods are rapidly convergent with no need for an additional mathematical acceleration.

Numerical methods for NLTE RT applications that provide a convergence much more rapid than the Jacobi's method are the G–S-type methods developed by Trujillo Bueno and Fabiani Bendicho (1995), and by Atanacković-Vukmanović (1991) and Atanacković-Vukmanović et al. (1997). These methods use the same preconditioning of the SE equations as in the MALI (Jacobi) method by Rybicki and Hummer (1991). However, in contrast to Jacobi's iterations, in the G–S methods the computation of the coefficients of (3.111) and the solution of the preconditioned SE equations (3.112) are performed along the iteration step, when sweeping point by point from the bottom up to the surface, using the up-to-date values as soon as available.

The generalization of the G–S and SOR[37] methods to the multilevel atom case is summarized in section 2 of Fabiani Bendicho et al. (1997). These methods were extended to the case of multilevel radiative transfer in spherical geometry by Asensio Ramos and Trujillo Bueno (2006), whereas Paletou and Léger (2007) made more explicit the generalization of the Gauss–Seidel method by Trujillo Bueno and Fabiani Bendicho (1995) to the multilevel RT.

Since mathematical techniques for additional acceleration (such as Ng acceleration, SOR, etc.) require a preliminary analysis to find the optimal values of the necessary parameters, numerical methods for RT problems fast enough to not need additional acceleration are highly desirable. The FBILI is such a method. The generalization of the FBILI method (see Section 3.7.10) to the multilevel line formation is described in general terms by Atanacković-Vukmanović (1991) and Atanacković-Vukmanović et al. (1997). Its implementation to the multilevel line transfer is made more explicit in the paper by Kuzmanovska et al. (2017), where the solution of the CaII lines formation problem in the solar atmosphere is compared with that obtained by the well-known code MULTI (Carlsson, 1986). It is shown that the extremely high convergence rate of the FBILI method is comparable to that of the global operator method, while its memory storage and CPU time per iteration are as small as that of a local operator method. The fact that the FBILI method with no additional acceleration is faster by a factor 5–6 than the Jacobi method may be of importance in the solution of other more complex (astro)physical problems.

Acknowledgements

I wish to thank the Instituto de Astrofisica de Canarias and especially the organizers of the XXIX Canary Islands Winter School of Astrophysics for the invitation and opportunity to give these lectures. I also wish to thank my teachers colleagues and friends, Eduardo Simonneau and Lucio Crivellari, who significantly contributed to the work presented in the lectures. My special thanks are extended to Lucio Crivellari for his careful reading, many useful suggestions and invaluable help in revising the manuscript. This work was also supported by the Ministry of Science, Education and Technological Development of Republic of Serbia through the Project 176004 'Stellar Physics'.

[37] The method of successive overrelaxation (SOR) is a variant of the G–S method, in which a current solution of the linear vector equation is given by $\mathbf{x}^{i+1} = \omega \mathbf{x}^{i+1} + (1-\omega)\mathbf{x}^{i}$, where ω is a prechosen real parameter. For an exhaustive treatment of the topic, see, e.g., Young (1971).

REFERENCES

Asensio Ramos, A. and Trujillo Bueno, J. 2006. *EAS Publ. Ser.* **18**, 25–48

Atanacković-Vukmanović, O. 1991. PhD Thesis, Department of Astronomy, Faculty of Mathematics, University of Belgrade

Atanacković-Vukmanović, O. 2003. *Serb. Astron. J.*, **167**, 27–34

Atanacković-Vukmanović, O. 2007. Pages 97–102 of: Demircan, O., Selam, S. O. and Albayrak, B. (eds.), *Solar and Stellar Physics through Eclipses.* ASP Conf. Ser., **370**, Astronomical Society of the Pacific

Atanacković-Vukmanović, O., and Simonneau, E. 1994. *JQSRT*, **51**, 525–543

Atanacković-Vukmanović, O., Crivellari, L. and Simonneau, E. 1997. *Astrophys. J.*, **487**, 735–746

Auer, L. 1973. *Astrophys. J.*, **180**, 469–472

Auer, L. 1976. *JQSRT*, **16**, 931–937

Auer, L. 2003. Pages 3-15 of: Hubený, I., Mihalas, D. and Werner, K. (eds.), *Stellar Atmosphere Modeling* ASP Conference Series, **288**, Astronomical Society of the Pacific

Auer, L., Fabiani Bendicho, P. and Trujillo Bueno, J. 1994. *Astron Astrophys*, **292**, 599–615

Auer, L. H. and Mihalas, D. 1969. *Astrophys. J.*, **158**, 641–655

Auer, L. H. and Mihalas, D. 1970. *MNRAS*, **149**, 65–74

Avrett, E. H. 1968. Pages 27–63 of: Athay, R. G., Mathis, J., and Skumanich, A., (eds.), *Resonance Lines in Astrophysics* NCAR

Avrett, E. H. and Loeser, R. 1969. *SAO Special Report*, **303**

Avrett, E. H. and Loeser, R. 1987. Pages 135–161 of: Kalkofen, W. (ed.), *Numerical Radiative Transfer* Cambridge University Press

Cannon, C. J. 1973. *JQSRT*, **13**, 627–633

Carlsson, M. 1986. *Uppsala Astron. Obs. Rep.*, **33**

Courant, R. and Hilbert, D. 1937. *Methods of Mathematical Physics Vol. I*, John Wiley and Sons

Crivellari, L. and Simonneau, E. 1991. *Astrophys. J.*, **367**, 612–618

Crivellari, L. and Simonneau, E. 1994. *Astrophys. J.*, **429**, 331–339

Crivellari, L. and Simonneau, E. 1995. *Astrophys. J.*, **451**, 328–334

Crivellari, L., Cardona, O. and Simonneau, E. 2002. *Astrophysics*, **45**, 480–488

Fabiani Bendicho, P., Trujillo Bueno, J. and Auer, L. 1997. *Astron. Astrophys.*, **324**, 161–176

Feautrier, P. 1964a. Pages 80–82 of: Avrett, E. H., Gingerich, O. and Whitney, C. A. (eds.), Proceedings of the First Harvard–Smithsonian Conference on Stellar Atmospheres, SAO Spec. Rep. No. 167, Cambridge

Feautrier, P. 1964b. *C. R. Acad. Sci.*, Paris, **258**, 3189

Feautrier, P. 1964c. Pages 108–110 of: Avrett, E. H., Gingerich, O. and Whitney, C. A. (eds.), Proceedings of the First Harvard–Smithsonian Conference on Stellar Atmospheres, SAO Spec. Rep. No. 167, Cambridge.

Fieldus, M. S., Lester, J. B. and Rogers, C. 1990. *Astron. Astrophys.*, **230**, 371–379

Füchtbauer, Ch. 1920. *Phys. Zeit.*, **21**, 322

Gros, M., Crivellari, L. and Simonneau, E. 1997. *Astrophys. J.*, **489**, 331–345

Hayek, W., Asplund, M., Carlsson, M., et al. 2010. *Astron. Astrophys.*, **517**, A49

Heinzel P. 1995. *Astron. Astrophys.*, **299**, 563–573

Hopf, E. 1928. *Zeitschrift für Phys.*, **46**, 374–382

Hubený, I. 2003. Pages 17–30 of: Hubený, I., Mihalas, D. and Werner, K. (eds.), *Stellar Atmosphere Modeling*, ASP Conference Series, **288**, Astronomical Society of the Pacific

Hubený, I. and Mihalas, D. 2015. *Theory of Stellar Atmospheres (An Introduction to Astrophysical Non-Equilibrium Quantitative Spectroscopic Analysis)*, Princeton University Press

Kalkofen, W. 1984. *Methods in Radiative Transfer*, Cambridge University Press

Kalkofen, W. 1987. *Numerical Radiative Transfer*, Cambridge University Press

Kuzmanovska-Barandovska, O. and Atanacković, O. 2010. *JQSRT*, **111**, 708–722

Kuzmanovska, O., Atanacković, O. and Faurobert, M. 2017. *JQSRT*, **196**, 230–241

Mihalas, D. 1978. *Stellar Atmospheres*, 2nd edition, W.H. Freeman and Co.

Mihalas, D. and Weibel Mihalas, B. 1984. *Foundations of Radiation Hydrodynamics*, Oxford University Press

Mihalas, D., Auer, L. H. and Mihalas, B. R. 1978. *Astrophys. J.*, **220**, 1001–1023

Milić, I. and Atanacković, O. 2014. *Adv. Space Res.*, **54**, 1297–1307

Ng, K.-C. 1974. *J. Chem. Phys.*, **61**, 2680–2689

Olson, G. L., Auer, L. H. and Buchler, J. R. 1986. *JQSRT* **35**, 431–442

Olson, G. L. and Kunasz, P. B. 1987. *JQSRT*, **38**, 325–336

Paletou, F. and Léger, L. 2007. *JQSRT*, **103**, 57–66

Pirković, I. and Atanacković, O. 2014. *Serb. Astron. J.*, **189**, 53–67

Puls, J. and Herrero, A. 1988. *Astron. Astrophys.*, **204**, 219–228

Rybicki, G. B. 1971. *JQSRT*, **11**, 589–595

Rybicki, G. B. 1972. Pages 145–165 of: Athay, R. G., House, L. L. and Newkirk, G. Jr. (eds), *Line Formation in the Presence of Magnetic Fields*, Manuscripts presented at a conference held in Boulder, Colorado, 1971. NCAR

Rybicki, G. B. 1991. Pages 1–8 of: Crivellari L., Hubený I. and Hummer D. G. (eds.), *Stellar Atmospeheres: Beyond Classical Models* NATO ASI Series C, 341

Rybicki, G. and Hummer, D. 1991. *Astron. Astrophys.*, **245**, 171–181

Scharmer, G. B. 1981. *Astrophys. J.*, **249**, 720–730

Scharmer, G. B. and Carlsson, M. 1985. *J. Comput. Phys.*, **59**, 56–80

Simonneau, E. and Atanacković-Vukmanović, O. 1991. Pages 105–110 of: Crivellari, L., Hubený, I. and Hummer, D. G. (eds.), *Stellar Atmospeheres: Beyond Classical Models* NATO Advanced Science Institutes (ASI) Series C, 341

Simonneau, E. and Crivellari, L. 1988. *Astrophys. J.*, **330**, 415–434

Simonneau, E. and Crivellari, L. 1993. *Astrophys. J.*, **409**, 830–840

Simonneau, E. and Crivellari, L. 1994. *Structural Algorithms to Solve Radiative Transfer Problems.* Research Project, Instituto de Astrofisica de Canarias

Simonneau, E., Cardona, O., Crivellari, L. 2012. *Astrophysics*, **55**, 110–126

Trujillo Bueno, J. and Fabiani Bendicho, P. 1995. *Astrophys. J.*, **455**, 646–657

Uitenbroek, H. 2001. *Astrophys. J.*, **557**, 389–398

van Noort, M., Hubený, I. and Lanz, T. 2002. *Astrophys. J.*, **568**, 1066–1094

Werner, K. and Husfeld, D. 1985. *Astron. Astophys.*, **148**, 417–422

Young, D. M. 1971. *Iterative Solution of Large Linear Systems*, Academic Press

4. Stellar Atmosphere Codes

MATS CARLSSON

Abstract

A description is given of stellar atmosphere codes – both codes for calculating the structure of the stellar atmosphere (i.e., including an energy equation) and codes for calculating the emergent spectrum from a given atmospheric structure. Emphasis has been given to codes that are either publicly available or in wide use by a large community. References are given for detailed code descriptions and for typical applications of the codes.

4.1 Introduction

Giving an overview of computer codes is always a very difficult task. A very large number of codes exist; some are known only by a few persons, some are widely used by a big community. There are codes that are kept under very tight control and there are others that are freely available.

The trend is to make codes more widely available by making them open source in some form or the other. Some journals have adopted a policy that all codes used to produce results reported on in a paper must be publicly available, and this is certainly reinforcing the trend towards open source. Free access to powerful computer codes would be a great asset to science in general, but there are also points of concern. A major problem is that minimizing the amount of misuse of a code implies a large effort in documentation and coding stringency – an effort that there is little tradition to fund. The size of this effort is always underestimated; this I personally painfully experienced when I decided to make my nonlocal thermodynamic equilibrium (NLTE) diagnostic code MULTI (Carlsson, 1986) public – the extra effort to make the code of maximum use for a wide community compared with keeping it as a personal tool is estimated to one year of work.

In the selection of codes to cover here, emphasis has been given to codes that are either publicly available or in wide use by a large community. For 3-D magnetohydrodynamic (MHD) codes, those used to produce publicly available grids of model atmospheres have been given preference. For the whole selection, my own bakground from cool stellar atmospheres and solar work has undoubtedly also played a role.

Given the title of the winter school, 'Applications of Radiative Transfer to Stellar and Planetary Atmospheres', we will put emphasis on codes that include radiative transfer. We will therefore not cover pure MHD codes or codes that only include simple radiative loss functions. The title of this contribution is 'Stellar Atmosphere Codes'. Hence, we will not attempt to cover codes that are used for planetary atmosphere simulations (except if the same code is also used for stellar atmospheres).

The structure of this chapter is as follows: In Section 4.2, we cover codes that calculate an emergent spectrum from a given atmosphere (diagnostic codes), and in Section 4.3 we cover codes that include an energy equation that gives the atmospheric structure. Some notes on radiation MHD codes are presented in Section 4.4.

A representative subset of the codes are described in Sections 4.5–4.9. These descriptions were mostly written by the main developers of the respective codes with only minor editing for consistency in style. They are all thanked for their responsiveness in this process. I apologize for possible errors or inaccuracies introduced in the editing. Code names in boldface in Sections 4.2–4.4 have descriptions in Sections 4.5–4.9 while code names in italics are not described there.

4.2 Diagnostic Codes

The *stellar atmosphere* is defined as the region of a star from which light escapes. Most of what we know about stars comes from analyzing the radiation they emit. Diagnostic codes here play an important role in enabling the calculation of the emergent spectrum from a given atmosphere. This entails solving the equation of radiative transfer. To make that possible, we need to know the opacity and the source function as function of frequency and as function of height in the atmosphere. Assuming local thermodynamic equilibrium (LTE) reduces the problem to finding the opacity as function of frequency for all spectral lines and continua that contribute to the total opacity in the spectral region of interest. There are many codes available for the synthesis of the spectrum from a given atmosphere under the assumption of LTE. They mainly differ in what formats of input atmosphere and line lists they use as input. **Synspec** (Hubený and Lanz, 2011, 2017) can read in model atmospheres from the structure codes **TLUSTY** or **ATLAS**. **Turbospectrum** (Alvarez and Plez, 1998; Plez, 2012) uses atmospheres from **Marcs**, **ATLAS** or a number of other formats. **SYNTHE** is the companion diagnostic code to the **ATLAS9** and **ATLAS12** structure codes. *SME* (Valenti and Piskunov, 1996; Piskunov and Valenti, 2017) can read atmospheres from **Marcs** and **ATLAS** and is often employed in spectroscopic diagnostics, both in individual studies of FGK stars and in surveys (e.g., the Gaia-ESO and Galah surveys). Another often used diagnostic code is *Moog* (Sneden et al., 2012). The major challenge for the improvement of diagnostic codes that assume LTE is in the completeness and accuracy of the line lists. See Kurucz (2014c) for a discussion.

LTE is a bad approximation when the radiation field is strong (e.g., in hot stars) or densities are low (e.g., in the line-forming region for strong spectral lines). The common approximation to employ is then statistical equilibrium – assuming that the population densities of the various energy levels do not vary in time.

Combining the statistical equilibrium and radiative transfer equations results in a nonlocal, nonlinear system of equations. Pioneering codes like *Linear-A* (Auer et al., 1972) and *Linear-B* (Rybicki, 1971) employed complete linearization (Auer and Mihalas, 1969) for the solution of the resulting equations while *Pandora* (Vernazza et al., 1973; Avrett and Loeser, 1992) employs an equivalent two-level atom formulation and iteration over transitions with special methods to ensure consistency of the solution.

Codes based on accelerated lambda iteration (ALI) started to dominate in the 1980s after the seminal paper by Scharmer (1981). **DETAIL/SURFACE** (Giddings, 1981; Butler and Giddings, 1985) in its original form used the Rybicki formulation of the complete linearization scheme (Rybicki, 1971) but later switched to the ALI scheme as formulated by Rybicki and Hummer (1991). **MULTI** (Carlsson, 1986) got rapid spread thanks to it adhering strictly to the FORTRAN 77 standard (thus being very portable) and being open source. It is still in widespread use 30 years later in spite of its limitations (standard version is restricted to 1-D plane-parallel atmospheres, assumes complete frequency redistribution (CRD) in all lines and cannot treat blends or polarization). In 1986, it took 63 seconds of compute-time to solve a Ca II six-level problem in an atmosphere with 45 depth-points on a VAX-11/750 computer. Today, the same problem with the same code takes 0.08 s on a MacBook Pro. This impressive speedup of a factor 790 is still far below the factor of 1M expected from Moore's law (doubling of computing power every 18 months), which shows that much of the increase in computing power is not in single-core performance but through parallelization.

Several codes have been made publicly available for solving non-LTE problems after the 1980s. A code that is often used as the non-LTE engine in inversions of observed spectral profiles, including polarization, is **Nicole** (Socas-Navarro et al., 2015). The code is restricted to one atom at a time and to 1D or 1.5-D (column by column as 1-D problems in a 3-D atmosphere) problems. Another freely available popular code is **RH** (Uitenbroek, 2000, 2001, 2003; Pereira and Uitenbroek, 2015). The code can treat partial frequency distribution; multiple atoms simultaneously; and 1-D (plane-parallel and spherical), 2-D and 3-D, geometries. In 3-D, there is no domain decomposition so one quickly runs out of memory. For 1.5-D problems, there is a special version parallelized with MPI. The code **PHOENIX** (see Section 4.8.3 for code references) is a powerful structure code in 1-D (see Section 4.3) that can also be used as a diagnostic code in 3-D, in full non-LTE, including planetary atmospheres. **Multi3d** (Leenaarts and Carlsson, 2009) is another alternative for full 3-D non-LTE diagnostic problems. The 1-D/1.5-D non-LTE codes mentioned (**Detail/Surface, Multi, Nicole, RH**) are all freely available while the 3-D non-LTE codes (**PHOENIX, Multi3d**) are only available through collaboration.

In addition to the codes that are included in the description chapters, there are a number of other codes (all proprietary) for calculating a synthetic spectrum for a given atmosphere, including various levels of sophistication for the physical description or geometry. *MUGA* (Trujillo Bueno and Fabiani Bendicho, 1995) is a non-LTE code that pioneered the use of Gauss–Seidel iterations, and multigrid in 3-D, *PORTA* (Štěpán and Trujillo Bueno, 2013) is a massively parallel code focusing on the detailed calculation of polarization in individual spectral lines, including scattering polarization and the Hanle and Zeeman effects. One should also keep in mind that many of the codes that have an energy equation to solve for the atmospheric structure (see Section 4.3) include modules to calculate the emergent spectrum.

4.3 Structure Codes

Diagnostic codes discussed in the previous section take as input a given atmospheric structure (normally temperature, electron density, density, line-of-sight velocity, microturbulence as functions of a depth variable – that can be height, column mass or standard optical depth). Some of the codes include the possibility to solve the equations of hydrostatic equilibrium and charge conservation to get the density and electron density consistent with the other variables. To also solve for the temperature structure, we need to include an energy equation. This equation should include all significant energy transport mechanisms – in stellar photospheres, certainly radiative transport but often also convective energy transport. In coronae, conduction is another important energy transport mechanism. In 1-D, convective energy transport has to be described by some approximate recipe – often with a mixing length theory. In a plane-parallel atmosphere, we get the equation

$$F_{\rm rad} + F_{\rm conv} = \sigma T_{\rm eff}^4, \tag{4.1}$$

where $F_{\rm rad}$ is the radiative flux, $F_{\rm conv}$ is the convective flux, σ is Stefan–Boltzmann's constant and $T_{\rm eff}$ is the effective temperature. Neglecting the convective flux for a moment, we can write the condition of radiative equilibrium as

$$\nabla F_{\rm rad} = 0 \tag{4.2}$$

Integrating the equation of radiative transfer over angle and frequency, we obtain

$$\int_0^\infty \chi_\nu (S_\nu - J_\nu) d\nu = 0, \tag{4.3}$$

where χ_ν is the monochromatic opacity, S_ν is the source function and J_ν is the mean intensity. Equation 4.3 shows that we need to know the opacity as a function of wavelength over the whole spectrum to include radiative equilibrium as part of the structure equations. The challenge is that the opacity varies orders of magnitude across spectral lines that are narrow in wavelength, which means that the opacity evaluation and the solution of the radiative transfer equation have to be performed for a very large number of frequency points to get an accurate spectrum. Realizing that it is not necessary to reproduce the spectrum but only to get the integral over frequency right relaxes the number of frequency points needed. Sampling the spectrum with enough points to get the effects on the integral from changes in the atmosphere in a statistical sense is called opacity sampling (OS). On the order of 10^5 frequency points are often used in OS methods.

Another simplification comes from the realization that we only need to sample high and low opacities in the right mixture over a small wavelength interval but not necessarily at the right wavelengths. Reordering the opacity in narrow wavelength intervals makes it possible to get a good statistical representation of high and low opacities with a far smaller number of frequency points; Figure 4.1 illustrates this reordering process. The method of reordered opacities over narrow wavelength regions is called opacity distribution function (ODF). One problem with the ODF method is that high opacities are always shifted to the same end of the wavelength interval. If high opacities come from different sources (e.g., different molecules) at different heights in the stellar atmosphere, this is not a good approximation. Another problem is that the full ODF tables are normally recalculated if one changes one abundance. This is not always necessary – there are methods to tabulate individual ODFs and sum them afterwards (Saxner and Gustafsson, 1984).

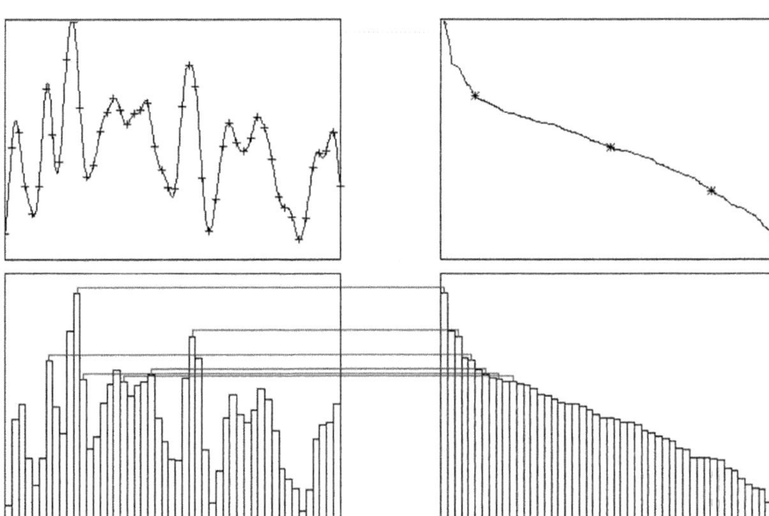

FIGURE 4.1. Wavelength-dependent opacity (upper left) and its ODF representation (upper right) using far fewer points. The lower panels show the same but in the form of histograms, illustrating more clearly the reordering scheme.

The code **ATLAS9** is a commonly used 1-D LTE atmospheric structure code that uses pretabulated ODFs to approximate the opacity as function of wavelength. The code **ATLAS12** is a version of the code that uses opacity sampling instead. Both codes are publicly available and have been used to calculate large grids of stellar atmosphere models.

Another 1-D LTE atmospheric structure code (only available through collaboration) that has been extensively used is **MARCS** (Gustafsson et al., 1975; Plez et al., 1992; Gustafsson et al., 2008). It also exists in both ODF and OS versions and has mostly been used for the modelling of cool star atmospheres, including detailed molecular opacities.

The Munich 1-D LTE code for cool atmosphere modelling, *MAFAGS* (Grupp, 2004a,b) exists both in ODF and OS versions and is widely used in the community.

For the modelling of hot stars, a number of stellar atmosphere codes exist that include departures from LTE. The most commonly used codes are **TLUSTY**, **POWR**, **PHOENIX**, **CMFGEN**, **WM-basic** and **FASTWIND**. They differ in how they treat NLTE line blanketing, winds, geometry and computing time. Puls (2009) gives an overview with more recent developments summarised by Sander (2017).

4.4 Radiation MHD Codes

The structure codes described in Section 4.3 all assume a time-independent state. In recent years, we have seen more and more an emphasis on the inadequacy of average models to describe the average of a time-dependent, dynamic atmosphere. Carlsson and Stein (1995) even questioned whether a nonmagnetic solar chromosphere exists – the average spectrum from a dynamic simulation that included acoustic waves travelling through the solar atmosphere needed a chromospheric temperature rise when modelled with a 1-D, static, semi-empirical atmosphere while the average temperature in the dynamic model did not exhibit a temperature rise. The modelling was performed with the 1-D hydrodynamic code **RADYN**, which solves the hydrodynamic equations together with population rate equations for models of hydrogen, calcium and helium.

Radiation hydrodynamical modelling in 3-D has also been possible under the assumption of LTE. The computational efforts make it prohibitive to even use the ODF approach, and Nordlund (1982) came up with an ingenious way to extend the ODF approach to the full spectrum with appropriately defined averages of the source function in addition to the opacity. Such a multigroup opacity approach can give a decent approximation to the nongrey radiative transfer with as few as four bins (although modern implementations often employ 10–12 bins for increased accuracy). Such 3-D hydrodynamic models are now a standard tool in spectral line synthesis from cool stars for the determination of stellar abundances. Detailed comparisons between such 3-D solar models and high-resolution spectra show a remarkable agreement without invoking large numbers of free parameters (Pereira et al., 2013a).

A number of codes have been developed to treat such 3-D dynamic problems, also including magnetic fields. Codes that have produced published grids of 3-D atmospheres include **CO5BOLD**, **Stagger** and **Bifrost**, where the latter includes a generalization of Nordlund's multigroup opacity method to include coherent scattering (Skartlien, 2000) and approximations for non-LTE radiative losses in the chromospere (Carlsson and Leenaarts, 2012) enabling more realistic modelling of chromospheric conditions. A comparison between the **CO5BOLD**, *MURaM* and **Stagger** codes was published by Beeck et al. (2012).

4.5 Code Descriptions: Diagnostics LTE

4.5.1 *SYNSPEC*

Main developer(s):
Ivan Hubený, Steward Observatory and Department of Astronomy, University of Arizona, 933 N. Cherry Ave., Tucson, AZ 85721, USA, *hubeny@as.arizona.edu*

Thierry Lanz, Observatoire de Cote d'Azur, Nice, France, *lanz@oca.eu*

Type of code: Calculates emergent spectrum for a given model atmosphere.

Geometry: 1-D plane-parallel

Main characteristics/capabilities: Computes a synthetic spectrum with any spectral resolution for a model atmosphere produced either by TLUSTY or for a Kurucz model. It requires an input line list, typically based on an updated Kurucz line list. If required, the code takes into account molecular lines.

Programming language: FORTRAN 77

Parallelization/architecture limitations: No parallelization implemented

Problem size possible, scaling, example(s) of running time for typical cases: The size is given by the input model atmosphere. The number of included atomic or molecular lines is only limited by the availability of input atomic data.

Literature references for code: Hubený and Lanz (2011, 2017)

Literature references for typical applications of code: Lanz and Hubený (2003); Lanz and Hubený (2007)

Availability: Code is publicly available for download, as specified in detail in Hubený and Lanz (2017).

4.5.2 *SYNTHE*

Main developer(s):
Robert Kurucz, Harvard-Smithsonian Center for Astrophysics, 60 Garden Street, Cambridge, MA, USA, *rkurucz@cfa.harvard.edu*

Type of code: Computes LTE spectrum from 1-D model atmosphere.

Geometry: 1-D plane-parallel

Main characteristics/capabilities: **SYNTHE** is a companion code to **ATLAS9** and **ATLAS12** (see Section 4.7.1) for the computation of detailed spectra that can be compared to high-signal-to-noise, high-resolution spectra.

Programming language: Fortran

Parallelization/architecture limitations: Serial

Literature references for code: Kurucz (2014a,c)

Availability: Freely available from http://kurucz.harvard.edu.
Fiorella Castelli maintains, distributes and uses Linux versions of the Kurucz programs; see wwwuser.oats.inaf.it/castelli.

4.5.3 *Turbospectrum*

Main developer(s):
Bertrand Plez, LUPM, CNRS, Université de Montpellier, 34095 Montpellier, France
bertrand.plez@umontpellier.fr

Type of code: Computes LTE spectrum from 1-D model atmosphere.

Geometry: 1-D, plane-parallel or spherical

Main characteristics/capabilities: Computes 1-D, LTE intensity or flux spectrum for given model atmosphere (MARCS, ATLAS and a number of specific formats). Shares a large number of routines and input physics with MARCS (equation of state, radiative transfer solver, etc.). Radiative transfer is solved along rays with a Feautrier scheme. Large number of lines can be included, with running time scaling accordingly. Source function includes continuum scattering. Possibility to include scattering for lines as well (ad hoc prescription). Appropriate for F-, G-, K-, M-, S- and C-type stars, dwarfs and giants. Limitation towards high and low temperatures due to equation of state. Includes all atoms and first ions, and about 600 molecular species.

Programming language: Fortran

Parallelization/architecture limitations: Serial

Problem size possible, scaling, example(s) of running time for typical cases: Running time scales linearly with number of wavelength points and number of lines. Dimension fixed by number of depth points times number of wavelength points. Fixed wavelength increment. Run time on a laptop for 625,000 wavelength points, 56 depth points, 7×10^6 lines: 8 minutes.

Literature references for code: Alvarez and Plez (1998); Plez (2012)

Literature references for typical applications of code: Plez et al. (2004); de Laverny et al. (2012); Davies et al. (2013); Jofré et al. (2017); Kunder et al. (2017); Pérez-Mesa et al. (2017)

Availability: Code is publicly available at www.pages-perso-bertrand-plez.univ-montp2.fr/ with companion molecular line lists. Additional developments are made in collaboration with the main developer, to allow public release.

4.6 Code Descriptions: Diagnostics NLTE

4.6.1 *DETAIL/SURFACE*

Main developer(s): Keith Butler, Universitätssternwarte, Ludwig-Maximilians-Universität, Scheinerstr. 1, 81679 Munich, Germany, *butler@usm.uni-muenchen.de*

Type of code: Non-LTE radiative transfer code to calculate spectrum from given atmosphere.

Geometry: 1-D plane-parallel

Main characteristics/capabilities: DETAIL provides a combined solution of radiative transfer-statistical equilibrium for fixed temperature-density structure for any number of atoms-ions. The structure is taken from a stellar atmosphere program (e.g., **ATLAS**) since DETAIL itself does not treat radiative/hydrostatic equilibrium. Models are data-driven, by keyword, making the program extremely flexible in the model atoms that can be dealt with. The original program used the Rybicki method for the solution, which scaled linearly with the number of frequencies. In the current version, the ALI iteration scheme devised by Rybicki and Hummer (1991) has been implemented. While the scaling is still linear in the number of frequencies, the solution overall is much more efficient.

Normally, Doppler profiles are used in DETAIL. SURFACE provides a formal solution with accurate line profiles for comparison with observation using the fixed structure and populations from DETAIL. When no populations are present, an LTE calculation is performed.

Programming language: Fortran 90

Parallelization/architecture limitations: None, but major parts are trivially parallel

Problem size possible, scaling, example(s) of running time for typical cases: Typically 70–90 depth points, x0000 frequencies, 100 s energy levels 1,000 s transitions. Run time is problem dependent but of order of minutes, not hours.

Literature references for code: Giddings (1981); Butler and Giddings (1985)

Literature references for typical applications of code: Becker and Butler (1995); Przybilla and Butler (2004); Morel et al. (2006); Przybilla et al. (2006, 2008, 2016); Bergemann et al. (2010, 2015, 2016, 2017); Mashonkina et al. (2011, 2017); Shi et al. (2014, 2018); Kupfer et al. (2017)

Availability: No specific license, program is generally available. No requirement of collaboration.

4.6.2 *Multi*

Main developer(s):
Mats Carlsson, Institute of Theoretical Astrophysics, University of Oslo, P.O. Box 1029 Blindern, NO-0315 Oslo, Norway, *mats.carlsson@astro.uio.no*

Type of code: Non-LTE radiative transfer code to calculate spectrum from given atmosphere.

Geometry: 1-D plane-parallel

Main characteristics/capabilities: Solves the statistical equilibrium equation together with the radiative transfer equation in 1-D. Assumes CRD. Only a single atom is treated in non-LTE; all others except hydrogen are assumed to be in LTE or non-LTE populations and can be read in from file. Hydrogen populations can be specified in an input file or set to LTE. Electron densities are either specified as input, assumed to be LTE or computed consistent with non-LTE hydrogen through inclusion of a charge conservation equation. Incoming radiation from the top boundary can be specified. Blends can only be treated iteratively. The user can choose between the global Scharmer operator (Scharmer, 1981; Scharmer and Carlsson, 1985), ensuring rapid convergence at the expense of a large memory footprint, or a local operator (Rybicki and Hummer, 1991).

Programming language: FORTRAN 77

Parallelization/architecture limitations: No parallelization

Problem size possible, scaling, example(s) of running time for typical cases: A six-level Ca II problem on 45 depth points takes 0.06 s on a MacBook Pro. Model atoms with hundreds of levels can be treated.

Literature references for code: Carlsson (1986)

Literature references for typical applications of code: Carlsson et al. (1992, 1994); Jorgensen et al. (1992); Carlsson and Judge (1993); Kiselman and Carlsson (1996)

Availability: The code is publicly available from http://folk.uio.no/matsc/mul23

4.6.3 *Multi3d*

Main developer(s):

Jorrit Leenaarts, Institute for Solar Physics, Department of Astronomy, Stockholm University, AlbaNova University Centre, SE-106 91 Stockholm, Sweden, *jorrit.leenaarts@astro.su.se*

Johan Pires Bjørgen, Institute for Solar Physics, Department of Astronomy, Stockholm University, AlbaNova University Centre, SE-106 91 Stockholm, Sweden, *johan.bjorgen@astro.su.se*

Andrii V. Sukhorukov, Instituto de Astrofísica de Canarias, E-38205 La Laguna, Tenerife, Spain, *andrii@iac.es*

Type of code: Non-LTE radiative transfer code to calculate spectrum from given atmosphere.

Geometry: The code requires a 3-D input atmosphere. Can compute the radiation field in the 1.5-D approximation (plane-parallel per column) or full 3-D.

Main characteristics/capabilities: Solves the statistical equilibrium equation together with the radiative transfer equation in full 3-D. Can compute lines in CRD or partial frequency distribution (PRD). Only a single atom is treated in non-LTE, all others except hydrogen are assumed to be in LTE. Hydrogen populations can be specified in an input file or set to LTE. Electron densities are either specified as input, assumed to be LTE or computed consistent with non-LTE hydrogen through inclusion of a charge conservation equation. Incoming radiation from the top boundary can be specified. Radiation emitted by the corona (if present in the atmosphere model) can be self-consistently included. Computations can be sped up using a multigrid method. Can compute the emergent radiation for any azimuth and inclination.

Programming language: Fortran 90

Parallelization/architecture limitations: MPI, parallelized over spatial domain and frequency

Problem size possible, scaling, example(s) of running time for typical cases: Scales well up to 4,096 cores. Has been run up to 8,192 cores. Atmospheres have been run up to $1536 \times 768 \times 768$ grid points. Running time depends on the problem and number of cores used. Typical CPU cost for a $504 \times 504 \times 504$ problem and hydrogen treated in 3-D non-LTE PRD is 200,000 CPU hours. CRD problems run a factor 5–10 faster.

Literature references for code: Leenaarts and Carlsson (2009)

Literature references for typical applications of code: Leenaarts et al. (2009, 2012, 2013b, 2016); Bjørgen and Leenaarts (2017); Sukhorukov and Leenaarts (2017)

Availability: The code is currently only available through collaboration with one of the main developers.

4.6.4 *NICOLE*

Main developer(s):
Hector Socas-Navarro, Instituto de Astrofísica de Canarias, Avda Vía Láctea S/N, La Laguna, 38205, Tenerife, Spain; Departamento de Astrofísica, Universidad de La Laguna, La Laguna, 38205, Tenerife, Spain, *hsocas@iac.es*

Type of code: NLTE radiative transfer for Stokes synthesis or inversion.

Geometry: 1.5-D plane-parallel (works with 3-D cubes, but each column is solved independently assuming infinite horizontal atmospheres).

Main characteristics/capabilities: Solves statistical equilibrium equations for given model atom assuming complete angle and frequency redistribution, field-free populations. Optionally, hydrostatic equilibrium and velocity field. Inversion mode assumes hydrostatic equilibrium for computation of z-scale and density stratification. Open source and massively parallel.

Programming language: Fortran 2003 (ANSI Standard)

Parallelization/architecture limitations: MPI over spatial columns

Problem size possible, scaling, example(s) of running time for typical cases: Typical setup: 50 depthpoints, six-level Ca II/III atom, 100 – 100 x,y map, computing time $\sim$1 hour per CPU. Scales linearly up to thousands of processors.

Literature references for code: Socas-Navarro et al. (2015)

Literature references for typical applications of code: Socas-Navarro (2011, 2015); de la Cruz Rodríguez et al. (2012); Henriques et al. (2017); Kuckein et al. (2017)

Availability: Open source. It is strongly advised that inversion mode be used through collaboration.

4.6.5 *PHOENIX*

See Section 4.8.3.

4.6.6 *RH*

Main developer(s):
Han Uitenbroek, National Solar Observatory, 3665 Discovery Drive, Boulder, CO 80303, USA, *huitenbroek@nso.edu*

Type of code: Non-LTE radiative transfer code to calculate spectrum from a given atmosphere.

Geometry: 1-, 2- and 3-D Cartesian geometry, spherical geometry. A derived 1.5-D version was created (see Pereira and Uitenbroek, 2015) for MPI-parallelized column-by-column transfer in a 3-D cube.

Main characteristics/capabilities: Solves the coupled equations of radiative transfer and statistical equilibrium in multiple atoms and molecules, including the effects of transitions overlapping in wavelength, partial frequency redistribution and Zeeman induced polarization. Hydrostatic equilibrium can be solved in 1-D. The effects of irradiation at the upper boundary (in 1-D), and upper and side boundaries (in 2- and 3-D) can be included. Parallelization over wavelength is supported with POSIX threads.

Programming language: C. A library of Interactive Data Language (IDL) routines for analyzing output data is included. The standard RH version reads and writes files in architecture-independent External Data Representation (XDR) format, while the 1.5-D version makes extensive use of the HDF5 library.

Parallelization/architecture limitations: The 1.5-D version employs parallelization through MPI

Problem size possible, scaling, example(s) of running time for typical cases: Typical run times in 1D geometry range from a few seconds for a six-level Ca II solution, including PRD, to minutes for complex atoms with hundreds of lines and levels. In full 3-D geometry, the number of atomic levels in LTE is severely limited, because of internal storage requirements. This limitation is not present for the 1.5-D version, which solves transfer column-by-column, neglecting horizontal transfer. A full 3-D NLTE solution with order of 10 levels and lines on a grid of several hundred squared can take weeks.

Literature references for code: Uitenbroek (2000, 2001, 2003); Pereira and Uitenbroek (2015)

Literature references for typical applications of code: Uitenbroek (1998, 2002); Miller-Ricci and Uitenbroek (2002); Uitenbroek et al. (2004); Uitenbroek and Tritschler (2006); Centeno et al. (2008); Uitenbroek and Criscuoli (2011); Rutten and Uitenbroek (2012); Leenaarts et al. (2013a,b); Pereira et al. (2013b); Maiorca et al. (2014); Holzreuter and Solanki (2015); Okamoto et al. (2015); Kowalski et al. (2017a)

Availability: The code is publicly available and can be obtained by sending an e-mail request to the primary developer.

4.7 Code Descriptions: Structure LTE

4.7.1 *ATLAS9, ATLAS12*

Main developer(s):
Robert Kurucz, Harvard-Smithsonian Center for Astrophysics, 60 Garden Street, Cambridge, MA, USA. *rkurucz@cfa.harvard.edu*

Type of code: Radiative-convective, hydrostatic equilibrium model atmosphere codes

Geometry: 1-D plane-parallel

Main characteristics/capabilities: ATLAS9 uses ODF line opacity, ATLAS12 uses OS line opacity. **SYNTHE** (see Section 4.5.2) is a companion program for calculating the emergent spectrum from a given model atmosphere.

Programming language: Fortran

Parallelization/architecture limitations: Serial

Problem size possible, scaling, example(s) of running time for typical cases: ATLAS9 can produce huge grids for fixed abundances because it pretabulates the opacities.
ATLAS12 computes the opacity for arbitrary abundances for each model, so it is much slower but more detailed.

Literature references for code: Kurucz (2014b,c)

Literature references for typical applications of code: Barban et al. (2003); Castelli and Kurucz (2004, 2006); Kirby (2011); Mészáros et al. (2012)

Availability: Freely available from http://kurucz.harvard.edu
Fiorella Castelli maintains, distributes and uses Linux versions of the Kurucz programs; see wwwuser.oats.inaf.it/castelli.

4.7.2 *MARCS*

Main developer(s):
Bengt Edvardsson, Kjell Eriksson, Bengt Gustafsson, Division of Astronomy and Space Physics, Department of Physics and Astronomy, Uppsala University, Box 516, 75120, Uppsala, Sweden, *bengt.edvardsson@physics.uu.se*,
kjell.eriksson@physics.uu.se, *bengt.gustafsson@physics.uu.se*

Bertrand Plez, LUPM, CNRS, Université de Montpellier, 34095 Montpellier, France *bertrand.plez@umontpellier.fr*

Type of code: Hydrostatic equilibrium, radiative+convective flux conservation, LTE

Geometry: 1-D, plane-parallel or spherical

Main characteristics/capabilities: Solves 1-D, LTE, hydrostatic, convection+ radiative flux conservation atmosphere problem. Radiative transfer is solved along rays with a Feautrier scheme for a large number of frequency points. Source function includes continuum scattering. Opacities are interpolated in tables precomputed for atomic and molecular lines. This makes it possible to include billions of lines while keeping running time low. Mixing-length treatment (MLT) of convection. Radiative and turbulent pressure included. Appropriate for F-, G-, K-, M-, S- and C-type stars, dwarfs and giants. Limitation towards high and low temperatures due to equation of state. Includes all atoms and first ions and about 600 molecular species.

Programming language: Fortran

Parallelization/architecture limitations: Serial. Double precision version available since 2017.

Problem size possible, scaling, example(s) of running time for typical cases: Solves easily with order of 100 depth points, on a tau scale from 10^{-6} to 100, 10^5 wavelengths points for opacity sampling. Typical models converge within five to 10 iterations, more for cool or low-gravity models. One iteration takes less than a minute on a laptop. Running time is about twice as long for spherical models. Running time scales linearly with number of wavelength points.

Literature references for code: Gustafsson et al. (1975, 2008); Plez et al. (1992, 2008)

Literature references for typical applications of code: Bell et al. (1976); Ekberg et al. (1986); Asplund et al. (1997); Bessell et al. (1998); Heiter and Eriksson (2006); Mészáros et al. (2012); Van Eck et al. (2017)

Availability: Code is currently only available through collaboration with one of the main developers. Grids, data and information are available on marcs.astro.uu.se.

4.8 Code Descriptions: Structure NLTE

4.8.1 *CMFGEN*

Main developer(s):
D. John Hillier, Department of Physics and Astronomy and Pittsburgh Particle Physics, Astrophysics, and Cosmology Center (PITT PACC), 3941 O'Hara Street, University of Pittsburgh, Pittsburgh, PA, 15260, USA, *hillier@pitt.edu*

Type of code: Non-LTE time-dependent stellar atmosphere code

Geometry: 1-D. Primarily spherical, but can also treat stellar atmospheres using a plane-parallel approach.

Main characteristics/capabilities: For stellar atmospheres, CMFGEN solves the radiative transfer equation, the non-LTE rate equations and the equation of radiative equilibrium. The structure below the sonic point is treated assuming hydrostatic equilibrium, while above the sonic point both the velocity law and mass-loss rate must be specified.

For supernovae, CMFGEN solves the time-dependent radiative transfer equation, the time-dependent non-LTE rate equations and the time-dependent energy equation. For the first model in a time sequence, we use the density, chemical, velocity and temperature structure arising from hydrodynamic explosion models.

In both cases, a separate code (CMF_FLUX) is used to compute observed spectra. Several diagnostic codes are also available with the CMFGEN distribution.

Programming language: Fortran 90 (compiles under Gfortran, PGI Fortran and Intel Fortran)

Parallelization/architecture limitations: Open Multiprocessing (OpenMP)

Problem size possible, scaling, example(s) of running time for typical cases: Optimal number of cores used with OpenMP (OMP) parallelization is 4 to 12. The code utilizes superlevels to reduce the number of unknown populations explicitly treated, but the number of superlevels and the assignments of levels to superlevels are easily changed. Small models (e.g., a pure H, He atmosphere) run in under an hour. A large atmosphere model (70 grid points with H, He, C, N, O, Ne, Ar, Si, S, P, Ca, Fe and Nk, and with $\sim 3,000$ superlevels) takes less than one day. Realistic supernova models (with ~ 100 grid points, $\sim 3,000$ super levels) take one to two days per time step.

Literature references for code: Hillier (1987, 1990, 2012); Hillier and Dessart (2012); Dessart et al. (2014)

Literature references for typical applications of code: Hillier and Miller (1999); Martins et al. (2002); Hillier et al. (2003); Bouret et al. (2012, 2013); Dessart et al. (2013a,b, 2016); Fierro-Santillán et al. (2017); Neugent et al. (2017); Zsargo et al. (2017); Wilk et al. (2018)

Availability: CMFGEN, auxiliary codes, atomic data and documentation are available online at www.pitt.edu/~hillier. Model atmosphere models are available at the same site and at several other sites (e.g., http://pollux.graal.univ-montp2.fr). Supernovae calculations are available through www-n.oca.eu/supernova/home.html and from the author. Researchers using the code for supernovae calculations might like to contact the author, as setting up the calculations can be difficult for first-time users. The author appreciates feedback on any issues that arise in using the code.

4.8.2 *FASTWIND*

Main developer(s):
Joachim Puls, LMU Munich, University Observatory, Scheinerstr. 1, D-81679 München, Germany, *uh101aw@usm.uni-muenchen.de*

Type of code: Unified, NLTE, stellar atmosphere/spectrum synthesis code

Geometry: 1-D spherical

Main characteristics/capabilities: Allows to calculate unified (photosphere + wind), NLTE, stellar atmospheres of hot, massive stars with winds, together with corresponding spectral energy distributions (SEDs) and spectral lines for specific elements (see described later in this section). Major applications are OB stars of all luminosity classes and early A-type supergiants, with detailed spectrum synthesis in the optical and infrared (IR). The temperature stratification is calculated from the electron thermal balance. In the standard version, line-blocking (from elements H to Zn) is calculated in an approximate, statistical way, to allow for a very fast performance. The user is responsible for providing the atomic models/line data for those elements that shall be analyzed in detail (models for H, He, N and Si are included by default). X-rays from wind-embedded shocks and microclumping can be accounted for.

A new version (thus far, for private use only) allows for an exact treatment of line-blocking, by means of a comoving-frame solution for 'all' lines from the elements H to Zn.

The standard distribution includes a package of IDL routines to display and analyze the results.

Programming language: Fortran 90

Parallelization/architecture limitations: None

Problem size possible, scaling, example(s) of running time for typical cases: The typical setup uses 55–75 radial grid points, automatically distributed in an optimum way. Typical computation times (atmosphere + spectral lines from H, He and N) are on the order of 20 minutes. The new version requires 120 minutes, including a detailed spectrum in the ranges from 900 to 2,000 and 3,400 to 7,000 Å.

Literature references for code: Santolaya-Rey et al. (1997); Puls et al. (2005); Rivero González et al. (2011); Carneiro et al. (2016, 2017); Puls (2017)

Literature references for typical applications of code: Repolust et al. (2004, 2005); Mokiem et al. (2005, 2007); Lefever et al. (2007); Markova and Puls (2008); Rivero González et al. (2012); Massey et al. (2013); Holgado et al. (2017); Schneider et al. (2018)

Availability: Available through collaboration and on request, if certain conditions are fulfilled.

4.8.3 *PHOENIX*

Main developer(s):

Peter Hauschildt, Hamburger Sternwarte, Gojenbergsweg 112, 21029 Hamburg, Germany, *yeti@hs.uni-hamburg.de*

Edward Baron, Homer L. Dodge Dept. of Physics and Astronomy, University of Oklahoma, 440 W. Brooks, Rm 100, Norman, OK 73019, USA, *baron@ou.edu*

Travis Barman, Lunar and Planetary Laboratory, The University of Arizona, 1629 E University Blvd, Tucson, AZ 85721-0092, USA, *barman@lpl.arizona.edu*

France Allard, C.R.A.L (UML 5574) Ecole Normale Superieure, 69364 Lyon Cedex 7, France, *fallard@ens-lyon.fr*

Type of code: NLTE stellar/planetary atmosphere modelling, including energy equation, velocity fields.

Geometry:
1-D: spherical, time dependent, Lagrange frame, special/general relativistic;
3-D: Cartesian, spherical or cylinder coordinate systems, time dependent, Euler or Lagrange frames, special/general relativistic

Main characteristics/capabilities: Full characteristics operator splitting radiative transfer solver in 1-D and 3-D with nonlocal operator splitting. Rate operator based multilevel NLTE solver for atoms, ions and the CO molecule.

1-D: computing LTE/NLTE atmosphere models in radiative/convective equilibrium for effective temperatures 200 K to 50 kK, as well as novae and supernova atmosphere models.

3-D: solving multilevel NLTE radiative transfer problems for given 3-D configurations, synthetic images and spectra for LTE and NLTE conditions, including irradiated planets.

Programming language: Fortran 2008, C++, C

Parallelization/architecture limitations: MPI+OpenMP+OpenCL+vectorization.

Multistage hierarchical parallelization with domain decompositions (space and wavelength, statistical equations).

Problem size possible, scaling, example(s) of running time for typical cases:
1-D: 64–256+ layers, > 2M wavelengths, LTE: > 1G lines (atomic, molecular), NLTE: 192 species currently available, > 10k levels per species (e.g., Fe II). Limited by available RAM per MPI process, run time variable from minutes to several hours depending on problem size. Weak and strong scaling to 10k MPI processes (depending on problem size).

3-D: 1M voxels, 2,500 NLTE level, 60K NLTE transitions, 650k wavelength points. Strong and weak scaling to 150k MPI processes with 16 OpenMP threads per process (depending on problem size).

Literature references for code: Hauschildt and Baron (2006, 2009, 2010, 2011, 2014); Baron and Hauschildt (2007); Chen et al. (2007); Baron et al. (2009, 2012); Knop et al. (2009); Seelmann et al. (2010); Jack et al. (2012a)

Literature references for typical applications of code: Baron et al. (2007, 2008); Hauschildt et al. (2008); Jack et al. (2009, 2012b, 2015); Berkner et al. (2013); Friesen et al. (2017)

Availability: Available through collaboration.

4.8.4 *PoWR*

Main developer(s):
Wolf-Rainer Hamann, Institut für Physik und Astronomie, Universität Potsdam, Germany *wrh@astro.physik.uni-potsdam.de*

Type of code: Non-LTE stellar atmosphere including wind

Geometry: 1-D spherical

Main characteristics/capabilities: The code solves consistently (1) the radiative transfer equation in the comoving frame; (2) the equations of statistical equilibrium (rate equations); and (3) the hydrostatic equation (in the quasistatic part of the photosphere), including the radiative force. The consistent solution is found by 'iteration with approximate lambda operators'. Consistency with the energy equation is obtained by a generalized Unsoeld–Lucy method. The emergent spectrum is finally calculated in the observer's frame ('formal integral').

Included physics: stellar wind with prespecified mass-loss rate and velocity law; complex model atoms with up to 1,000 levels; iron-group elements treated in the superlevel approach; millions of iron-group lines from the Kurucz database; wind inhomogeneities in the microclumping approximation; and embedded X-ray sources (optional). Additional physics in the 'formal integral': refined model atoms (e.g., multiplet splitting); pressure line broadening; radius-dependent microturbulence; macroclumping (optional); and wind rotation (optional).

The code comes together with a visualization tool (WRplot).

Programming language: Fortran 90 (100,000 lines); various shell scripts and tools

Parallelization/architecture limitations: Not parallel

Problem size possible, scaling, example(s) of running time for typical cases: Typical setup: 50 depth points, 55 impact parameter points, 200,000 frequency points, 500 explicit non-LTE levels; time depends very much on the model and is greatly facilitated if one can start from a similar previous model; computation may converge within a few hours up to one week on a modern workstation.

Literature references for code: Hamann (1985); Hamann and Koesterke (1998); Gräfener et al. (2002); Hamann and Gräfener (2003); Šurlan et al. (2013); Sander et al. (2017)

Literature references for typical applications of code: Sander et al. (2012); Hainich et al. (2015); Ramachandran et al. (2018)

Many more papers with application of the PoWR code can be found via
www.astro.physik.uni-potsdam.de/~www/research/pub_astro_1_de.html

Availability: Code is currently only available through collaboration with one of the developers; a public release is planned.

Grids of models for WR-stars and O-stars at different metallicities are available via a web interface at
www.astro.physik.uni-potsdam.de/PoWR

4.8.5 *TLUSTY*

Main developer(s):

Ivan Hubený, Steward Observatory and Department of Astronomy, University of Arizona, 933 N. Cherry Ave., Tucson, AZ 85721, USA, *hubeny@as.arizona.edu*

Thierry Lanz, Observatoire de Cote d'Azur, Nice, France, *lanz@oca.eu*

Type of code: Self-consistent model stellar atmospheres or accretion disks

Geometry: 1-D plane-parallel

Main characteristics/capabilities: Program produces plane-parallel, horizontally homogeneous model stellar atmospheres in hydrostatic and radiative (or radiative + convective) equilibrium, by solving simultaneously these two equilibrium conditions together with the radiative transfer equation and the set of kinetic equilibrium equations. Departures from LTE are allowed for a set of occupation numbers of selected atomic and ionic energy levels. For cool objects for which LTE models are satisfactory, the code can use precalculated opacity tables that include all important atomic and molecular opacity sources. The program can also compute a vertical structure of an annulus of an (azimuthally symmetric) accretion disk, replacing the radiative equilibrium equation by an appropriate energy balance equation that accounts for viscous energy dissipation.

Programming language: FORTRAN 77

Parallelization/architecture limitations: No parallelization implemented

Problem size possible, scaling, example(s) of running time for typical cases: The program is fully data oriented as far as the choice of atomic species, ions, energy levels, transitions and opacity sources to be considered explicitly. The typical setup for a modern metal line-blanketed atmospheres: 50–100 depth points, 1,000–1,500 NLTE levels, 200,000–400,000 frequency points. This requires a processor with at least 2GB of memory.

Literature references for code: Hubený (1988); Hubený and Lanz (1995, 2017)

Literature references for typical applications of code: Lanz and Hubený (2003); Lanz and Hubený (2007); Wade and Hubený (1998); Hubený et al. (2000, 2001); Cunha et al. (2006); Davis and Hubený (2006)

Availability: Code is publicly available for download, as specified in detail in Hubený and Lanz (2017).

4.8.6 *WM-basic*

Main developer(s):

Adalbert W. A. Pauldrach, University Observatory Munich (USM), Ludwig-Maximilians-Universität München, Scheinerstraße 1, D-81679 München, Germany, *uh10107@usm.uni-muenchen.de*
Tadziu L. Hoffmann, University Observatory Munich (USM), Ludwig-Maximilians-Universität München, Scheinerstraße 1, D-81679 München, Germany, *hoffmann@usm.uni-muenchen.de*
Johann A. Weber, University Observatory Munich (USM), Ludwig-Maximilians-Universität München, Scheinerstraße 1, D-81679 München, Germany, *jweber@usm.uni-muenchen.de*

Type of code: Radiation hydrodynamics, nonequilibrium thermodynamics

Geometry: 1-D spherical with an extension to a 3-D grid code

Main characteristics/capabilities: With our method, we perform calculations of model atmospheres of hot stars with line-driven winds that consist of (a) a solution of the hydrodynamics describing velocity and density structures of the outflow, based on radiative acceleration by line, continuum and Thomson absorption and scattering; (b) the computation of the occupation numbers from a solution of the rate equations containing all important radiative and collisional processes, using sophisticated model atoms and corresponding line lists; (c) a calculation of the radiation field from a detailed radiative transfer solution taking into account the Doppler-shifted line opacities and emissivities along with the continuum radiative transfer; and (d) a computation of the temperature from the requirement of radiative (absorption/emission) and thermal (heating/cooling) balance. An accelerated Lambda iteration (ALImI, explained in detail by Pauldrach et al., 2014) procedure is used to achieve consistency of occupation numbers, radiative transfer and temperature. If required, an updated radiative acceleration can be computed from the converged model, and the process iteratively repeated. In addition, secondary effects such as the production of extreme ultraviolet (EUV) and X-ray radiation in the cooling zones of shocks embedded in the wind and arising from the nonstationary, unstable behaviour of radiation-driven winds can, together with K-shell absorption, be optionally considered. The radiative transfer and the computation of the abundances and energy balance can also be applied to supernovae and H II regions. An extension of the program implements the numerical modelling of both 3-D radiative transfer and time-dependent effects in H II regions, allowing for several sources of ionization with different spectral energy distributions.

Programming language: Fortran 2003 and C++

Parallelization/architecture limitations: Optimization for vectorized computation and parallelization of the 3-D radiative transfer code with OpenMP

Problem size possible, scaling, example(s) of running time for typical cases: Typical setup: 41 depth points, up to five ionization stages of 26 elements, up to 50 sophisticatedly packed excitation levels per ion for stellar atmosphere models. For H II region models, the number of grid points is typically increased to 101 in the spherically symmetric case, whereas we have used up to 10,000 grid cells in the 3-D case. The execution time of a simulation run depends on the resolution and the object being modelled: approximately 10–15 minutes for a stellar atmosphere model or a spherically symmetric H II region model on a 2GHz machine (memory usage below 100MB), on the order of one day for a time-dependent 3-D simulation with 10,000 grid points (memory usage approximately 7GB).

Literature references for code: Pauldrach (1987); Pauldrach et al. (1990, 1993, 1994, 1998, 2001, 2004, 2012, 2014); Taresch et al. (1997); Haser et al. (1998); Sauer et al. (2006); Weber et al. (2013, 2015); Hoffmann et al. (2014)

Literature references for typical applications of code: Pauldrach and Puls (1990); Pauldrach et al. (1990, 1994, 2001, 2004, 2012, 2014); Sellmaier et al. (1996); Giveon et al. (2002); Sternberg et al. (2003); Rubin et al. (2007, 2008, 2016); Leitherer et al. (2010); Kaschinski et al. (2012); Hoffmann et al. (2012, 2014, 2016); Weber et al. (2015)

Availability: A 32-bit Windows executable is available from www.usm.lmu.de/~adi/Programs/Programs.html. The source code is currently only available through collaboration with one of the main developers.

4.9 Code Descriptions: Radiation MHD

4.9.1 *CO5BOLD*

Main developer(s):
Bernd Freytag, Division of Astronomy and Space Physics, Uppsala University, Regementsvägen 1, SE-75120 Uppsala, Sweden, *Bernd.Freytag@physics.uu.se*
Matthias Steffen, Leibniz-Institut fr Astrophysik Potsdam (AIP), An der Sternwarte 16, D-14482 Potsdam, Germany, *msteffen@aip.de*

Type of code:
Radiation hydrodynamics
Optionally: magnetic fields in the ideal MHD approximation
Optionally: dust

Geometry:
2-D or 3-D with a Cartesian grid

Main characteristics/capabilities:
CO5BOLD is used to model surface convection, pulsations, waves, shocks and winds on the Sun or stellar and even substellar objects. The code solves the coupled equations of

compressible (magneto-)hydrodynamics in the presence of an external gravity field and nonlocal frequency-dependent radiation transport.

There are two different setups: one is used to produce local 'box-in-a-star' models of small patches of the surface of stars such as the Sun assuming constant gravity and employing periodic side boundaries. The other is able to produce global 'star-in-a-box' models of an entire star (minus the tiny core region) such as Betelgeuse, employing a central gravitational potential and open boundary conditions on all outer surfaces of the computational domain.

Programming language: Fortran 90

Parallelization/architecture limitations: OpenMP for all modules. Hybrid (MPI + OpenMP) for the modules required for global radiation hydrodynamics (RHD) models.

Problem size possible, scaling, example(s) of running time for typical cases: Typical model sizes are 256^3–500^3 grid points. The run time depends on the number of grid points, the number of angles and opacity bins in the radiation-transport solvers, the stellar time to cover, the numerical time step, etc., and varies a lot between different types of models. The typical run time lies between node weeks and node months.

Literature references for code: Caffau et al. (2011); Freytag (2013, 2017)

Literature references for typical applications of code: Freytag and Steffen (2004); Wedemeyer et al. (2004); Wedemeyer-Böhm et al. (2005); Freytag and Höfner (2008); Chiavassa et al. (2009); Ludwig et al. (2009); Freytag et al. (2002, 2010, 2017); Sbordone et al. (2010); Caffau et al. (2011); Ludwig and Kučinskas (2012); Spite et al. (2013); Tremblay et al. (2013a,b); Steiner et al. (2014); Sonoi et al. (2015); Steffen et al. (2015); Vasilyev et al. (2017)

Availability: The code is available through collaboration with one of the approved users.

4.9.2 *Radyn*

Main developer(s):

Mats Carlsson, Institute of Theoretical Astrophysics, University of Oslo, P.O. Box 1029 Blindern, NO-0315 Oslo, Norway, *mats.carlsson@astro.uio.no*

Joel Allred, NASA/Goddard Space Flight Center, Code 671, Greenbelt, MD 20771, USA, *joel.c.allred@nasa.gov*

Adam Kowalski, National Solar Observatory, University of Colorado Boulder, 3665 Discovery Drive, Boulder, CO 80303, USA, *Adam.F.Kowalski@colorado.edu*

Type of code: Radiation hydrodynamics

Geometry: 1-D plane-parallel but with possibility of including loop geometry (varying acceleration of gravity as function of height)

Main characteristics/capabilities: Solves the 1-D equations of hydrodynamics (conservation of mass, energy, momentum and charge) on an adaptive grid (Dorfi and Drury, 1987) together with non-LTE rate equations for main species (typically hydrogen, calcium and helium). Extra energy input may be in the form of an energy term as a function of height and time, a piston at the lower boundary or an electron beam with total energy, energy spectrum index and lower cutoff energy as a function of time. Heating by incident coronal radiation and optically thin radiative losses in the corona is also included.

Programming language: Fortran 90

Parallelization/architecture limitations: Parallelized with MPI over radiative transitions

Problem size possible, scaling, example(s) of running time for typical cases: Typical setup: 300 depth points, six-level hydrogen atom, six-level calcium atom and nine-level helium atom. Almost linear scaling up to 28 cores but not further (because of parallelization only over radiative transitions). Running time depends on the problem: one hour of solar time for piston-driven studies of acoustic wave propagation takes order hours to run, flare simulations from order hours for low-energy input to weeks for large energy input.

Literature references for code: Carlsson and Stein (1992); Allred et al. (2015)

Literature references for typical applications of code: Carlsson and Stein (1992, 1995, 1997, 2002); Allred et al. (2005, 2006); Kennedy et al. (2015); Kowalski et al. (2017b); Kuridze et al. (2016); Reid et al. (2017); Rubio da Costa and Kleint (2017); Simões et al. (2017)

Availability: Code is currently only available through collaboration with one of the main developers, but a public release is planned for late 2020.

4.9.3 *Bifrost*

Main developer(s):
Boris Gudiksen, Mats Carlsson, Viggo Hansteen, Institute of Theoretical Astrophysics, University of Oslo, P.O. Box 1029 Blindern, NO-0315 Oslo, Norway, *boris.gudiksen@astro.uio.no*, *mats.carlsson@astro.uio.no*,
viggo.hansteen@astro.uio.no

Jorrit Leenarts, Institute for Solar Physics, Department of Astronomy, Stockholm University, AlbaNova University Centre, SE-106 91 Stockholm, Sweden,
jorrit.leenaarts@astro.su.se

Juan Martínez-Sykora, Lockheed Martin Solar and Astrophysics Lab, Org. ADBS, Bldg. 252, 3251 Hanover Street Palo Alto, CA 94304 USA, *jms@lmsal.com*

Type of code: Radiation magnetohydrodynamics

Geometry: Fully 3-D

Main characteristics/capabilities: Able to include the upper convective zone and lower corona, including optically thick radiation with scattering, thermal conduction and description for the radiative processes in the upper chromosphere. On top of these capabilities, Bifrost is able to easily include several other physical effects, such as hydrogen and helium out of statistical equilibrium and the effect on the energy equation, and metals out of statistical equilibrium and the generalized Ohm's law.

Programming language: Fortran 90

Parallelization/architecture limitations: Vectorized MPI

Problem size possible, scaling, example(s) of running time for typical cases: Typical problem size is 768^3 gridpoints, with a resolution of roughly 30 km giving horizontal size of 24 Mm, while the vertical scale is usually from 2 Mm below the photosphere to 14 Mm above. There is in principle no maximum scale, as Bifrost has been able to scale well to 68,000 cores.

Literature references for code: Gudiksen et al. (2011)

Literature references for typical applications of code: Leenaarts et al. (2012); Olluri et al. (2013); Carlsson et al. (2016); Golding et al. (2016); Hansteen et al. (2017); Martínez-Sykora et al. (2017a,b)

Availability: Code is currently only available through collaboration with the main developers.

4.9.4 *Stagger*

Main developer(s):

Åke Nordlund, Centre for Star and Planet Formation, Niels Bohr Institute, Natural History Museum of Denmark, University of Copenhagen, Øster Voldgade 5-7, 1350 Copenhagen K, Denmark, *aake@nbi.dk*

Klaus Galsgaard, Niels Bohr Institute, Natural History Museum of Denmark, University of Copenhagen, Øster Voldgade 5-7, 1350 Copenhagen K, Denmark, *kg@nbi.dk*

Remo Collet, Department of Physics and Astronomy, Aarhus University, Ny Munkegade 120, DK-8000 Aarhus C, Denmark, *remo@phys.au.dk*

Robert F. Stein, Department of Physics and Astronomy, Michigan State University, East Lansing, MI 48824, USA, *steinr@msu.edu*

Type of code: Radiation magnetohydrodynamics

Geometry: 3-D

Main characteristics/capabilities: Solves the time-dependent, 3-D conservation equations of standard MHD coupled with radiative transfer. The MHD equations are discretized using a high-order finite-difference scheme (sixth-order derivatives, fifth-order interpolations) and solved on a Cartesian mesh. The radiative transfer equation is solved along sets of parallel rays uniformly distributed across the simulation domain at various inclinations using a long-characteristics scheme. The wavelength dependence of radiative transfer quantities (extinction coefficient and source function) is modelled through opacity binning. The radiative transfer solution is used to calculate the necessary heating rates for the energy conservation equation. The solution of the MHD equations is advanced in time using a third-order Runge–Kutta method.

Applications of the code include solar and stellar surface convection simulations and modelling of supersonic turbulence in the interstellar medium and fragmentation of giant molecular clouds.

Programming language: Fortran 90

Parallelization/architecture limitations: Parallelization with MPI: domain decomposition for magnetohydrodynamics; ray reconstruction and parallelization over ray bundles for radiative transfer

Problem size possible, scaling, example(s) of running time for typical cases: Typical setup for solar and stellar convection simulations: numerical resolution from 200^3 to $2\,000^3$, radiative transfer with typically 12 inclined rays plus vertical, and four to 12 opacity bins. With this setup, scaling is linear up to about 256 cores.

Running time depends on the problem; as an example, one hour of solar surface convection simulation in a box of $\sim 6 \times 6 \times 4\,\mathrm{Mm}$, without magnetic fields, with a mesh with 240^3 numerical resolution and radiative transfer with 12 opacity bins and 13 rays takes about 21 hours on 192 Intel Xeon Sandy Bridge cores with InfiniBand interconnect.

Literature references for code: Nordlund et al. (1994); www.astro.ku.dk/~kg/Papers/MHD_code.ps.gz

Literature references for typical applications of code: Galsgaard and Nordlund (1996); Padoan et al. (2007); Nordlund et al. (2009); Collet et al. (2011, 2017); Beeck et al. (2012); Stein and Nordlund (2012); Magic et al. (2013)

Availability: Code is currently only available through collaboration with one of the developers.

Acknowledgements

This research was supported by the Research Council of Norway through its Centres of Excellence scheme, project number 262622, and through grants of computing time from the Programme for Supercomputing.

REFERENCES

Allred, J. C., Hawley, S. L., Abbett, W. P. and Carlsson, M. 2005. Radiative Hydrodynamic Models of the Optical and Ultraviolet Emission from Solar Flares. *ApJ*, **630**, 573–586

Allred, J. C., Hawley, S. L., Abbett, W. P. and Carlsson, M. 2006. Radiative Hydrodynamic Models of Optical and Ultraviolet Emission from M Dwarf Flares. *ApJ*, **644**, 484–496

Allred, J. C., Kowalski, A. F. and Carlsson, M. 2015. A Unified Computational Model for Solar and Stellar Flares. *ApJ*, **809**, 104

Alvarez, R., and Plez, B. 1998. Near-Infrared Narrow-Band Photometry of M-Giant and Mira Stars: Models Meet Observations. *A&A*, **330**, 1109–1119

Asplund, M., Gustafsson, B., Kiselman, D. and Eriksson, K. 1997. Line-Blanketed Model Atmospheres for R Coronae Borealis Stars and Hydrogen-Deficient Carbon Stars. *A&A*, **318**, 521–534

Auer, L. H. and Mihalas, D. 1969. Non-LTE Model Atmospheres. III. A Complete-Linearization Method. *ApJ*, **158**, 641

Auer, L. H., Heasley, J. N. and Milkey, R. W. 1972. A Computational Program for the Solution of Non-LTE Transfer Problems by the Complete Linearization Method. *Contributions from the Kitt Peak National Observatory*, **555**(2), 193–214

Avrett, E. H. and Loeser, R. 1992. The PANDORA Atmosphere Program (Invited Review). Page 489 of: Giampapa, M. S. and Bookbinder, J. A. (eds), *Cool Stars, Stellar Systems, and the Sun.* Astronomical Society of the Pacific Conference Series, vol. 26

Barban, C., Goupil, M. J., Van't Veer-Menneret, C., Garrido, R., Kupka, F. and Heiter, U. 2003. New Grids of ATLAS9 Atmospheres. II. Limb-Darkening Coefficients for the Strömgren Photometric System for A-F Stars. *A&A*, **405**, 1095–1105

Baron, E. and Hauschildt, P. H. 2007. A 3D Radiative Transfer Framework. II. Line Transfer Problems. *A&A*, **468**, 255–261

Baron, E., Branch, D. and Hauschildt, P. H. 2007. Reddening, Abundances, and Line Formation in SNe II. *ApJ*, **662**, 1148–1155

Baron, E., Hauschildt, P. H., and Chen, B. 2009. A 3D Radiative Transfer Framework. V. Homologous Flows. *A&A*, **498**, 987–992

Baron, E., Hauschildt, P. H., Chen, B. and Knop, S. 2012. A 3D Radiative Transfer Framework. X. Arbitrary Velocity Fields in the Comoving Frame. *A&A*, **548**, A67

Baron, E., Jeffery, D. J., Branch, D., Bravo, E., García-Senz, D. and Hauschildt, P. H. 2008. Detailed Spectral Modeling of a Three-Dimensional Pulsating Reverse Detonation Model: Too Much Nickel. *ApJ*, **672**, 1038–1042

Becker, S. R. and Butler, K. 1995. Non-LTE Line Formation for Iron Group Elements. III. Fe IV. *A&A*, **301**, 187

Beeck, B., Collet, R., Steffen, M., et al. 2012. Simulations of the Solar Near-Surface Layers with the CO5BOLD, MURaM, and Stagger Codes. *A&A*, **539**, A121

Bell, R. A., Eriksson, K., Gustafsson, B. and Nordlund, A. 1976. A Grid of Model Atmospheres for Metal-Deficient Giant Stars. II. *A&AS*, **23**, 37–95

Bergemann, M., Collet, R., Amarsi, A. M., Kovalev, M., Ruchti, G. and Magic, Z. 2017. Non-Local Thermodynamic Equilibrium Stellar Spectroscopy with 1D and <3D> Models. I. Methods and Application to Magnesium Abundances in Standard Stars. *ApJ*, **847** (September), 15

Bergemann, M., Kudritzki, R.-P., Gazak, Z., Davies, B. and Plez, B. 2015. Red Supergiant Stars as Cosmic Abundance Probes. III. NLTE Effects in J-Band Magnesium Lines. *ApJ*, **804**(May), 113

Bergemann, M., Pickering, J. C. and Gehren, T. 2010. NLTE Analysis of CoI/CoII Lines in Spectra of Cool Stars with New Laboratory Hyperfine Splitting Constants. *MNRAS*, **401**(January), 1334–1346

Bergemann, M., Serenelli, A., Schönrich, R., et al. 2016. The Gaia-ESO Survey: Hydrogen Lines in Red Giants Directly Trace Stellar Mass. *A&A*, **594**(October), A120

Berkner, A., Hauschildt, P. H. and Baron, E. 2013. 3D Radiative Transfer Effects in Parametrized Starspots. *A&A*, **550**, A104

Bessell, M. S., Castelli, F. and Plez, B. 1998. Model Atmospheres Broad-Band Colors, Bolometric Corrections and Temperature Calibrations for O-M Stars. *A&A*, **333**, 231–250

Bjørgen, J. P. and Leenaarts, J. 2017. Numerical Non-LTE 3D Radiative Transfer Using a Multigrid Method. *A&A*, **599**, A118

Bouret, J.-C., Hillier, D. J., Lanz, T. and Fullerton, A. W. 2012. Properties of Galactic Early-Type O-Supergiants: A Combined FUV-UV and Optical Analysis. *A&A*, **544**, A67

Bouret, J.-C., Lanz, T., Martins, F., et al. 2013. Massive Stars at Low Metallicity: Evolution and Surface Abundances of O Dwarfs in the SMC. *A&A*, **555**, A1

Butler, K. and Giddings, J. 1985. *Newsletter on Analysis of Astronomical Spectra, Daresbury Laboratory*, **9**.

Caffau, E., Ludwig, H.-G., Steffen, M., Freytag, B. and Bonifacio, P. 2011. Solar Chemical Abundances Determined with a CO5BOLD 3D Model Atmosphere. *Sol. Phys.*, **268**, 255–269

Carlsson, M. 1986. *A Computer Program for Solving Multi-Level Non-LTE Radiative Transfer Problems in Moving or Static Atmospheres.* Uppsala Astronomical Observatory Reports, **33**.

Carlsson, M. and Judge, P. G. 1993. On I Lines in the Sun and Stars. I Understanding the Resonance Lines. *ApJ*, **402**, 344–357

Carlsson, M. and Leenaarts, J. 2012. Approximations for Radiative Cooling and Heating in the Solar Chromosphere. *A&A*, **539**, A39

Carlsson, M. and Stein, R. F. 1992. Non-LTE Radiating Acoustic Shocks and CA II K2V Bright Points. *ApJ*, **397**, L59–L62

Carlsson, M. and Stein, R. F. 1995. Does a Nonmagnetic Solar Chromosphere Exist? *ApJ*, **440**, L29–L32

Carlsson, M. and Stein, R. F. 1997. Formation of Solar Calcium H and K Bright Grains. *ApJ*, **481**, 500–514

Carlsson, M. and Stein, R. F. 2002. Dynamic Hydrogen Ionization. *ApJ*, **572**, 626–635

Carlsson, M., Hansteen, V. H., Gudiksen, B. V., Leenaarts, J. and De Pontieu, B. 2016. A Publicly Available Simulation of an Enhanced Network Region of the Sun. *A&A*, **585**, A4

Carlsson, M., Rutten, R. J. and Shchukina, N. G. 1992. The Formation of the MG I Emission Features Near 12 Microns. *A&A*, **253**, 567–585

Carlsson, M., Rutten, R. J., Bruls, J. H. M. J. and Shchukina, N. G. 1994. The Non-LTE Formation of Li I Lines in Cool Stars. *A&A*, **288**, 860–882

Carneiro, L. P., Puls, J. and Hoffmann, T. L. 2017. Carbon Line Formation and Spectroscopy in O-Type Stars. *ArXiv e-prints*

Carneiro, L. P., Puls, J., Sundqvist, J. O. and Hoffmann, T. L. 2016. Atmospheric NLTE Models for the Spectroscopic Analysis of Blue Stars with Winds. III. X-Ray Emission from Wind-Embedded Shocks. *A&A*, **590**, A88

Castelli, F. and Kurucz, R. L. 2004. New Grids of ATLAS9 Model Atmospheres. *ArXiv Astrophysics e-prints*

Castelli, F. and Kurucz, R. L. 2006. Computed H_{beta} Indices from ATLAS9 Model Atmospheres. *A&A*, **454**, 333–340

Centeno, R., Trujillo Bueno, J., Uitenbroek, H. and Collados, M. 2008. The Influence of Coronal EUV Irradiance on the Emission in the He I 10830 Å and D_3 Multiplets. *ApJ*, **677**, 742–750

Chen, B., Kantowski, R., Baron, E., Knop, S. and Hauschildt, P. H. 2007. Steps for Solving the Radiative Transfer Equation for Arbitrary Flows in Stationary Space-Times. *MNRAS*, **380**, 104–112

Chiavassa, A., Plez, B., Josselin, E. and Freytag, B. 2009. Radiative Hydrodynamics Simulations of Red Supergiant Stars. I. Interpretation of Interferometric Observations. *A&A*, **506**, 1351–1365

Collet, R., Hayek, W., Asplund, M., Nordlund, Å., Trampedach, R. and Gudiksen, B. 2011. Three-Dimensional Surface Convection Simulations of Metal-Poor Stars. The Effect of Scattering on the Photospheric Temperature Stratification. *A&A*, **528**, A32

Collet, R., Nordlund, Å., Asplund, M., Hayek, W. and Trampedach, R. 2017. The Benchmark Halo Giant HD 122563: CNO Abundances Revisited with Three-Dimensional Hydrodynamic Model Stellar Atmospheres. *ArXiv e-prints*

Cunha, K., Hubený, I. and Lanz, T. 2006. Neon Abundances in B Stars of the Orion Association: Solving the Solar Model Problem? *ApJ*, **647**, L143–L146

Davies, B., Kudritzki, R.-P., Plez, B., et al. 2013. The Temperatures of Red Supergiants. *ApJ*, **767**, 3

Davis, S. W. and Hubený, I. 2006. A Grid of Relativistic, Non-LTE Accretion Disk Models for Spectral Fitting of Black Hole Binaries. *ApJS*, **164**, 530–535

de la Cruz Rodríguez, J., Socas-Navarro, H., Carlsson, M. and Leenaarts, J. 2012. Non-Local Thermodynamic Equilibrium Inversions from a 3D Magnetohydrodynamic Chromospheric Model. *A&A*, **543**, A34

de Laverny, P., Recio-Blanco, A., Worley, C. C. and Plez, B. 2012. The AMBRE Project: A New Synthetic Grid of High-Resolution FGKM Stellar Spectra. *A&A*, **544**, A126

Dessart, L., Hillier, D. J., Blondin, S. and Khokhlov, A. 2014. Critical Ingredients of Type Ia Supernova Radiative-Transfer Modelling. *MNRAS*, **441**, 3249–3270

Dessart, L., Hillier, D. J., Waldman, R. and Livne, E. 2013b. Type II-Plateau Supernova Radiation: Dependences on Progenitor and Explosion Properties. *MNRAS*, **433**, 1745–1763.

Dessart, L., Hillier, D. J., Woosley, et al. 2016. Inferring Supernova IIb/Ib/Ic Ejecta Properties from Light Curves and Spectra: Correlations from Radiative-Transfer Models. *MNRAS*, **458**, 1618–1635

Dessart, L., Waldman, R., Livne, E., Hillier, D. J. and Blondin, S. 2013a. Radiative Properties of Pair-Instability Supernova Explosions. *MNRAS*, **428**, 3227–3251

Dorfi, E. A. and Drury, L. O. 1987. Simple Adaptive Grids for 1-D Initial Value Problems. *Journal of Computational Physics*, **69**, 175–195

Ekberg, U., Eriksson, K. and Gustafsson, B. 1986. Line Blanketing in Model Atmospheres of Carbon Stars. *A&A*, **167**, 304–310

Fierro-Santillán, C., Zsargó, J., Klapp, J., Díaz-Azuara, S. A., Arrieta, A. and Arias, L. 2017. Analysis and Fit of Stellar Spectra Using a Mega-Database of CMFGEN Models. Pages 397–397 of: Eldridge, J. J., Bray, J. C., McClelland, L. A. S. and Xiao, L. (eds.), *The Lives and Death-Throes of Massive Stars*. IAU Symposium, vol. 329.

Freytag, B. 2013. Advances in the Hydrodynamics Solver of CO5BOLD. *Memorie della Societa Astronomica Italiana Supplementi*, **24**, 26.

Freytag, B. 2017. Boundary Conditions in CO5BOLD. *Mem. Soc. Astron. Italiana*, **88**, 12.

Freytag, B. and Höfner, S. 2008. Three-Dimensional Simulations of the Atmosphere of an AGB Star. *A&A*, **483**, 571–583

Freytag, B. and Steffen, M. 2004. Numerical Simulations of Convection in A-Stars. Pages 139–147 of: Zverko, J., Ziznovsky, J., Adelman, S. J. and Weiss, W. W. (eds.), *The A-Star Puzzle*. IAU Symposium, vol. 224

Freytag, B., Allard, F., Ludwig, H.-G., Homeier, D. and Steffen, M. 2010. The Role of Convection, Overshoot, and Gravity Waves for the Transport of Dust in M Dwarf and Brown Dwarf Atmospheres. *A&A*, **513**, A19

Freytag, B., Liljegren, S. and Höfner, S. 2017. Global 3D Radiation-Hydrodynamics Models of AGB Stars: Effects of Convection and Radial Pulsations on Atmospheric Structures. *A&A*, **600**, A137

Freytag, B., Steffen, M. and Dorch, B. 2002. Spots on the Surface of Betelgeuse – Results from New 3D Stellar Convection Models. *Astronomische Nachrichten*, **323**, 213–219

Friesen, B., Baron, E., Parrent, J. T., et al. 2017. Optical and Ultraviolet Spectroscopic Analysis of SN 2011fe at Late Times. *MNRAS*, **467**, 2392–2411

Galsgaard, K. and Nordlund, Å. 1996. Heating and Activity of the Solar Corona 1. Boundary Shearing of an Initially Homogeneous Magnetic Field. *J. Geophys. Res.*, **101**, 13445–13460

Giddings, J. R. 1981. PhD thesis, University of London

Giveon, U., Sternberg, A., Lutz, D., Feuchtgruber, H. and Pauldrach, A. W. A. 2002. The Excitation and Metallicity of Galactic H II Regions from Infrared Space Observatory SWS Observations of Mid-Infrared Fine-Structure Lines. *ApJ*, **566**, 880–897

Golding, T. P., Leenaarts, J. and Carlsson, M. 2016. Non-Equilibrium Helium Ionization in an MHD Simulation of the Solar Atmosphere. *ApJ*, **817**, 125

Gräfener, G., Koesterke, L. and Hamann, W.-R. 2002. Line-Blanketed Model Atmospheres for WR Stars. *A&A*, **387**, 244–257.

Grupp, F. 2004a. MAFAGS-OS: New Opacity Sampling Model Atmospheres for A, F and G Stars. I. The Model and the Solar Flux. *A&A*, **420**(June), 289–305

Grupp, F. 2004b. MAFAGS-OS: New Opacity Sampling Model Atmospheres for A, F and G Stars. II. Temperature Determination and Three 'Standard' Stars. *A&A*, **426**(October), 309–322

Gudiksen, B. V., Carlsson, M., Hansteen, V. H., Hayek, W., Leenaarts, J. and Martínez-Sykora, J. 2011. The Stellar Atmosphere Simulation Code Bifrost: Code Description and Validation. *A&A*, **531**, A154.

Gustafsson, B., Bell, R. A., Eriksson, K. and Nordlund, A. 1975. A Grid of Model Atmospheres for Metal-Deficient Giant Stars. I. *A&A*, **42**, 407–432

Gustafsson, B., Edvardsson, B., Eriksson, K., Jørgensen, U. G., Nordlund, Å. and Plez, B. 2008. A Grid of MARCS Model Atmospheres for Late-Type Stars. I. Methods and General Properties. *A&A*, **486**, 951–970

Hainich, R., Pasemann, D., Todt, H., Shenar, T., Sander, A. and Hamann, W.-R. 2015. Wolf–Rayet Stars in the Small Magellanic Cloud. I. Analysis of the Single WN Stars. *A&A*, **581**, A21

Hamann, W.-R. 1985. Line Formation in Expanding Atmospheres: Accurate Solution Using Approximate Lambda Operators. *A&A*, **148**, 364–368

Hamann, W.-R. and Gräfener, G. 2003. A Temperature Correction Method for Expanding Atmospheres. *A&A*, **410**, 993–1000

Hamann, W.-R. and Koesterke, L. 1998. Spectrum Formation in Clumped Stellar Winds: Consequences for the Analyses of Wolf–Rayet Spectra. *A&A*, **335**, 1003–1008

Hansteen, V. H., Archontis, V., Pereira, T. M. D., Carlsson, M., Rouppe van der Voort, L. and Leenaarts, J. 2017. Bombs and Flares at the Surface and Lower Atmosphere of the Sun. *ApJ*, **839**, 22

Haser, S. M., Pauldrach, A. W. A., Lennon, D. J., et al. 1998. Quantitative UV Spectroscopy of Early O Stars in the Magellanic Clouds: The Determination of the Stellar Metallicities. *A&A*, **330**, 285–305

Hauschildt, P. H. and Baron, E. 2006. A 3D Radiative Transfer Framework. I. Non-Local Operator Splitting and Continuum Scattering Problems. *A&A*, **451**, 273–284

Hauschildt, P. H. and Baron, E. 2009. A 3D Radiative Transfer Framework. IV. Spherical and Cylindrical Coordinate Systems. *A&A*, **498**, 981–985

Hauschildt, P. H. and Baron, E. 2010. A 3D Radiative Transfer Framework. VI. PHOENIX/3D Example Applications. *A&A*, **509**, A36

Hauschildt, P. H. and Baron, E. 2011. A 3D Radiative Transfer Framework. VIII. OpenCL Implementation. *A&A*, **533**, A127

Hauschildt, P. H. and Baron, E. 2014. A 3D Radiative Transfer Framework. XI. Multi-Level NLTE. *A&A*, **566**, A89

Hauschildt, P. H., Barman, T. and Baron, E. 2008. Irradiated Planets. *Physica Scripta Volume T*, **130**(1), 014033

Heiter, U. and Eriksson, K. 2006. Geometry of Giant Star Model Atmospheres: A Consistency Test. *A&A*, **452**, 1039–1048

Henriques, V. M. J., Mathioudakis, M., Socas-Navarro, H. and de la Cruz Rodríguez, J. 2017. A Hot Downflowing Model Atmosphere for Umbral Flashes and the Physical Properties of Their Dark Fibrils. *ApJ*, **845**, 102

Hillier, D. J. 1987. Modeling the Extended Atmospheres of WN Stars. *ApJS*, **63**, 947–964

Hillier, D. J. 1990. An Iterative Method for the Solution of the Statistical and Radiative Equilibrium Equations in Expanding Atmospheres. *A&A*, **231**, 116–124

Hillier, D. J. 2012. Hot Stars with Winds: The CMFGEN Code. Pages 229–234 of: Richards, M. T. and Hubený, I. (eds), *From Interacting Binaries to Exoplanets: Essential Modeling Tools*. IAU Symposium, vol. 282

Hillier, D. J. and Dessart, L. 2012. Time-Dependent Radiative Transfer Calculations for Supernovae. *MNRAS*, **424**, 252–271

Hillier, D. J. and Miller, D. L. 1999. Constraints on the Evolution of Massive Stars through Spectral Analysis. I. The WC5 Star HD 165763. *ApJ*, **519**, 354–371

Hillier, D. J., Lanz, T., Heap, S. R., et al. 2003. A Tale of Two Stars: The Extreme O7 Iaf+ Supergiant AV 83 and the OC7.5 III((f)) Star AV 69. *ApJ*, **588**, 1039–1063

Hoffmann, T. L., Lieb, S., Pauldrach, A. W. A., Lesch, H., Hultzsch, P. J. N. and Birk, G. T. 2012. Numerical Models for the Diffuse Ionized Gas in Galaxies. I. Synthetic Spectra of Thermally Excited Gas with Turbulent Magnetic Reconnection as Energy Source. *A&A*, **544**, A57

Hoffmann, T. L., Pauldrach, A. W. A. and Kaschinski, C. B. 2016. Radiation-Driven Winds of Hot Luminous Stars. XVIII. The Unreliability of Stellar and Wind Parameter Determinations from Optical vs. UV Spectral Analysis of Selected Central Stars of Planetary Nebulae and the Possibility of Some CSPNs as Single-Star Supernova Ia Progenitors. *A&A*, **592**, A158

Hoffmann, T. L., Sauer, D. N., Pauldrach, A. W. A. and Hultzsch, P. J. N. 2014. Non-LTE Models for Synthetic Spectra of Type Ia Supernovae. IV. A Modified Feautrier Scheme for Opacity-Sampled Pseudo-Continua at High Expansion Velocities and Application to Synthetic SN Ia Spectra. *A&A*, **569**, A62

Holgado, G., Simón-Díaz, S., Barbá, R. H., et al. 2017. The IACOB Project. V. Spectroscopic Parameters of the O-Type Stars in the Modern Grid of Standards for Spectral Classification. *ArXiv e-prints*

Holzreuter, R. and Solanki, S. K. 2015. Three-Dimensional Non-LTE Radiative Transfer Effects in Fe i Lines. III. Line Formation in Magneto-Hydrodynamic Atmospheres. *A&A*, **582**, A101

Hubený, I. 1988. A Computer Program for Calculating Non-LTE Model Stellar Atmospheres. *Computer Physics Communications*, **52**, 103–132

Hubený, I., Agol, E., Blaes, O. and Krolik, J. H. 2000. Non-LTE Models and Theoretical Spectra of Accretion Disks in Active Galactic Nuclei. III. Integrated Spectra for Hydrogen-Helium Disks. *ApJ*, **533**, 710–728

Hubený, I. and Lanz, T. 1995. Non-LTE Line-Blanketed Model Atmospheres of Hot Stars. 1: Hybrid Complete Linearization/Accelerated Lambda Iteration Method. *ApJ*, **439**, 875–904

Hubený, I. and Lanz, T. 2011. *Synspec: General Spectrum Synthesis Program.*

Hubený, I. and Lanz, T. 2017. A Brief Introductory Guide to TLUSTY and SYNSPEC. *ArXiv e-prints*

Hubený, I., Blaes, O., Krolik, J. H. and Agol, E. 2001. Non-LTE Models and Theoretical Spectra of Accretion Disks in Active Galactic Nuclei. IV. Effects of Compton Scattering and Metal Opacities. *ApJ*, **559**, 680–702

Jack, D., Baron, E. and Hauschildt, P. H. 2015. Identification of the Feature that Causes the I-band Secondary Maximum of a Type Ia Supernova. *MNRAS*, **449**, 3581–3586

Jack, D., Hauschildt, P. H. and Baron, E. 2009. Time-Dependent Radiative Transfer with PHOENIX. *A&A*, **502**, 1043–1049

Jack, D., Hauschildt, P. H. and Baron, E. 2012a. A 3D Radiative Transfer Framework. IX. Time Dependence. *A&A*, **546**, A39

Jack, D., Hauschildt, P. H. and Baron, E. 2012b. Near-Infrared Light Curves of Type Ia Supernovae. *A&A*, **538**, A132

Jofré, P., Heiter, U., Worley, C. C., et al. 2017. Gaia FGK Benchmark Stars: Opening the Black Box of Stellar Element Abundance Determination. *A&A*, **601**, A38

Jorgensen, U. G., Carlsson, M. and Johnson, H. R. 1992. The Calcium Infrared Triplet Lines in Stellar Spectra. *A&A*, **254**, 258

Kaschinski, C. B., Pauldrach, A. W. A. and Hoffmann, T. L. 2012. Radiation-Driven Winds of Hot Luminous Stars. XVII. Parameters of Selected Central Stars of PN from Consistent Optical and UV Spectral Analysis and the Universality of the Mass–Luminosity Relation. *A&A*, **542**, A45

Kennedy, M. B., Milligan, R. O., Allred, J. C., Mathioudakis, M. and Keenan, F. P. 2015. Radiative Hydrodynamic Modelling and Observations of the X-Class Solar Flare on 2011 March 9. *A&A*, **578**, A72

Kirby, E. N. 2011. Grids of ATLAS9 Model Atmospheres and MOOG Synthetic Spectra. *PASP*, **123**, 531

Kiselman, D. and Carlsson, M. 1996. The NLTE Formation of Neutral-Boron Lines in Cool Stars. *A&A*, **311**, 680–689

Knop, S., Hauschildt, P. H. and Baron, E. 2009. Comoving-Frame Radiative Transfer in Arbitrary Velocity Fields. II. Large Scale Applications. *A&A*, **501**, 813–820

Kowalski, A. F., Allred, J. C., Daw, A., Cauzzi, G. and Carlsson, M. 2017b. The Atmospheric Response to High Nonthermal Electron Beam Fluxes in Solar Flares. I. Modeling the Brightest NUV Footpoints in the X1 Solar Flare of 2014 March 29. *ApJ*, **836**, 12

Kowalski, A. F., Allred, J. C., Uitenbroek, H., et al. 2017a. Hydrogen Balmer Line Broadening in Solar and Stellar Flares. *ApJ*, **837**, 125

Kuckein, C., Diercke, A., González Manrique, S. J., et al. 2017. Ca II 8542 Å Brightenings Induced by a Solar Microflare. *A&A*, **608**, A117

Kunder, A., Kordopatis, G., Steinmetz, M., et al. 2017. The Radial Velocity Experiment (RAVE): Fifth Data Release. *AJ*, **153**, 75

Kupfer, T., Przybilla, N., Heber, U., Jeffery, C. S., Behara, N. T. and Butler, K. 2017. Quantitative Spectroscopy of Extreme Helium Stars: Model Atmospheres and a Non-LTE Abundance Analysis of BD+10deg2179. *MNRAS*, **471**, 877–890

Kuridze, D., Mathioudakis, M., Christian, D. J., et al. 2016. Observations and Simulations of the Na I D_1 Line Profiles in an M-class Solar Flare. *ApJ*, **832**, 147

Kurucz, R. L. 2014a. How to Build a Model of the Atmosphere and Spectrum. Pages 25–37 of: Niemczura, E., Smalley, B. and Pych, W. (eds.), *Determination of Atmospheric Parameters of B-, A-, F- and G-Type Stars*. Series: GeoPlanet: Earth and Planetary Sciences. Springer International Publishing.

Kurucz, R. L. 2014b. Model Atmosphere Codes: ATLAS12 and ATLAS9. Pages 39–51 of: Niemczura, E., Smalley, B. and Pych, W. (eds.), *Determination of Atmospheric Parameters of B-, A-, F- and G-Type Stars*. Series: GeoPlanet: Earth and Planetary Sciences. Springer International Publishing.

Kurucz, R. L. 2014c. Problems with Atomic and Molecular Data: Including All the Lines. Pages 63–73 of: Niemczura, E., Smalley, B. and Pych, W. (eds.), *Determination of Atmospheric Parameters of B-, A-, F- and G-Type Stars*. Series: GeoPlanet: Earth and Planetary Sciences. Springer International Publishing.

Lanz, T. and Hubený, I. 1995. Non-LTE Line-Blanketed Model Atmospheres of Hot Stars. 2: Hot, Metal-Rich White Dwarfs. *ApJ*, **439**, 905–916

Lanz, T. and Hubený, I. 2003. A Grid of Non-LTE Line-Blanketed Model Atmospheres of O-Type Stars. *ApJS*, **146**, 417–441

Lanz, T. and Hubený, I. 2007. A Grid of NLTE Line-Blanketed Model Atmospheres of Early B-Type Stars. *ApJS*, **169**, 83–104

Leenaarts, J., and Carlsson, M. 2009. MULTI3D: A Domain-Decomposed 3D Radiative Transfer Code. Page 87 of: Lites, B., Cheung, M., Magara, T., Mariska, J. and Reeves, K. (eds.), *The Second Hinode Science Meeting: Beyond Discovery – Toward Understanding*. Astronomical Society of the Pacific Conference Series, vol. 415.

Leenaarts, J., Carlsson, M., and Rouppe van der Voort, L. 2012. The Formation of the Hα Line in the Solar Chromosphere. *ApJ*, **749**, 136

Leenaarts, J., Carlsson, M., Hansteen, V. and Rouppe van der Voort, L. 2009. Three-Dimensional Non-LTE Radiative Transfer Computation of the CA 8542 Infrared Line from a Radiation-MHD Simulation. *ApJ*, **694**, L128–L131

Leenaarts, J., Pereira, T. M. D., Carlsson, M., Uitenbroek, H. and De Pontieu, B. 2013a. The Formation of IRIS Diagnostics. I. A Quintessential Model Atom of Mg II and General Formation Properties of the Mg II h&k Lines. *ApJ*, **772**, 89

Leenaarts, J., Pereira, T. M. D., Carlsson, M., Uitenbroek, H., and De Pontieu, B. 2013b. The Formation of IRIS Diagnostics. II. The Formation of the Mg II h&k Lines in the Solar Atmosphere. *ApJ*, **772**, 90

Leenaarts, J., Golding, T., Carlsson, M., Libbrecht, T. and Joshi, J. 2016. The Cause of Spatial Structure in Solar He I 1083 nm Multiplet Images. *A&A*, **594**, A104

Lefever, K., Puls, J. and Aerts, C. 2007. Statistical Properties of a Sample of Periodically Variable B-Type Supergiants: Evidence for Opacity-Driven Gravity-Mode Oscillations. *A&A*, **463**, 1093–1109

Leitherer, C., Ortiz Otálvaro, P. A., Bresolin, F., et al. 2010. A Library of Theoretical Ultraviolet Spectra of Massive, Hot Stars for Evolutionary Synthesis. *ApJS*, **189**, 309–335

Ludwig, H.-G. and Kučinskas, A. 2012. Three-Dimensional Hydrodynamical CO^{5}BOLD Model Atmospheres of Red Giant Stars. I. Atmospheric Structure of a Giant Located Near the RGB Tip. *A&A*, **547**, A118

Ludwig, H.-G., Caffau, E., Steffen, M., Freytag, B., Bonifacio, P. and Kučinskas, A. 2009. The CIFIST 3D Model Atmosphere Grid. *Mem. Soc. Astron. Italiana*, **80**, 711

Magic, Z., Collet, R., Asplund, M., et al. 2013. The Stagger-Grid: A Grid of 3D Stellar Atmosphere Models. I. Methods and General Properties. *A&A*, **557**, A26

Maiorca, E., Uitenbroek, H., Uttenthaler, S., Randich, S., Busso, M. and Magrini, L. 2014. A New Solar Fluorine Abundance and a Fluorine Determination in the Two Open Clusters M67 and NGC 6404. *ApJ*, **788**, 149

Markova, N. and Puls, J. 2008. Bright OB Stars in the Galaxy. IV. Stellar and Wind Parameters of Early to Late B Supergiants. *A&A*, **478**, 823–842

Martínez-Sykora, J., De Pontieu, B., Carlsson, M., Hansteen, V. H., Nóbrega-Siverio, D. and Gudiksen, B. V. 2017b. Two-Dimensional Radiative Magnetohydrodynamic Simulations of Partial Ionization in the Chromosphere. II. Dynamics and Energetics of the Low Solar Atmosphere. *ApJ*, **847**, 36

Martínez-Sykora, J., De Pontieu, B., Hansteen, V. H., Rouppe van der Voort, L., Carlsson, M. and Pereira, T. M. D. 2017a. On the Generation of Solar Spicules and Alfvénic Waves. *Science*, **356**, 1269–1272

Martins, F., Schaerer, D. and Hillier, D. J. 2002. On the Effective Temperature Scale of O Stars. *A&A*, **382**, 999–1004

Mashonkina, L., Gehren, T., Shi, J.-R., Korn, A. J. and Grupp, F. 2011. A Non-LTE Study of Neutral and Singly-Ionized Iron Line Spectra in 1D Models of the Sun and Selected Late-Type Stars. *A&A*, **528**(April), A87

Mashonkina, L., Sitnova, T. and Belyaev, A. K. 2017. Influence of Inelastic Collisions with Hydrogen Atoms on the Non-LTE Modelling of Ca I and Ca II Lines in Late-Type Stars. *A&A*, **605**(September), A53

Massey, P., Neugent, K. F., Hillier, D. J. and Puls, J. 2013. A Bake-Off between CMFGEN and FASTWIND: Modeling the Physical Properties of SMC and LMC O-Type Stars. *ApJ*, **768**, 6

Mészáros, S., Allende Prieto, C., Edvardsson, B., et al. 2012. New ATLAS9 and MARCS Model Atmosphere Grids for the Apache Point Observatory Galactic Evolution Experiment (APOGEE). *AJ*, **144**, 120

Miller-Ricci, E. and Uitenbroek, H. 2002. Improving the Numerical Modeling of the O I Resonance Triplet in the Solar Spectrum. *ApJ*, **566**, 500–504

Mokiem, M. R., de Koter, A., Puls, J., Herrero, A., Najarro, F., and Villamariz, M. R. 2005. Spectral Analysis of Early-Type Stars Using a Genetic Algorithm Based Fitting Method. *A&A*, **441**, 711–733

Mokiem, M. R., de Koter, A., Vink, J. S., et al. 2007. The Empirical Metallicity Dependence of the Mass-Loss Rate of O- and Early B-Type Stars. *A&A*, **473**, 603–614

Morel, T., Butler, K., Aerts, C., Neiner, C. and Briquet, M. 2006. Abundance Analysis of Prime B-Type Targets for Asteroseismology. I. Nitrogen Excess in Slowly-Rotating β Cephei Stars. *A&A*, **457**, 651–663

Neugent, K. F., Massey, P., Hillier, D. J. and Morrell, N. 2017. The Evolution and Physical Parameters of WN3/O3s: A New Type of Wolf–Rayet Star. *ApJ*, **841**, 20

Nordlund, A. 1982. Numerical Simulations of the Solar Granulation. I Basic Equations and Methods. *A&A*, **107**, 1–10

Nordlund, Å., Galsgaard, K. and Stein, R. F. 1994. Magnetoconvection and Magnetoturbulence. Page 471 of: Rutten, R. J. and Schrijver, C. J. (eds.), *NATO Advanced Science Institutes (ASI) Series C*. NATO Advanced Science Institutes (ASI) Series C, vol. 433

Nordlund, Å., Stein, R. F. and Asplund, M. 2009. Solar Surface Convection. *Living Reviews in Solar Physics*, **6**, 2

Okamoto, T. J., Antolin, P., De Pontieu, B., Uitenbroek, H., Van Doorsselaere, T. and Yokoyama, T. 2015. Resonant Absorption of Transverse Oscillations and Associated Heating in a Solar Prominence. I. Observational Aspects. *ApJ*, **809**, 71

Olluri, K., Gudiksen, B. V. and Hansteen, V. H. 2013. Non-Equilibrium Ionization in the Bifrost Stellar Atmosphere Code. *AJ*, **145**, 72

Padoan, P., Nordlund, Å., Kritsuk, A. G., Norman, M. L. and Li, P. S. 2007. Two Regimes of Turbulent Fragmentation and the Stellar Initial Mass Function from Primordial to Present-Day Star Formation. *ApJ*, **661**, 972–981

Pauldrach, A. 1987. Radiation Driven Winds of Hot Luminous Stars. III Detailed Statistical Equilibrium Calculations for Hydrogen to Zinc. *A&A*, **183**, 295–313

Pauldrach, A. W. A. and Puls, J. 1990. Radiation-Driven Winds of Hot Stars. VIII The Bistable Wind of the Luminous Blue Variable P Cygni (B1 Ia/+/). *A&A*, **237**, 409–424

Pauldrach, A. W. A., Feldmeier, A., Puls, J. and Kudritzki, R. P. 1993. Radiation Driven Winds of Hot Stars: Theory of O-Star Atmospheres as a Spectroscopic Tool. *Space Sci. Rev.*, **66**, 105–125

Pauldrach, A. W. A., Hoffmann, T. L. and Hultzsch, P. J. N. 2014. Non-LTE Models for Synthetic Spectra of Type Ia Supernovae. III. An Accelerated Lambda-Iteration Procedure for the Mutual Interaction of Strong Spectral Lines in SN Ia Models with and without Energy Deposition. *A&A*, **569**, A61

Pauldrach, A. W. A., Hoffmann, T. L. and Lennon, M. 2001. Radiation-Driven Winds of Hot Luminous Stars. XIII. A Description of NLTE Line Blocking and Blanketing towards Realistic Models for Expanding Atmospheres. *A&A*, **375**, 161–195

Pauldrach, A. W. A., Hoffmann, T. L. and Méndez, R. H. 2004. Radiation-Driven Winds of Hot Luminous Stars. XV. Constraints on the Mass-Luminosity Relation of Central Stars of Planetary Nebulae. *A&A*, **419**, 1111–1122

Pauldrach, A. W. A., Kudritzki, R. P., Puls, J. and Butler, K. 1990. Radiation Driven Winds of Hot Luminous Stars. VII The Evolution of Massive Stars and the Morphology of Stellar Wind Spectra. *A&A*, **228**, 125–154

Pauldrach, A. W. A., Kudritzki, R. P., Puls, J., Butler, K. and Hunsinger, J. 1994. Radiation-Driven Winds of Hot Luminous Stars. 12: A First Step towards Detailed UV-Line Diagnostics of O-Stars. *A&A*, **283**, 525–560

Pauldrach, A. W. A., Lennon, M., Hoffmann, T. L., Sellmaier, F., Kudritzki, R.-P. and Puls, J. 1998. Realistic Models for Expanding Atmospheres. Page 258 of: Howarth, I. (ed.), *Properties of Hot Luminous Stars*. Astronomical Society of the Pacific Conference Series, vol. 131.

Pauldrach, A. W. A., Vanbeveren, D. and Hoffmann, T. L. 2012. Radiation-Driven Winds of Hot Luminous Stars XVI. Expanding Atmospheres of Massive and Very Massive Stars and the Evolution of Dense Stellar Clusters. *A&A*, **538**, A75

Pereira, T. M. D. and Uitenbroek, H. 2015. RH 1.5D: A Massively Parallel Code for Multi-Level Radiative Transfer with Partial Frequency Redistribution and Zeeman Polarisation. *A&A*, **574**, A3

Pereira, T. M. D., Asplund, M., Collet, R., Thaler, I., Trampedach, R. and Leenaarts, J. 2013a. How Realistic Are Solar Model Atmospheres? *A&A*, **554**, A118

Pereira, T. M. D., Leenaarts, J., De Pontieu, B., Carlsson, M. and Uitenbroek, H. 2013b. The Formation of IRIS Diagnostics. III. Near-Ultraviolet Spectra and Images. *ApJ*, **778**, 143

Pérez-Mesa, V., Zamora, O., García-Hernández, D. A., et al. 2017. Rubidium and Zirconium Abundances in Massive Galactic Asymptotic Giant Branch Stars Revisited. *A&A*, **606**, A20

Piskunov, N. and Valenti, J. A. 2017. Spectroscopy Made Easy: Evolution. *A&A*, **597**(January), A16

Plez, B. 2008. MARCS Model Atmospheres. *Physica Scripta Volume T*, **133**(1), 014003

Plez, B. 2012. *Turbospectrum: Code for Spectral Synthesis*

Plez, B., Brett, J. M. and Nordlund, A. 1992. Spherical Opacity Sampling Model Atmospheres for M-Giants. I Techniques, Data and Discussion. *A&A*, **256**, 551–571

Plez, B., Hill, V., Cayrel, R., et al. 2004. Lead Abundance in the Uranium Star CS 31082-001. *A&A*, **428**, L9–L12

Przybilla, N. and Butler, K. 2004. Non-LTE Line Formation for Hydrogen Revisited. *ApJ*, **609**, 1181–1191

Przybilla, N., Butler, K., Becker, S. R. and Kudritzki, R. P. 2006. Quantitative Spectroscopy of BA-Type Supergiants. *A&A*, **445**, 1099–1126

Przybilla, N., Fossati, L., Hubrig, S., et al. 2016. B Fields in OB Stars (BOB): Detection of a Magnetic Field in the He-Strong Star CPD -57deg 3509. *A&A*, **587**, A7

Przybilla, N., Nieva, M.-F. and Butler, K. 2008. A Cosmic Abundance Standard: Chemical Homogeneity of the Solar Neighborhood and the ISM Dust-Phase Composition. *ApJ*, **688**, L103

Puls, J. 2009. Modeling the Atmospheres of Massive Stars. *Communications in Asteroseismology*, **158**, 113

Puls, J. 2017. FASTWIND Reloaded: Complete Comoving Frame Transfer for 'All' Contributing Lines. Pages 435–435 of: Eldridge, J. J., Bray, J. C., McClelland, L. A. S. and Xiao, L. (eds.), *The Lives and Death-Throes of Massive Stars*. IAU Symposium, vol. 329.

Puls, J., Urbaneja, M. A., Venero, R., et al. 2005. Atmospheric NLTE-Models for the Spectroscopic Analysis of Blue Stars with Winds. II. Line-Blanketed Models. *A&A*, **435**, 669–698

Ramachandran, V., Hainich, R., Hamann, W.-R., et al. 2018. Stellar Population of the Superbubble N 206 in the LMC. I. Analysis of the Of-Type Stars. *A&A*, **609**, A7

Reid, A., Mathioudakis, M., Kowalski, A., Doyle, J. G. and Allred, J. C. 2017. Solar Ellerman Bombs in 1D Radiative Hydrodynamics. *ApJ*, **835**, L37

Repolust, T., Puls, J. and Herrero, A. 2004. Stellar and Wind Parameters of Galactic O-Stars: The Influence of Line-Blocking/Blanketing. *A&A*, **415**, 349–376

Repolust, T., Puls, J., Hanson, M. M., Kudritzki, R.-P. and Mokiem, M. R. 2005. Quantitative H and K Band Spectroscopy of Galactic OB-Stars at Medium Resolution. *A&A*, **440**, 261–286

Rivero González, J. G., Puls, J. and Najarro, F. 2011. Nitrogen Line Spectroscopy of O-Stars. I. Nitrogen III Emission Line Formation Revisited. *A&A*, **536**, A58

Rivero González, J. G., Puls, J., Massey, P. and Najarro, F. 2012. Nitrogen Line Spectroscopy in O-Stars. III. The Earliest O-Stars. *A&A*, **543**, A95

Rubin, R. H., Simpson, J. P., Colgan, et al. 2007. Spitzer Observations of M83 and the Hot Star, HII Region Connection. *MNRAS*, **377**, 1407–1418

Rubin, R. H., Simpson, J. P., Colgan, S. W. J., et al. 2008. Spitzer Observations of M33 and the Hot Star, HII Region Connection. *MNRAS*, **387**, 45–62

Rubin, R. H., Simpson, J. P., Colgan, S. W. J., et al. 2016. Spitzer Observations of Extragalactic H II Regions. III. NGC 6822 and the Hot Star, H II Region Connection. *MNRAS*, **459**, 1875–1891

Rubio da Costa, F. and Kleint, L. 2017. A Parameter Study for Modeling Mg II h and k Emission during Solar Flares. *ApJ*, **842**, 82

Rutten, R. J. and Uitenbroek, H. 2012. Chromospheric Backradiation in Ultraviolet Continua and Hα. *A&A*, **540**, A86

Rybicki, G. B. 1971. A Modified Feautrier Method. *J. Quant. Spec. Radiat. Transf.*, **11**, 589–595

Rybicki, G. B. and Hummer, D. G. 1991. An Accelerated Lambda Iteration Method for Multilevel Radiative Transfer. I Non-Overlapping Lines with Background Continuum. *A&A*, **245**, 171–181

Sander, A., Hamann, W.-R. and Todt, H. 2012. The Galactic WC Stars: Stellar Parameters from Spectral Analyses Indicate a New Evolutionary Sequence. *A&A*, **540**, A144

Sander, A. A. C. 2017. Recent Advances in Non-LTE Stellar Atmosphere Models. Pages 215–222 of: Eldridge, J. J., Bray, J. C., McClelland, L. A. S. and Xiao, L. (eds.), *The Lives and Death-Throes of Massive Stars.* IAU Symposium, vol. 329

Sander, A. A. C., Hamann, W.-R., Todt, H., Hainich, R. and Shenar, T. 2017. Coupling Hydrodynamics with Comoving Frame Radiative Transfer. I. A Unified Approach for OB and WR Stars. *A&A*, **603**, A86

Santolaya-Rey, A. E., Puls, J. and Herrero, A. 1997. Atmospheric NLTE-Models for the Spectroscopic Analysis of Luminous Blue Stars with Winds. *A&A*, **323**, 488–512

Sauer, D. N., Hoffmann, T. L. and Pauldrach, A. W. A. 2006. Non-LTE Models for Synthetic Spectra of Type Ia Supernovae/Hot Stars with Extremely Extended Atmospheres. II. Improved Lower Boundary Conditions for the Numerical Solution of the Radiative Transfer. *A&A*, **459**, 229–240

Saxner, M. and Gustafsson, B. 1984. A Method for Adding Opacity Distribution Functions. *A&A*, **140**, 334–340

Sbordone, L., Bonifacio, P., Caffau, E., et al. 2010. The Metal-Poor End of the Spite Plateau. I. Stellar Parameters, Metallicities, and Lithium Abundances. *A&A*, **522**, A26

Scharmer, G. B. 1981. Solutions to Radiative Transfer Problems Using Approximate Lambda Operators. *ApJ*, **249**, 720–730

Scharmer, G. B. and Carlsson, M. 1985. A New Approach to Multi-Level Non-LTE Radiative Transfer Problems. *Journal of Computational Physics*, **59**, 56–80

Schneider, F. R. N., Sana, H., Evans, C. J., et al. 2018. An Excess of Massive Stars in the Local 30 Doradus Starburst. *ArXiv e-prints*

Seelmann, A. M., Hauschildt, P. H. and Baron, E. 2010. A 3D Radiative Transfer Framework. VII. Arbitrary Velocity Fields in the Eulerian Frame. *A&A*, **522**, A102

Sellmaier, F. H., Yamamoto, T., Pauldrach, A. W. A. and Rubin, R. H. 1996. A Possible Solution for the [Ne III] Problem in HII Regions. *A&A*, **305**, L37

Shi, J. R., Gehren, T., Zeng, J. L., Mashonkina, L. and Zhao, G. 2014. Statistical Equilibrium of Copper in the Solar Atmosphere. *ApJ*, **782**(February), 80

Shi, J. R., Yan, H. L., Zhou, Z. M. and Zhao, G. 2018. NLTE Analysis of Copper Lines in Different Stellar Populations. *ApJ*, **862**(July), 71

Simões, P. J. A., Kerr, G. S., Fletcher, L., Hudson, H. S., Giménez de Castro, C. G. and Penn, M. 2017. Formation of the Thermal Infrared Continuum in Solar Flares. *A&A*, **605**, A125

Skartlien, R. 2000. A Multigroup Method for Radiation with Scattering in Three-Dimensional Hydrodynamic Simulations. *ApJ*, **536**, 465–480

Sneden, C., Bean, J., Ivans, I., Lucatello, S. and Sobeck, J. 2012. *MOOG: LTE Line Analysis and Spectrum Synthesis*. Astrophysics Source Code Library

Socas-Navarro, H. 2011. A High-Resolution Three-Dimensional Model of the Solar Photosphere Derived from Hinode Observations. *A&A*, **529**, A37

Socas-Navarro, H. 2015. The Solar Oxygen Abundance from an Empirical Three-Dimensional Model. *A&A*, **577**, A25

Socas-Navarro, H., de la Cruz Rodríguez, J., Asensio Ramos, A., Trujillo Bueno, J. and Ruiz Cobo, B. 2015. An Open-Source, Massively Parallel Code for Non-LTE Synthesis and Inversion of Spectral Lines and Zeeman-Induced Stokes Profiles. *A&A*, **577**, A7

Sonoi, T., Samadi, R., Belkacem, K., Ludwig, H.-G., Caffau, E. and Mosser, B. 2015. Surface-Effect Corrections for Solar-Like Oscillations Using 3D Hydrodynamical Simulations. I. Adiabatic Oscillations. *A&A*, **583**, A112

Spite, M., Caffau, E., Bonifacio, P., et al. 2013. Carbon-Enhanced Metal-Poor Stars: The Most Pristine Objects? *A&A*, **552**, A107

Steffen, M., Prakapavičius, D., Caffau, E., et al. 2015. The Photospheric Solar Oxygen Project. IV. 3D-NLTE Investigation of the 777 nm Triplet Lines. *A&A*, **583**, A57

Stein, R. F. and Nordlund, Å. 2012. On the Formation of Active Regions. *ApJ*, **753**, L13

Steiner, O., Salhab, R., Freytag, B., Rajaguru, P., Schaffenberger, W. and Steffen, M. 2014. Properties of Small-Scale Magnetism of Stellar Atmospheres. *PASJ*, **66**, S5

Štěpán, J. and Trujillo Bueno, J. 2013. PORTA: A Three-Dimensional Multilevel Radiative Transfer Code for Modeling the Intensity and Polarization of Spectral Lines with Massively Parallel Computers. *A&A*, **557**, A143

Sternberg, A., Hoffmann, T. L. and Pauldrach, A. W. A. 2003. Ionizing Photon Emission Rates from O- and Early B-Type Stars and Clusters. *ApJ*, **599**, 1333–1343

Sukhorukov, A. V. and Leenaarts, J. 2017. Partial Redistribution in 3D Non-LTE Radiative Transfer in Solar-Atmosphere Models. *A&A*, **597**, A46

Šurlan, B., Hamann, W.-R., Aret, A., Kubát, J., Oskinova, L. M. and Torres, A. F. 2013. Macroclumping as Solution of the Discrepancy between Hα and P v Mass Loss Diagnostics for O-Type Stars. *A&A*, **559**, A130

Taresch, G., Kudritzki, R. P., Hurwitz, M., et al. 1997. Quantitative Analysis of the FUV, UV and Optical Spectrum of the O3 Star HD 93129A. *A&A*, **321**, 531–548

Tremblay, P.-E., Ludwig, H.-G., Freytag, B., Steffen, M. and Caffau, E. 2013a. Granulation Properties of Giants, Dwarfs, and White Dwarfs from the CIFIST 3D Model Atmosphere Grid. *A&A*, **557**, A7

Tremblay, P.-E., Ludwig, H.-G., Steffen, M. and Freytag, B. 2013b. Spectroscopic Analysis of DA White Dwarfs with 3D Model Atmospheres. *A&A*, **559**, A104

Trujillo Bueno, J. and Fabiani Bendicho, P. 1995. A Novel Iterative Scheme for the Very Fast and Accurate Solution of Non-LTE Radiative Transfer Problems. *ApJ*, **455**, 646

Uitenbroek, H. 1998. The Effect of Photospheric Granulation on the Determination of the Lithium Abundance in Solar-Type Stars. *ApJ*, **498**, 427–440

Uitenbroek, H. 2000. The CO Fundamental Vibration-Rotation Lines in the Solar Spectrum. II. Non-LTE Transfer Modeling in Static and Dynamic Atmospheres. *ApJ*, **536**, 481–493

Uitenbroek, H. 2001. Multilevel Radiative Transfer with Partial Frequency Redistribution. *ApJ*, **557**, 389–398

Uitenbroek, H. 2002. The Effect of Coherent Scattering on Radiative Losses in the Solar Ca II K Line. *ApJ*, **565**, 1312–1322

Uitenbroek, H. 2003. The Accuracy of the Center-of-Gravity Method for Measuring Velocity and Magnetic Field Strength in the Solar Photosphere. *ApJ*, **592**, 1225–1233

Uitenbroek, H. and Criscuoli, S. 2011. Why One-Dimensional Models Fail in the Diagnosis of Average Spectra from Inhomogeneous Stellar Atmospheres. *ApJ*, **736**, 69

Uitenbroek, H. and Tritschler, A. 2006. The Contrast of Magnetic Elements in Synthetic CH- and CN-Band Images of Solar Magnetoconvection. *ApJ*, **639**, 525–533

Uitenbroek, H., Miller-Ricci, E., Asensio Ramos, A. and Trujillo Bueno, J. 2004. The Zeeman Effect in the G Band. *ApJ*, **604**, 960–968

Valenti, J. A. and Piskunov, N. 1996. Spectroscopy Made Easy: A New Tool for Fitting Observations with Synthetic Spectra. *A&AS*, **118**(September), 595–603

Van Eck, S., Neyskens, P., Jorissen, A., et al. 2017. A Grid of MARCS Model Atmospheres for Late-Type Stars. II. S Stars and Their Properties. *A&A*, **601**, A10

Vasilyev, V., Ludwig, H.-G., Freytag, B., Lemasle, B. and Marconi, M. 2017. Spectroscopic Properties of a Two-Dimensional Time-Dependent Cepheid Model. I. Description and Validation of the Model. *A&A*, **606**, A140

Vernazza, J. E., Avrett, E. H. and Loeser, R. 1973. Structure of the Solar Chromosphere: Basic Computations and Summary of the Results. *ApJ*, **184**, 605–632

Wade, R. A. and Hubený, I. 1998. Detailed Mid- and Far-Ultraviolet Model Spectra for Accretion Disks in Cataclysmic Binaries. *ApJ*, **509**, 350–361

Weber, J. A., Pauldrach, A. W. A. and Hoffmann, T. L. 2015. Three-Dimensional Modeling of Ionized Gas. II. Spectral Energy Distributions of Massive and Very Massive Stars in Stationary and Time-Dependent Modeling of the Ionization of Metals in H II Regions. *A&A*, **583**, A63

Weber, J. A., Pauldrach, A. W. A., Knogl, J. S. and Hoffmann, T. L. 2013. Three-Dimensional Modeling of Ionized Gas. I. Did Very Massive Stars of Different Metallicities Drive the Second Cosmic Reionization? *A&A*, **555**, A35

Wedemeyer, S., Freytag, B., Steffen, M., Ludwig, H.-G. and Holweger, H. 2004. Numerical Simulation of the Three-Dimensional Structure and Dynamics of the Non-Magnetic Solar Chromosphere. *A&A*, **414**, 1121–1137

Wedemeyer-Böhm, S., Kamp, I., Bruls, J. and Freytag, B. 2005. Carbon Monoxide in the Solar Atmosphere. I. Numerical Method and Two-Dimensional Models. *A&A*, **438**, 1043–1057

Wilk, K. D., Hillier, D. J. and Dessart, L. 2018. Ejecta Mass Diagnostics of Type Ia Supernovae. *MNRAS*, **474**, 3187–3211

Zsargo, J., Arrieta, A., Fierro, C., et al. 2017. A Mega-Grid of CMFGEN Model Atmospheres for Rapid Analysis of Stellar Spectra. Page 407 of: Miroshnichenko, A., Zharikov, S., Korčáková, D. and Wolf, M. (eds.), *The B[e] Phenomenon: Forty Years of Studies*. Astronomical Society of the Pacific Conference Series, vol. 508.

5. Radiative Transfer in the (Expanding) Atmospheres of Early-Type Stars, and Related Problems

JOACHIM PULS

Abstract

In many cases, the quantitative spectroscopy of early-type stars requires to account for their line-driven winds, and theoretical models of such winds are based on a consistent calculation of the radiative line acceleration. Both topics ask for a thorough understanding of radiative transfer in expanding atmospheres. In this chapter, we concentrate on three issues, and compare, when possible, with corresponding results for plane-parallel, hydrostatic conditions: First, we investigate how sphericity alone affects the radiation field in those cases where Doppler shifts can be neglected (continua). Subsequently, we consider the impact of velocity fields on the line transfer, both by applying the so-called Sobolev approximation, and by presenting the more exact comoving-frame approach. Restrictions and extensions of both methods are discussed. Finally, we concentrate on the coupling between radiation field and occupation numbers via the NLTE rate equations. We illustrate the basic problem within the conventional lambda iteration, which is then solved by means of the so-called Accelerated Lambda Iteration (ALI), and by a 'preconditioning' of the rate equations.

5.1 (Very Brief) Introduction

One of the most striking observational features of early-type stars are their quasi-stationary UV P Cygni profiles (Figure 5.1), which indicate fast outflows, and, together with other diagnostics, only small variability of global quantities such as mass-loss rate, $\dot{M}$, and terminal velocity, v_∞.

These winds and their characteristic quantities have to be explained, diagnostic tools have to be developed, and predictions have to be given. All these tasks are comprised in the *theory of expanding atmospheres.*[1]

Beginning with the theoretical work by Lucy and Solomon (1970) and Castor et al. (1975, "CAK"), it turned out that the winds from early-type stars are driven by radiative line acceleration, and subsequent diagnostics revealed that typical mass-loss rates lie in the range $10^{-7} \ldots 10^{-5} M_\odot \mathrm{yr}^{-1}$, with v_∞ between 200 and 3,000 $\mathrm{km\,s^{-1}}$, fairly proportional to the corresponding photospheric escape speeds.[2]

In order to account for the presence of these winds when synthesizing theoretical spectral energy distributions (SEDs) (quantitative spectroscopy!), and to enable the calculation of the line acceleration required to set up theoretical models, the *radiative transfer in expanding media* needs to be formulated and understood, which is the topic of the following chapter. Particularly, there are two effects that give rise to major differences compared to plane-parallel, hydrostatic calculations used, e.g., for the analysis of late-type

[1] In addition to the references provided in the following, we also recommend the textbooks by Mihalas (1978) and Hubený and Mihalas (2014).

[2] For specific reviews on the topic of line-driven winds, see Kudritzki and Puls (2000) and Puls et al. (2008).

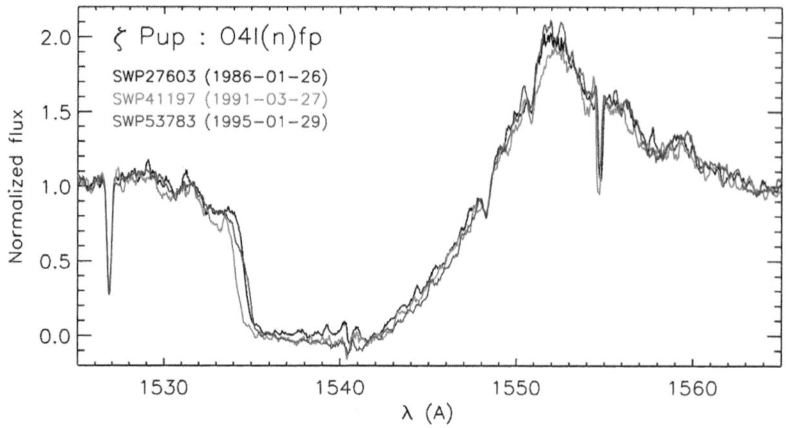

FIGURE 5.1. Three UV P Cgyni profiles of the C IV resonance line from the O4 supergiant ζ Pup obtained (several years apart) with the International Ultraviolet Explorer (IUE) space mission. Note how the overall shape of the spectrum (indicating a terminal velocity, v_∞, of roughly 2,500 $\mathrm{km\,s^{-1}}$) remains fairly constant.

stars (see Chapters 2 and 6): *sphericity*, which affects the radiation field and (wind-) density, to be covered in Sections 5.2 and 5.3, and *velocity fields*, which mostly affect the line transfer, via the induced Doppler shifts, to be discussed in Section 5.4.

5.2 From p-p Symmetry to Spherical Atmospheres with Velocity Fields

As long as $\Delta r/R_* \ll 1$, with Δr the vertical extent of the atmosphere and R_* the stellar radius, plane-parallel (p-p) symmetry can be assumed, at least in a 1-D treatment. Such an approach is valid, e.g., for the solar photosphere, when refraining from a precise description of convection. Since the curvature of the stellar atmosphere is neglected in a p-p approach, the angle between a photon's path and the isocontours of important quantities such as density and temperature remains constant throughout the atmosphere. On the other hand, when $\Delta r/R_* \gtrsim 1$, as in the solar corona or in the winds of early-type stars or red giants/supergiants, at least spherical symmetry needs to be adopted, but in any case the aforementioned angle changes drastically when propagating from the bottom to the top of the atmosphere.

5.2.1 *Coordinate Systems and Symmetries*

When using a cartesian coordinate system, a vector $\mathbf{r}$ is expressed via $\mathbf{r} = x\mathbf{e}_x + y\mathbf{e}_y + z\mathbf{e}_z$, while in a spherical coordinate system $\mathbf{r} = \Theta\mathbf{e}_\Theta + \Phi\mathbf{e}_\Phi + r\mathbf{e}_r$, where $\mathbf{e}_x$, $\mathbf{e}_y$, $\mathbf{e}_z$ and $\mathbf{e}_\Theta$, $\mathbf{e}_\Phi$, $\mathbf{e}_r$ form a right-handed, orthonormal base. In such systems, the specific intensity depends on $I(x, y, z, t; \mathbf{n}, \nu)$ and $I(\Theta, \Phi, r, t; \mathbf{n}, \nu)$, respectively, where $\mathbf{n}$ is the direction vector, ν the frequency, and t the time. Related symmetries are the plane-parallel one, where all physical quantities depend only on z, e.g., $I(\mathbf{r}, t; \mathbf{n}, \nu) \rightarrow I(z, t; \mathbf{n}, \nu)$, and the spherical symmetry, with physical quantities depending only on r, e.g., $I(\mathbf{r}, t; \mathbf{n}, \nu) \rightarrow I(r, t; \mathbf{n}, \nu)$.

Since the specific intensity has direction $\mathbf{n}$ into $\mathrm{d}\Omega$, additional angles θ, ϕ with respect to $(\mathbf{e}_x, \mathbf{e}_y, \mathbf{e}_z)$ or $(\mathbf{e}_\Theta, \mathbf{e}_\Phi, \mathbf{e}_r)$ are required, with polar angle, $\theta = \sphericalangle(\mathbf{e}_z, \mathbf{n})$ or $\theta = \sphericalangle(\mathbf{e}_r, \mathbf{n})$, respectively (see Figure 5.2). Thus, I_ν can be expressed as $I_\nu(x, y, z, t; \theta, \phi)$ or $I_\nu(r, \Theta, \Phi, t; \theta, \phi)$.

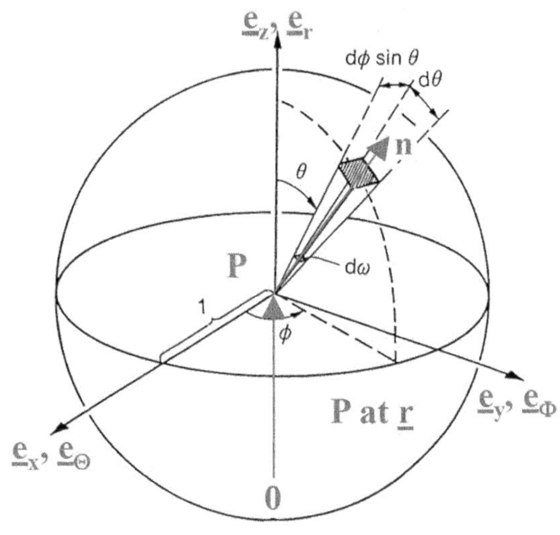

FIGURE 5.2. Directional angles, θ, ϕ, and solid angle element $d\Omega = d\phi \sin\theta \times d\theta$, as used to calculate the specific intensity $I_\nu(\mathbf{r}, \mathbf{n}, t)$ at point P, for both a cartesian and a spherical coordinate system (see text).

Both in plane-parallel and spherical symmetry, the intensity does not dependent on azimuthal direction, ϕ (again Figure 5.2), and we finally obtain $I_\nu \rightarrow I_\nu(z, t; \theta)$ or $\rightarrow I_\nu(r, t; \theta)$, respectively.

5.2.2 *Hydrostatic Equilibrium*

In plane-parallel atmospheres without winds (e.g., Kurucz atmospheres), but also in atmospheric models aiming at a description of early-type stars with thin winds (e.g., TLUSTY or DETAIL/SURFACE; see Appendix A); the pressure/density stratification is conventionally prescribed assuming hydrostatic equilibrium, namely

$$\frac{\partial P}{\partial z} = \rho(z)\left(-g_{\rm grav} + g_{\rm rad}(z)\right), \tag{5.1}$$

where $g_{\rm grav} = GM_*/R_*^2$ and again $\Delta z(\text{photosphere}) \ll R_*$. Integration of (5.1) gives either

$$P_{\rm tot}(z) = g_{\rm grav} \cdot m,$$

where $P_{\rm tot} = P_{\rm gas} + P_{\rm rad}$ and the mass column density is defined as $m \equiv \int_z^\infty \rho(z)dz$, or, neglecting $g_{\rm rad}$ and adopting a constant surface temperature T_*,

$$\rho(z) \approx \rho(z=0)\; e^{-z/H},$$

with photospheric scale height

$$H = \frac{k_{\rm B} T_*}{\mu\, m_{\rm H}\, g_{\rm grav}} = \frac{2\, v_{\rm sound}^2(T_*)}{v_{\rm esc}^2}\, R_*.$$

Here $v_{\rm sound} = \sqrt{k_{\rm B}T/\mu\, m_{\rm H}}$ is the isothermal speed of sound (of order of few $\rm km\,s^{-1}$), μ the mean molecular weight, and $v_{\rm esc} = \sqrt{2GM_*/R_*}$ the photospheric escape velocity (usually of order of several 100 $\rm km\,s^{-1}$). Alternatively, neglecting again $g_{\rm rad}$,

$$\rho(m) \approx \frac{1}{H} m, \quad \text{i.e.,} \qquad \log\rho = \log m - \log H. \tag{5.2}$$

When velocity fields are taken into account, conservation of mass leads to the *equation of continuity*

$$\frac{\partial \rho}{\partial t} + \nabla \cdot (\rho \mathbf{v}) = 0,$$

which for a steady one-dimensional spherical flow reduces to

$$4\pi\, r^2 \rho\, v = \text{const } = \dot{M}, \tag{5.3}$$

where $\dot{M}$ is the (constant) mass-loss rate through a spherical surface.

From the conservation of momentum, one obtains Euler's equation

$$\frac{\partial \rho}{\partial t} + \nabla \cdot (\rho \mathbf{v}\mathbf{v}) = -\nabla P + \rho\, \mathbf{g}^{\text{ext}}. \tag{5.4}$$

By $\mathbf{vv}$ we denote the dyadic product, and $\mathbf{g}^{\text{ext}}$ the total external acceleration. From vector calculus it holds that $\nabla \cdot (\rho \mathbf{vv}) = \mathbf{v}\,[\nabla \cdot (\rho \mathbf{v})] + [\rho \mathbf{v} \cdot \nabla]\,\mathbf{v}$. For a one-dimensional spherical flow, (5.4) reduces to the *equation of motion*

$$\rho\, v\, \frac{\partial v}{\partial r} = -\frac{\partial P}{\partial r} + \rho\, g_r^{\text{ext}}. \tag{5.5}$$

The LHS of (5.5) is the advection term due to inertia. The comparison of (5.5) – where gravity and radiative acceleration are taken into account – with (5.1), namely

$$\frac{\partial P}{\partial r} = \rho(r) \left(-\frac{G M_*}{r^2} + g_{\text{rad}}(r) \right) - \rho(r) v(r)\, \frac{\partial v}{\partial r}$$

and

$$\frac{\partial P}{\partial z} = \rho(z) \left(-\frac{G M_*}{R_*^2} + g_{\text{rad}}(z) \right),$$

shows the importance of the advection term.

5.2.3 *When Is a (Quasi-)Hydrostatic Approach Justified?*

By using the equation of state $P = (k_{\text{B}} T / \mu m_{\text{H}}) \rho = v_{\text{sound}}^2\, \rho$ and the equation of continuity (5.3), the equations of motion and of hydrostatic equilibrium can be rewritten as follows:

$$\left(v_{\text{sound}}^2(r) - v^2(r)\right) \frac{\partial \rho}{\partial r} = -\rho(r) \left(g_{\text{grav}}(r) - g_{\text{rad}}(r) + \frac{\text{d}v_{\text{sound}}^2}{\text{d}r} - \frac{2 v^2(r)}{r} \right) \text{[hydrodyn.]}$$

$$v_{\text{sound}}^2(z) \frac{\partial \rho}{\partial z} = -\rho(z) \left(g_{\text{grav}}(R_*) - g_{\text{rad}}(z) + \frac{\text{d}v_{\text{sound}}^2}{\text{d}z} \right) \quad \text{[hydrostatic]}.$$

By comparing both equations, we note that the 'only' difference is an additional term $\propto v^2$ both on the left and right side of the equation of motion, and we conclude that for $v \ll v_{\text{sound}}$ – i.e., in deeper photospheric regions, well below the sonic point where $v(r_{\text{S}}) = v_{\text{sound}}$ – the hydrodynamic density stratification approaches the (quasi-)hydrostatic one.

Thus, p-p atmospheres using hydrostatic equilibrium yield reasonable results *even in the presence of winds*, as long as the studied features (continua, lines) are formed below the sonic point (see also the following subsection).

5.2.4 *Unified Atmospheres*

The concept of 'unified atmospheres' (= wind + photosphere) was founded by Gabler et al. (1989). Nowadays, two flavors of such a description are present:

(a) The complete stratification is adapted from theoretical wind models based on the (modified) CAK theory (Friend and Abbott, 1986; Pauldrach et al., 1986), such that either $\dot{M}$ and v_∞ of the models agree with the required input values, or the stratification results from a self-consistent calculation w.r.t. $g_{\rm rad}$ (without the possibility to choose arbitrary combinations of wind and stellar parameters as input). Both methods are used within the atmosphere code WM-basic (Pauldrach et al., 2001). The disadvantage of this approach is that it is difficult (or even impossible) to manipulate the density/velocity stratification in case the theory is not applicable or too simplified.

(b) A quasistatic photosphere is combined with an empirical wind structure (PoWR, CMFGEN, PHOENIX, FASTWIND; see Appendix A), with the disadvantage that the transition region is somewhat ill defined. Specifically, in deep layers $\rho(r)$ is calculated from (quasi-)hydrostatic equilibrium (5.1) (with R_* replaced by r), and the corresponding velocity is derived via

$$v(r) = \frac{\dot{M}}{4\pi r^2 \rho(r)} \qquad \text{for } v \ll v_{\rm sound} \qquad (\text{roughly: } v < 0.1 v_{\rm sound}).$$

In the outer layers, at first $v(r)$ is defined using the semi-empirical 'beta velocity law' for radiation driven winds (e.g., Pauldrach et al. 1986, and Figure 5.3),

$$v(r) = v_\infty \left(1 - \frac{bR_*}{r}\right)^\beta, \tag{5.6}$$

with $0.5 < \beta \lesssim 2\ldots 3$, and b derived from the transition velocity. In this regime, then, the density results from

$$\rho(r) = \frac{\dot{M}}{4\pi r^2 v(r)}.$$

Finally, a certain transition zone is defined to ensure a smooth transition from the deeper to the outer layers. This unified description is quite flexible, and the corresponding input/fit parameters are $\dot{M}$, v_∞, β, and the transition velocity. A comparison of a hydrostatic and unified atmospheric structure is presented in Figure 5.4. We stress that at the same $\tau_{\rm Ross}$ or m, the wind density (for $v \gtrsim v_{\rm sound}$) is *lower* than the hydrostatic one.

5.2.5 *Plane-Parallel or Unified Atmospheres?*

Since the calculation of unified atmospheres plus corresponding SEDs is much more time consuming than the calculation of plane-parallel ones, it is reasonable to check beforehand which approach is required. Accounting for the formation region of optical lines (see Figure 5.4), unified models become vital if $\tau_{\rm Ross} \gtrsim 10^{-2}$ at the transition between photosphere and wind (roughly located at $0.1 v_{\rm sound}$). Using a typical velocity law ($\beta = 1$), as a rule of thumb

$$\dot{M}_{\rm max} = \dot{M}(\tau_{\rm Ross} = 10^{-2} \text{ at } 0.1 v_{\rm sound}) \approx 6 \cdot 10^{-8} M_\odot {\rm yr}^{-1} \cdot \frac{R_*}{10\,R_\odot} \cdot \frac{v_\infty}{1{,}000\,{\rm km\,s^{-1}}}.$$

If the actual $\dot{M} < \dot{M}_{\rm max}$ for the considered object, most diagnostic features are formed in the quasihydrostatic part of the atmosphere, and plane-parallel models can be used.

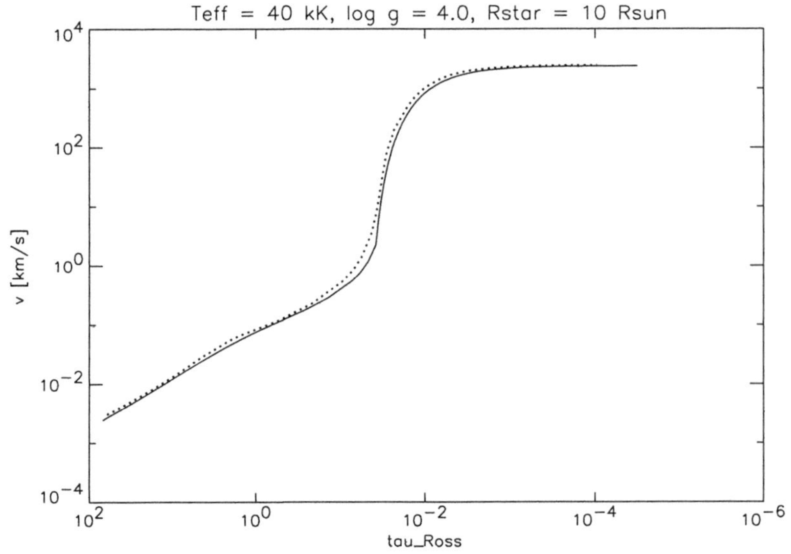

FIGURE 5.3. Velocity fields for unified O-star models with a comparatively thin wind. Dotted: hydrodynamic solution following Pauldrach et al. (1986); solid: analytical velocity law (5.6) with similar terminal velocity and $\beta = 0.8$, extended towards larger depths using a quasihydrostatic approach.

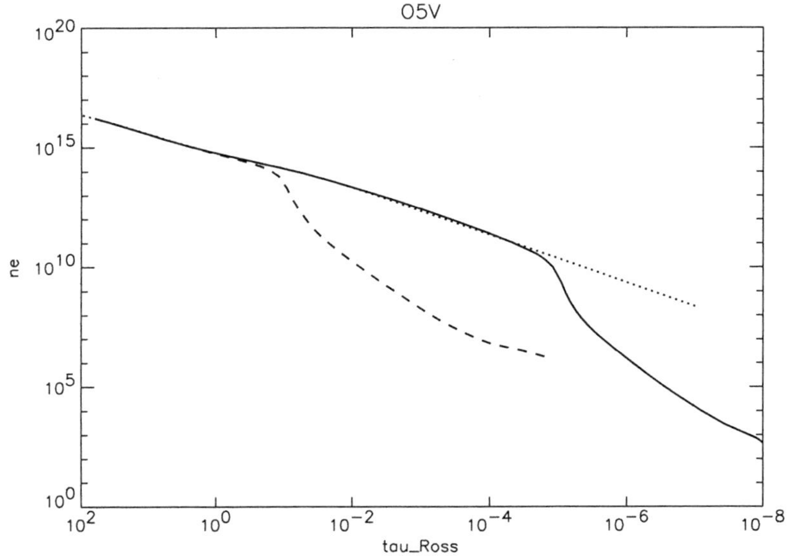

FIGURE 5.4. Electron density as a function of $\tau_{\rm Ross}$, for different atmospheric models of an O5-dwarf. Dotted: hydrostatic model atmosphere, cf. (5.2); solid, dashed: unified models with a thin and a moderately dense wind, respectively. In case of the denser wind, the cores of the optical lines ($\tau_{\rm Ross} \approx 10^{-1} - 10^{-2}$) are formed at significantly different densities than in the hydrostatic model, whereas the unified, thin-wind model and the hydrostatic one would lead to similar results.

Typically, this refers to the optical spectroscopy of late O-dwarfs and B-stars up to luminosity class II (for early subtypes) or Ib (mid/late subtypes).

5.3 Radiative Transfer: From p-p to Spherical Symmetry

5.3.1 *Basic Considerations*

In the following, we will mostly restrict ourselves to 1-D problems, since multi-D problems are beyond the scope of this overview. At first we will summarize the major changes in the description/properties of the radiation field when switching from a plane-parallel to a spherically symmetric situation.

Basically, the specific intensity and its moments are similarly defined when proceeding from the p-p height coordinate, z, to the radial distance, r.

$$I(z,\mu) \rightarrow I(r,\mu) \text{ with } \mu = \cos\theta \text{ and } \theta = \sphericalangle(\mathbf{e}_r, \mathbf{n}),$$

where here and in the following notation, the ν and t dependence has been suppressed. From the adopted symmetry (independence from the azimuthal direction, e.g., Figure 5.2), the nth moment of the specific intensity, namely

$$M_n = \frac{1}{2}\int_{-1}^{+1} I(r,\mu)\mu^n \mathrm{d}\mu,$$

is equally defined as in the p-p case when $z \rightarrow r$. For n = 0,1,2, we obtain the mean intensity, the Eddington flux and the second moment, $J(r)$, $H(r)$ and $K(r)$, respectively. The flux(-density) vector,

$$\mathcal{F} = \left(0, 0, 4\pi H\right)^T,$$

has only an r-component different from zero, which is proportional to the Eddington flux.

Regarding the radiation stress tensor, $\mathbf{P}$, only the diagonal elements are different from zero (as in the p-p case), and the only difference thus far refers to the divergence of the stress tensor (which is related to the radiation force; see (5.11)). While in p-p symmetry, only its z-component is different from zero, and

$$\left(\nabla\cdot\mathbf{P}\right)_z = \frac{\partial p_\mathrm{R}}{\partial z}, \text{ with } p_\mathrm{R} \text{ (radiation pressure scalar) } = \frac{4\pi}{c}K(z),$$

in spherical symmetry only the r-component is different from zero, and

$$\left(\nabla\cdot\mathbf{P}\right)_r = \frac{\partial p_\mathrm{R}}{\partial r} + \frac{3p_\mathrm{R} - u}{r}, \text{ with } u \text{ (radiation energy density) } = \frac{4\pi}{c}J(r).$$

Behaviour at large distances from the surface: optically thin envelopes. An important difference between p-p and spherically symmetric configurations relates to the behaviour of the radiation field at large distances from the stellar surface, which in case of spherical symmetry is affected by geometrical dilution. To estimate corresponding effects, let's assume an optically thin envelope, i.e., $I_\nu(r)$:= const for a specific ray, and that the radiation field leaving the effective photosphere, R_eff, is isotropic: $I_\nu^{+,\mathrm{phot}}(R_\mathrm{eff},\mu)$:= const $= I_\nu^+(R_\mathrm{eff})$:

$$\Rightarrow M_n = \frac{1}{2}\int_{-1}^{+1} I_\nu(\mu)\mu^n \mathrm{d}\mu \quad \rightarrow \quad \frac{1}{2}\int_{\mu_*}^{+1} I_\nu^+(R_\mathrm{eff})\mu^n \mathrm{d}\mu = \frac{1}{2}I_\nu^+(R_\mathrm{eff})\frac{\left(1-\mu_*^{n+1}\right)}{n+1}.$$

In this case, for the 0th moment we find $J_\nu \approx W I_\nu^+(R_{\rm eff})$, with dilution factor

$$W = \frac{1}{2}(1-\mu_*) \text{ and } \mu_* = \sqrt{1-\left(\frac{R_{\rm eff}}{r}\right)^2} \tag{5.7}$$

where μ_* is the cosine of the (half) cone angle subtended by stellar disk, θ_*, which can be calculated via $\sin\theta_* = R_{\rm eff}/r$. Now, for $r \gg R_{\rm eff}$,

$$\mu_*^{n+1} \to \left(1 - \frac{n+1}{2}\left(\frac{R_{\rm eff}}{r}\right)^2\right),$$

and *any* moment

$$J_\nu = H_\nu = K_\nu = \ldots . \to \quad \frac{1}{4} I_\nu^+(R_{\rm eff})\left(\frac{R_{\rm eff}}{r}\right)^2.$$

In other words, all moments become equal, and the Eddington factors (ratios of moments) converge to unity for $r \gg R_{\rm eff}$. This is specific for (spherical) envelopes at large distances, and different from corresponding plane-parallel results. (Exercise: perform the same calculation for plane-parallel conditions and large z.)

The equation of radiative transfer (RTE). Independent from any coordinate system and discarding general relativity (GR) effects, the RTE reads

$$\left(\frac{1}{\mathrm{c}}\frac{\partial}{\partial t} + \mathbf{n}\cdot\nabla\right) I_\nu(\mathbf{r},\mathbf{n},t) = \eta_\nu(\mathbf{r},\mathbf{n},t) - \chi_\nu(\mathbf{r},\mathbf{n},t) I_\nu(\mathbf{r},\mathbf{n},t), \tag{5.8}$$

where η_ν is the total emissivity, χ_ν the total opacity, and $\mathbf{n}\cdot\nabla$ the directional derivative $\triangleq \frac{\mathrm{d}}{\mathrm{d}s}$ along path s.

In plane-parallel geometry, $\mathbf{n}\cdot\nabla \to \mu \mathrm{d}/\mathrm{d}z$, since the actual path is longer than the height difference, $\mathrm{d}s = \mathrm{d}z/\mu$, and we obtain, when discarding the time-dependence for stationary conditions,

$$\mu\frac{\mathrm{d}}{\mathrm{d}z} I_\nu(z,\mu) = \eta_\nu(z,\mu) - \chi_\nu(z,\mu) I_\nu(z,\mu) \quad \text{(plane-parallel, stationary).}$$

In spherical geometry, μ is no longer constant along a certain direction $\mathbf{n}$. Restricting ourselves to spherical symmetry,

$$\mathbf{n}\cdot\nabla \Rightarrow \mu\frac{\partial}{\partial r} + \frac{(1-\mu^2)}{r}\frac{\partial}{\partial\mu},$$

which can be shown by using the so-called p-z geometry (see the following section). For stationary processes, we then have

$$\left(\mu\frac{\partial}{\partial r} + \frac{(1-\mu^2)}{r}\frac{\partial}{\partial\mu}\right) I_\nu(r,\mu) = \eta_\nu(r,\mu) - \chi_\nu(r,\mu) I_\nu(r,\mu) \quad \text{(sph. symm., stationary).}$$

Moments of the RTE. The zero- and first-order-moment equations are obtained by integrating the RTE over $\mathrm{d}\Omega$ or by multiplying with $\mathbf{n}/c$ and integrating over $\mathrm{d}\Omega$, respectively, and are very useful and insightful for many problems/applications.

In the general case, the zero-order-moment equation reads

$$\frac{4\pi}{\mathrm{c}}\frac{\partial}{\partial t} J_\nu + \nabla\cdot\mathcal{F}_\nu = \oint (\eta_\nu - \chi_\nu I_\nu)\,\mathrm{d}\Omega. \tag{5.9}$$

After integrating over frequency, the RHS of this equation becomes zero ('radiative equilibrium'), as long as only radiation energy is transported. For time-independent problems, this then refers to 'flux-conservation', $\nabla \cdot \mathcal{F} = 0$, with $\mathcal{F}$ the *total* flux.

For plane-parallel, stationary and static conditions, (5.9) collapses to

$$\frac{\mathrm{d}H_\nu}{\mathrm{d}z} = \eta_\nu - \chi_\nu J_\nu,$$

whereas for spherically symmetric, stationary, and (quasi-)static conditions, it reads

$$\frac{1}{r^2}\frac{\partial(r^2 H_\nu)}{\partial r} = \eta_\nu - \chi_\nu J_\nu. \tag{5.10}$$

We stress that the two preceding equations are valid only in case of *(quasi-)static* atmospheres, since otherwise the opacities become angle dependent, due to the apparent Doppler shifts (see Section 5.4), and cannot be put in front of the angular integrals. Thus, the latter two equations cannot be used in case of stellar winds, and the more general formulation of the RHS of (5.9) has to be accounted for. In an approximate way, though, these equations might still be applied for pure continuum problems in the presence of velocity fields, if an exact treatment of ionization edges plays a minor role.

The general first-order-moment equation is given by

$$\frac{1}{c^2}\frac{\partial}{\partial t}\mathcal{F}_\nu + \nabla \cdot \mathbf{P}_\nu = \frac{1}{c}\oint (\eta_\nu - \chi_\nu I_\nu)\,\mathbf{n}\mathrm{d}\Omega, \tag{5.11}$$

where now the frequency-integrated RHS is just the negative of the total radiation force, $-\mathbf{f}_{\mathrm{rad}} = -\rho\mathbf{g}_{\mathrm{rad}}$ (force exerted by radiation field onto the material).

The limit for plane-parallel, stationary and static conditions reads

$$\frac{\mathrm{d}K_\nu}{\mathrm{d}z} = -\chi_\nu H_\nu,$$

while for spherically symmetric, stationary and (quasi-)static conditions we find

$$\frac{\partial K_\nu}{\partial r} + \frac{3K_\nu - J_\nu}{r} = -\chi_\nu H_\nu. \tag{5.12}$$

The same caveats concerning (quasi-)static conditions as such as the preceding apply also here. We note as well that for static conditions the emissivity contribution to the radiation force vanishes, if the emission is isotropic as assumed here (since there are no Doppler shifts in this case).

5.3.2 *Solution Methods*

Ray-by-ray solution – p-z geometry. The following elegant method (based on Hummer and Rybicki, 1971) to solve the RTE for spherical atmospheres can be *only* applied to spherically symmetric problems, and for conditions where Doppler shifts do not play a decisive role, i.e., where opacities and emissivities can be assumed as isotropic (e.g., continuum formation in winds, if interactions of edges with other processes do not play a role). In brief, the methods works as follows (compare with Figure 5.5):

- Define p-rays (with impact parameter p) tangential to each discrete radial shell.
- Augment those with a bunch of (equidistant) p-rays resolving the core.

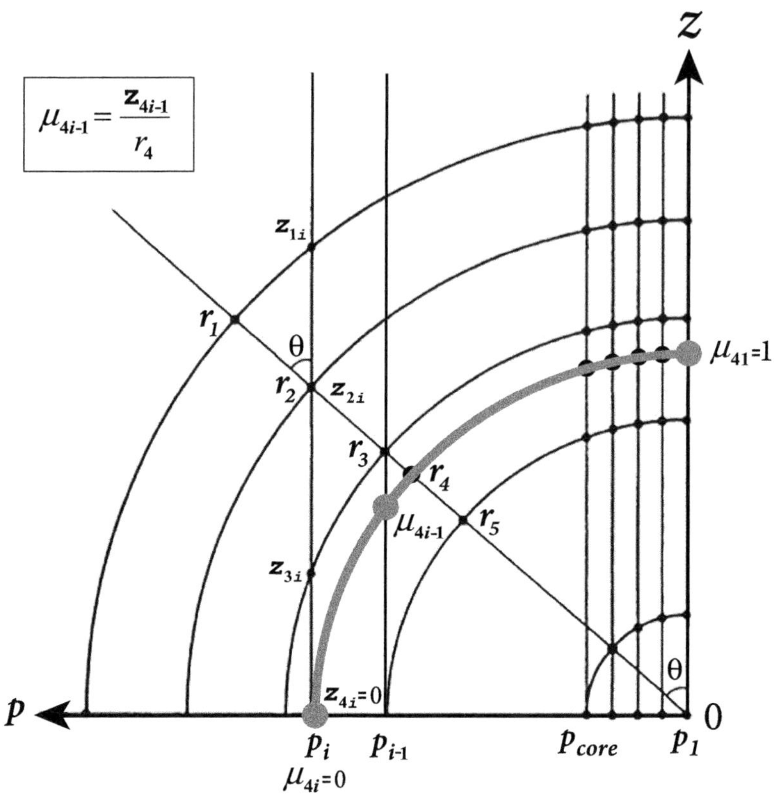

FIGURE 5.5. Sketch of p-z geometry (adapted from Mihalas, 1978). See text.

- Use only the forward hemisphere, i.e.,

$$z_{di} = \sqrt{r_d^2 - p_i^2} \quad \text{with} \quad z_{di} \geq 0.$$

  In this way, all points $z_{di}, i = 1, \mathrm{NP}$, are located on the same r_d-shell, i.e., have the same physical parameters, in particular emissivities and opacities (due to spherical symmetry and neglect of Doppler shifts).

Now one solves the RTE along each p-ray. From first principles,

$$\pm \frac{\mathrm{d}I_\nu^\pm(z, p_i)}{\mathrm{d}z} = \eta_\nu(r) - \chi_\nu(r) I_\nu^\pm(z, p_i)$$

with + for $\mu > 0$ and − for $\mu < 0$, using appropriate boundary conditions (core vs. noncore rays) and standard methods (finite differences, etc.). After being calculated, $I_\nu^\pm(z_{di}(r_d), p_i), i = 1, \mathrm{NP}$, samples the specific intensity at the same radius, r_d, but at different angles,

$$\pm\mu_{di} = \frac{z_{di}}{r_d},$$

starting at $|\mu_{di}| = 1$ for $i = 1$ and $d = 1, \mathrm{NZ}$ (central ray, $p_i = 0$) until $\mu_{di} = 0$ (tangent ray, where $p_i = r_d$ and thus $z_{di} = 0$). In other words, along individual r_d-shells, the specific intensities $I_\nu^\pm(r_d, \mu) = I_\nu^\pm(z_d, \mu)$ are sampled for all relevant μ, and corresponding moments can be calculated by integration.

Feautrier variables. In fact, the RTE is not solved for $I_\nu^\pm$ separately, but for a linear combination of I_ν^+ and I_ν^-, using the so-called Feautrier variables, u_ν and v_ν, which allows to construct a second-order scheme (higher accuracy, diffusion limit for large optical depths can be easily represented), similar as in the plane-parallel case:

$$u_\nu(z,p) = \frac{1}{2}(I_\nu^+(z,p) + I_\nu^-(z,p)) \qquad \text{mean intensity like}$$

$$v_\nu(z,p) = \frac{1}{2}(I_\nu^+(z,p) - I_\nu^-(z,p)) \qquad \text{flux like}$$

$$\Rightarrow \frac{\partial v_\nu}{\partial z} = \chi_\nu(S_\nu - u_\nu), \qquad \frac{\partial u_\nu}{\partial z} = -\chi_\nu v_\nu$$

$$\Rightarrow \frac{\partial^2 u_\nu}{\partial \tau_\nu^2} = u_\nu - S_\nu \qquad \text{(2nd order, with } \mathrm{d}\tau_\nu = -\chi_\nu \mathrm{d}z)$$

The source function S_ν is defined in the customary way as η_ν/χ_ν. Corresponding boundary conditions have to be provided, of course. For the inner boundary and for core rays, mostly a first-order condition using the diffusion approximation is applied, while for noncore rays, a second-order condition is formulated, using symmetry arguments. For the outer boundary, either $I_\nu^-(z_{\max}, p) = 0$ is set, or higher-order terms need to be accounted for, in case of optically thick conditions (e.g., at and bluewards of the He II edge). For atmospheres illuminated by companions, etc., this needs to be adapted.

As it turns out, this *formal solution* for $I_\nu(\mu)$ (or $u_\nu(\mu)$ and $v_\nu(\mu)$) and corresponding angle-averaged quantities (moments) is (partly strongly) affected by inaccuracies, due to the specific way of discretization within the p-z grid. However, the *ratios* of such moments (= Eddington factors) remain much more precise, since the aforementioned errors cancel to a major part.

The variable Eddington factor method. Thus, the conventional method to solve the RTE in spherically symmetric atmospheres (again: no Doppler shifts!) is to consider the moments equations (only radius-dependent), and to use the Eddington factors from the (previously described) formal solution to close the relations. This procedure ensures high accuracy (because of direct solution for angle-averaged quantities and second-order scheme), while the Eddington factors (from the formal solution) quickly stabilize in the course of global iterations. One additional advantage of using the moments equations is that the optimum diagonal accelerated lambda operator (see Section 5.5.2) can be easily calculated in parallel with the solution (and without major computational effort). Using the zero-order and first-order moment of the RTE ((5.10) and (5.12)), and the conventional Eddington factor $f_\nu = K_\nu/J_\nu$, we obtain

$$\frac{\partial(r^2 H_\nu)}{\partial \tau_\nu} = r^2(J_\nu - S_\nu) \quad \text{and} \quad \frac{\partial(f_\nu J_\nu)}{\partial \tau_\nu} - \frac{(3f_\nu - 1)J_\nu}{\chi_\nu r} = H_\nu,$$

now with $\mathrm{d}\tau_\nu = -\chi_\nu \mathrm{d}r$. Introducing a sphericality factor q_ν via

$$\ln(r^2 q_\nu) = \int_{r_{\text{core}}}^{r} \left[\frac{3f_\nu - 1}{r' f_\nu}\right] \mathrm{d}r' + \ln(r_{\text{core}}^2), \tag{5.13}$$

the second equation becomes

$$\frac{\partial(f_\nu q_\nu r^2 J_\nu)}{\partial \tau_\nu} = q_\nu r^2 H_\nu,$$

and can be combined with the first one to yield a second-order scheme for $r^2 J_\nu$,

$$\frac{\partial^2(f_\nu q_\nu r^2 J_\nu)}{\partial X_\nu^2} = \frac{1}{q_\nu} r^2 (J_\nu - S_\nu), \qquad \text{with } \mathrm{d}X_\nu = q_\nu \mathrm{d}\tau_\nu.$$

For comparison, the corresponding equation in p-p symmetry is given by

$$\frac{\partial^2(f_\nu J_\nu)}{\partial \tau_\nu^2} = (J_\nu - S_\nu),$$

and is just the limit of the spherically symmetric case, for $q_\nu \to 1$ and $r^2 \to R_*^2$.

5.4 Line Transfer in (Rapidly) Expanding Atmospheres

The basic problem for *line* transfer in rapidly expanding (or accreting) atmospheres is the Doppler shift (discarded in Section 5.3) that affects both opacities and emissivities, giving rise to an intricate coupling of location, frequency and angle. As detailed later, a very high resolution in the radial grid ($\Delta v = \mathrm{O}(v_{\mathrm{th}}/3)$) is required when standard (observer's frame) RT methods are applied,[3] with v_{th} the thermal speed of the considered ion. E.g., for $v_\infty = 2{,}000 \text{ km s}^{-1}$, and $v_{\mathrm{th}} = 8 \text{ km s}^{-1}$ (representative for CNO-elements in a hot star wind), this leads to $\approx$750 radial grid points.[4]

In such cases, only the RTE for the specific intensity should be solved (maybe cast to 'Rybicki form' if a separation into scattering and thermal part is possible), while the use of the aforementioned variable Eddington factor method is prohibitive, since it does not account for Doppler shifts.

In the following, we mostly consider the pure line case (except when stated differently), assuming that the continuum is optically thin (which is not so wrong for 'normal' OB-star winds, but invalid, e.g., for WR-star winds with much larger mass-loss rates).

Moreover, we assume pure Doppler broadening, which captures the essential broadening effect when calculating NLTE occupation numbers, etc., by means of scattering integrals, $\bar{J}$ (see (5.15)). For the calculation of emergent profiles, however, other broadening functions that describe also the line wings in a realistic manner (e.g., Stark and Voigt profiles) should be used if necessary.

5.4.1 *Notation: Line Opacity and Profile Function*

The inclusion of Doppler shifts leads to complications and possible confusion. Thus, before tackling the actual problem, we must define the notation we are going to use.[5]

The line opacity, as a function of radial distance r and frequency ν, can be expressed as

$$\chi_\nu(r) = \bar{\chi}_{\mathrm{L}}(r)\phi(\nu, r) \text{ , with } \phi(\nu, r) = \frac{1}{\Delta\nu_{\mathrm{D}}(r)\sqrt{\pi}} \exp\left[-\left(\frac{\nu - \tilde{\nu}}{\Delta\nu_{\mathrm{D}}(r)}\right)^2\right]$$

$$\text{and} \quad \Delta\nu_{\mathrm{D}}(r) = \frac{\tilde{\nu} v_{\mathrm{th}}(r)}{c} \quad \text{for a Doppler profile.}$$

Here the profile function $\phi(\nu, r)$ is normalized with respect to frequency, and has dimensions $[\phi] = T$; $\tilde{\nu}$ is the line-center frequency, and v_{th} includes any kind of microturbulence (if present). The line opacity integrated over frequency is given by

[3] Many such methods also require a very high resolution in μ.

[4] This problem becomes mitigated when a large "microturbulence" of order 100 km s^{-1}(due to an inhomogeneous wind structure) is accounted for (e.g., Hamann, 1980; Puls et al., 1993).

[5] Further specifications will be given in Appendix B.

$$\bar{\chi}_{\rm L}(r) = \frac{\pi e^2}{m_e c} f_{lu} \left(n_l - n_u \frac{g_l}{g_u} \right),$$

where l and u denote the lower and upper levels of the transition, f_{lu} the oscillator strength, n_l and n_u the occupation numbers of the levels, and g_l and g_u the corresponding statistical weights. We stress that $[\bar{\chi}_L] = L^{-1}T^{-1}$, while $[\chi_\nu] = L^{-1}$.

When the material in the atmosphere is in motion with respect to the frame of an external observer at rest, matter particles "see" the radiation field at frequencies corresponding to their own *comoving frame* (CMF), and opacity and emissivity become *angle dependent* in the observer's frame, due to Doppler shifts. In fact, the atoms absorb and emit photons at frequency

$$\nu_{\rm CMF} = \nu - (\tilde{\nu}/c)\, \mathbf{n} \cdot \mathbf{v}(\mathbf{r}),$$

where ν is the frequency in the observer's frame, and nonrelativistic velocities are assumed. In spherical geometry, it holds that $\mathbf{n} \cdot \mathbf{v}(\mathbf{r}) = \mu v(r)$. The profile function, evaluated at CMF frequencies, is then

$$\phi(\nu_{\rm CMF}, r) = \frac{1}{\Delta\nu_{\rm D}(r)\sqrt{\pi}} \exp\left[- \left(\frac{\nu - \tilde{\nu} - \mu\, \tilde{\nu}\, v(r)/c}{\Delta\nu_{\rm D}(r)} \right)^2 \right].$$

For simplicity's sake, in the following we will assume that $v_{\rm th}$ is spatially constant[6] and define, in the observer's frame, the frequency shift measured in Doppler units as

$$x \equiv \frac{\nu - \tilde{\nu}}{\Delta\nu_{\rm D}} \quad \text{with} \quad \Delta\nu_{\rm D} = \frac{\tilde{\nu} v_{\rm th}}{c}.$$

With the preceding choice, the transformation between observer's frame and CMF is

$$x_{\rm CMF} = x - \mu v'(r), \quad \text{with} \quad v'(r) = \frac{v(r)}{v_{\rm th}} \quad \in \left(0, \frac{v_\infty}{v_{\rm th}} >> 1\right),$$

so that

$$\phi_\nu(x_{\rm CMF}, r) = \phi_\nu(x - \mu v', r) = \frac{1}{\Delta\nu_{\rm D}\sqrt{\pi}} \exp\left[- \left(x - \mu v'(r)\right)^2 \right],$$

The preceding profile function, whose dimension is still T, depends primarily on $x_{\rm CMF}$.

In order to simplify the following discussion, it is convenient to include the factor $(\Delta\nu_{\rm D})^{-1}$ into the opacity, so that the profile function, in units of Doppler shift, is now dimensionless (and normalized with respect to x), whilst $[\bar{\chi}_{\rm L}(r)/\Delta\nu_{\rm D}] = L^{-1}$. We then have

$$\chi_\nu(x_{\rm CMF}, r) = \frac{\bar{\chi}_{\rm L}(r)}{\Delta\nu_{\rm D}} \phi(x_{\rm CMF}, r)\,, \text{ with } \phi(x_{\rm CMF}, r) = \frac{1}{\sqrt{\pi}} \exp\left[- \left(x - \mu v'(r)\right)^2 \right],$$

$$\text{and} \quad \frac{\bar{\chi}_{\rm L}(r)}{\Delta\nu_{\rm D}} = \frac{\bar{\chi}_{\rm L}(r)\tilde{\lambda}}{v_{\rm th}}.$$

Since $\mu v'(r) \in [-v_\infty/v_{\rm th}], +v_\infty/v_{\rm th}]$, x must vary within the same range (essentially, $x \in [-\infty, +\infty]$), and not only within a range of a few thermal Doppler widths[7].

[6] The generalization to a depth-dependent $v_{\rm th}$ will be considered in Appendix B.1.
[7] Several integrals involving $\phi(x)$ are presented in Appendix B.2.

5.4.2 *Sobolev Theory*

The resonance zone. Since $\mu v'(r)$ enters into the argument of ϕ, we must know (for instance, for computing the optical depth) the variation of the former quantity along a path $\mathrm{d}s$, i.e., $\mathrm{d}\mu v'(r)/\mathrm{d}s$. (Recall that $\mathbf{n}\cdot\nabla = \mathrm{d}/\mathrm{d}s$.) We consider again a $p-z$ geometry, in which the z-axis shall be parallel to $\mathbf{n}$, and obtain

$$\frac{\mathrm{d}\mu v'(r)}{\mathrm{d}s} \rightarrow \left.\frac{\mathrm{d}\mu v'(r)}{\mathrm{d}z}\right|_p = \mu \left.\frac{\mathrm{d}v'}{\mathrm{d}r}\frac{\mathrm{d}r}{\mathrm{d}z}\right|_p + \left.\frac{\mathrm{d}\mu}{\mathrm{d}z}\right|_p v' = \mu^2\frac{\mathrm{d}v'}{\mathrm{d}r} + (1-\mu^2)\frac{v'}{r}.$$

Note that contrasted to Section 5.3.2, $\mu < 0$ implies here that $z < 0$, so that for negative angles we consider the back hemisphere of the $p-z$ system. Moreover it holds that

$$\mu^2\frac{\mathrm{d}v'}{\mathrm{d}r} + (1-\mu^2)\frac{v'}{r} > 0 \quad \text{for} \quad v' > 0 \quad \text{and} \quad \frac{\mathrm{d}v'}{\mathrm{d}r} > 0.$$

Thus, in spherical symmetry $\mu v'(r)$ increases monotonically along *any* given direction $\mathbf{n}$, as long as $v'(r) > 0$ is monotonically increasing.

Now, as the optical depth is defined by

$$\tau_x(z) = \int_{z_{\min}}^{z} \frac{\bar{\chi}_\mathrm{L}(z')}{\Delta\nu_\mathrm{D}}\phi\left(x - [\mu v'](z'), z'\right)\mathrm{d}z',$$

it depends on the argument of the profile function, and it is clear that line processes are only effective in a (very) localized region, the so-called *resonance zone*, whenever $\phi(x_\mathrm{CMF})$ is nonnegligible, i.e., when $(x - \mu v') \in [-\Delta x_\mathrm{Dop}, +\Delta x_\mathrm{Dop}] \approx [-3, 3]$ (see Figure 5.6).

In order to achieve a proper representation of the line transfer process, both frequencies x and projected velocities $\mu v'(z)$ must be highly resolved, on scales corresponding to v_Dop. If, on the one hand, the $\mu v'(z)$-spacing were too coarse, the resonance zones would be missed or not resolved, the intensities would remain constant (or too large), and the quantity $\bar{I}$ (related to the scattering integral, and required later on; see Figure 5.6) would become too large. With regard to Figure 5.6, this would mean that in the most extreme case, all three curves (intensities) would remain constant, at a value equal to I_0, resulting in a dramatic overestimate of $\bar{I}$. If, on the other hand, the x-spacing was too coarse, the variation of $I(x)$ (from 'neighboring' resonance zones) would be insufficiently sampled. For our example, this could mean that the left and/or right curves were absent (due to missing frequencies), and the middle curve would not be centered, since there might be no frequency where $x - \mu v'$ is *exactly* zero.

In spherical geometry, the first point is a specific problem, since the general spacing refers to the radial grid (and not to specific p-rays), and a high resolution in $v'(r)$ does not guarantee a high resolution in $\mu v'(z)$. In models using cartesian coordinates (μ=const along a specific ray), the first point leads to the condition that $\Delta\mu = \Delta x/v'_\mathrm{max}$, i.e., an intricate coupling of frequency and angle.

'Standard' Sobolev theory. As discussed in the previous paragraph, line processes (contrasted to continuum ones) occur in a very localized region within a rapidly expanding medium. V. Sobolev (1960; but work done already during World War II) was the first to obtain a completely local approximation that is quite accurate (and can be extended to become even more precise). The following reasoning follows (in part) Owocki and Puls (1996); for an alternative and very insightful derivation, see Rybicki and Hummer (1978). For simplicity, in this reasoning, we do the following:

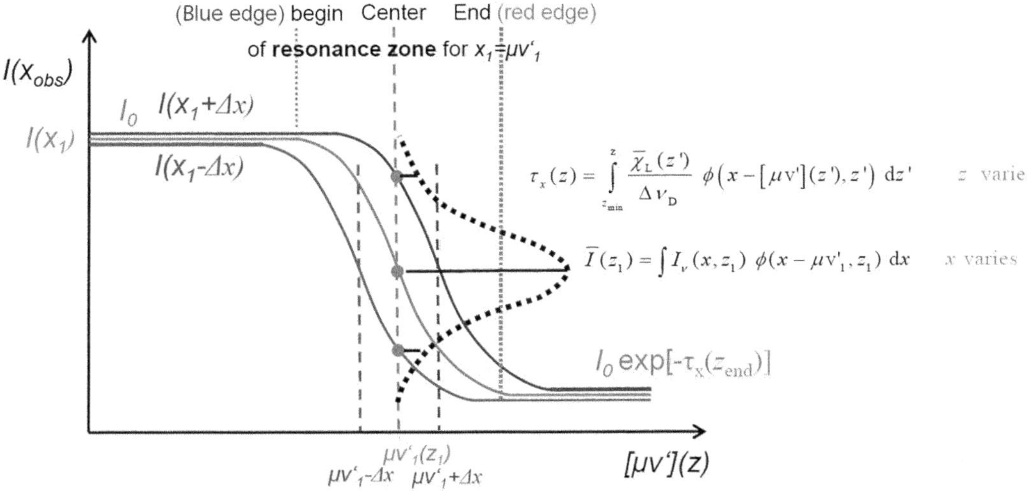

FIGURE 5.6. Radiative line transfer in expanding atmospheres: resonance zones and related aspects, for the case of pure absorption. Displayed is the specific intensity along an arbitrary ray, as a function of $[\mu v'](z) = \mu(z)v'(r(z))$ ($v' > 0$ for outflows). The three curves show the variation of the intensity when crossing the corresponding resonance zones, for observer's frame frequencies x_1 (central curve), $x_1 + \Delta x$ (rightmost curve), and $x_1 - \Delta x$ (leftmost curve), respectively. The centers of the resonance zones are marked by dashed vertical lines. The evaluation of the quantity $\bar{I}(z_1)$ (see insert) is indicated as well. For this quantity, the intensities from different frequencies contribute as follows: at the considered location z_1, $I(x_1 + \Delta x)$ (rightmost curve) has just entered its own resonance zone, $I(x_1)$ needs to be evaluated at the center of the resonance zone corresponding to $\mu v'(z_1) = x_1$, and $I(x_1 - \Delta x)$ (leftmost curve) has almost passed its resonance zone. Obviously, the largest contribution is provided by $I(x_1, z_1)$. Note the intricate coupling between location and frequency.

- Concentrate on outflows, i.e., $v(r) > 0$ (but $dv/dr < 0$, as occurring, e.g., in flows with embedded shocks, is not excluded).
- Adopt, as before, a spatially constant thermal speed, $v_{\rm th}(r) := v_{\rm th}$
- Define $\chi_{\rm l}(r) = \bar{\chi}_{\rm L}(r)/\Delta\nu_{\rm D}$.

Under such conditions, the optical depth difference between two points z_1 and z_2 (along impact parameter p) is given by

$$t(x, p, z_1, z_2,) = \int_{\min(z_1,z_2)}^{\max(z_1,z_2)} \chi_{\rm l}(r')\,\phi(x - \mu' v'(r'))\,dz' \qquad (5.14)$$

with (as usual) $\mu' = z'/r'$, and $r' = \sqrt{z'^2 + p^2}$. Then, without any approximation,

$$I_\nu(x, p, z)$$

$$= \underbrace{I_{\rm core}\, e^{-t(x,p,z,z_B)}}_{\text{direct component, only present for } \mu>0 \text{ and } p\le R_*} + \underbrace{\int_0^{t(x,p,z,z_B)} S(r') e^{-t(x,p,z,z')} dt(x, p, z, z')}_{\text{diffuse component (radiation scattered/emitted in the wind)}}$$

with

$$z_B = \begin{cases} z_* \text{ for } z > 0,\ p \le R_* \\ -\infty \text{ else} \end{cases}$$

This equation is valid for both outwards ($\mu \geq 0$) and inwards ($\mu < 0$) rays, depending on the sign of z. (Here, we use a p-z geometry extending over both hemispheres, with $z > 0$ for the front, and $z < 0$ for the back hemisphere.)

To calculate the scattering integrals, required to couple with the rate equations, we first integrate over $\phi(x)\mathrm{d}x$,

$$\bar{I}(\mu, r) = \int_{-\infty}^{+\infty} I_\nu(x, \mu, r)\phi\left(x - \mu v'(r)\right)\mathrm{d}x, \qquad \text{and then over } \mathrm{d}\mu,$$

$$\bar{J}(r) = \frac{1}{2}\int_{-1}^{+1} \bar{I}(\mu, r)\,\mathrm{d}\mu \tag{5.15}$$

Now we consider that the integrands provide a contribution only if $x \approx \mu' v'(r')$ (5.14) or $x \approx \mu v'(r)$ (5.15), respectively, due to the behaviour of ϕ. For the optical depth difference, this means that

$$t(x, p, z, z_B,) = \int_{\min(z,z_B)}^{\max(z,z_B)} \chi_\mathrm{l}(r')\phi\left(x - \mu' v'(r')\right)\mathrm{d}z' \approx \chi_\mathrm{l}(r_0)\int_{z_\mathrm{min}}^{z_\mathrm{max}} \phi\left(x - \mu' v'(r')\right)\mathrm{d}z', \tag{5.16}$$

where r_0 is the position of the corresponding resonance zone, which (at least in principle) needs to be calculated from

$$\left[\mu' v'\right](r_0) = x, \qquad \text{i.e.,} \pm\sqrt{1 - \frac{p^2}{r_0^2}}v'(r_0) = x \qquad \text{(nonlinear eq.)},$$

which has a unique solution for strictly monotonic flows (otherwise there is more than one resonance zone). The RHS of (5.16) constitutes the heart of the *Sobolev approximation*: line opacities (and source functions, discussed later in this section) are assumed to be constant over the resonance zones!

Furthermore, we switch from an integration over $\mathrm{d}z'$ to an integration over CMF frequency, $\mathrm{d}x_\mathrm{CMF} = \mathrm{d}(x - \mu' v'(r'))$,

$$\left.\frac{\mathrm{d}x_\mathrm{CMF}}{\mathrm{d}z}\right|_p = -\left.\frac{\mathrm{d}(\mu v')}{\mathrm{d}z}\right|_p = \text{(see Section 5.4.2)} = -\left(\mu^2\frac{\mathrm{d}v'}{\mathrm{d}r} + (1-\mu^2)\frac{v'}{r}\right) =: -Q(r, \mu).$$

For $Q > 0$, we have the following situation: by considering the boundaries, $x_\mathrm{CMF}(z) = x - \mu v'(r)$, $x_\mathrm{CMF}(z_B) \to \infty$ (bluewards of blue edge of resonance zone), and by putting $Q(r', \mu')$ in front of the integral (using the same argument as before), we arrive at

$$t(x, p, z, z_B) \approx \chi_\mathrm{l}(r_0)\int_{x_\mathrm{CMF}(z_B)}^{x_\mathrm{CMF}(z)} \frac{-1}{Q(r', \mu')}\phi(x_\mathrm{CMF})\,\mathrm{d}x_\mathrm{CMF} \approx$$

$$\approx \frac{\chi_\mathrm{l}(r_0)}{Q(r_0, \mu_0)}\int_{x-\mu v'(z)}^{\infty} \phi(\xi)\,\mathrm{d}\xi = \tau_\mathrm{S}(r_0, \mu_0)\;\Phi(x - \mu v'(r)) \tag{5.17}$$

This result can be generalized to also include negative values of Q, if we define

$$\tau_S(r_0, \mu_0) = \frac{\chi_\mathrm{l}(r_0)}{|Q(r_0, \mu_0)|} = \frac{\bar{\chi}_\mathrm{L}(r_0)}{\Delta\nu_\mathrm{D}\left|\mu^2\frac{\mathrm{d}v'}{\mathrm{d}r} + (1-\mu^2)\frac{v'}{r}\right|_{r_0,\mu_0}} \tag{5.18}$$

as the Sobolev optical depth, evaluated at the resonance zone. In the most general case, Q is the directional derivative of the velocity in direction $\mathbf{n}$, i.e.,

$$|Q| = |\mathbf{n} \cdot \nabla(\mathbf{n} \cdot \mathbf{v}')| = \left|\frac{\mathrm{d}v_l'}{dl}\right|, \text{ if } l \text{ has direction } \mathbf{n}.$$

For a further understanding of Equations (5.17) and (5.18), a few comments might be relevant:

- $\Phi(\infty) = 0$ (blue – starwards – side of resonance zone), and $\Phi(-\infty) = 1$ (red side of resonance zone), as long as $v > 0$. Thus:
  $t(x, p, z, z_B) \to 0$ for z 'before' the resonance zone, and $I(z) = I_{\mathrm{core}}$.
  $t(x, p, z, z_B) \to \tau_S$ for z 'behind' the resonance zone, and $I(z) \approx I_{\mathrm{core}} \exp(-\tau_{\mathrm{S}})$ (without emission, compare with Figure 5.6).
- For pure Doppler-profiles, $\Phi(x) = \frac{1}{2}\mathrm{erfc}(x)$.
- Since v' is the velocity in units of the thermal speed, and since $\Delta\nu_{\mathrm{D}} = v_{\mathrm{th}}/\tilde{\lambda}$, we can alternatively write

$$\tau_S(r_0, \mu_0) = \frac{\bar{\chi}_{\mathrm{L}}(r_0)\, \tilde{\lambda}}{\left|\mu^2 \frac{\mathrm{d}v}{\mathrm{d}r} + (1 - \mu^2)\frac{v}{r}\right|_{r_0, \mu_0}},$$

  when v and r are measured in actual units (then v/r has units of s^{-1}).

Since also the integrand of the diffuse component contributes only for $x \approx \mu' v'$,

$$\int_0^{t(x,p,z,z_B)} S(r')\, \mathrm{e}^{-t(x,p,z,z')}\, \mathrm{d}t(x, p, z, z') \to S(r_0) \int_0^t \mathrm{e}^{-t'}\mathrm{d}t' = S(r_0)\left(1 - \mathrm{e}^{-t}\right),$$

the specific intensity can be approximated by

$$I_\nu(x, p, z) \approx I_{\mathrm{core}}(p)\mathrm{e}^{-\tau_{\mathrm{S}}(r_0,\mu_0)\Phi(x_{\mathrm{CMF}})} + S(r_0)\left(1 - \mathrm{e}^{-\tau_{\mathrm{S}}(r_0,\mu_0)\Phi(x_{\mathrm{CMF}})}\right). \tag{5.19}$$

This means that behind the resonance zone (where $\Phi(x_{\mathrm{CMF}}) = 1$),

$$I_\nu(x, p, z_{\mathrm{behind}}) \approx I_{\mathrm{core}}(p)\mathrm{e}^{-\tau_{\mathrm{S}}(r_0,\mu_0)} + S(r_0)\left(1 - \mathrm{e}^{-\tau_{\mathrm{S}}(r_0,\mu_0)}\right) = \text{const},$$

while before the resonance zone (where $\Phi(x_{\mathrm{CMF}}) = 0$),

$$I_\nu(x, p, z_{\mathrm{before}}) \approx \begin{cases} I_{\mathrm{core}}(p) = \text{const for } p \le R_* \\ 0 \quad \text{else} \end{cases}$$

Only inside the resonance zone, the optical depth increases and the intensity varies accordingly (again, compare with Figure 5.6).

We stress that to calculate the *specific intensity* in Sobolev approximation (required, e.g., for the emergent profile), the location of the resonance zone has to be evaluated for each frequency and impact parameter!

Now comes the second 'trick'. As already outlined, we first calculate

$$\bar{I}(r, \mu) = \int_{-\infty}^{+\infty} \left[I_{\mathrm{core}}(p)\mathrm{e}^{-\tau_{\mathrm{S}}(r_0,\mu_0)\Phi(x_{\mathrm{CMF}})} + S(r_0)\left(1 - \mathrm{e}^{-\tau_{\mathrm{S}}(r_0,\mu_0)\Phi(x_{\mathrm{CMF}})}\right)\right] \phi(x_{\mathrm{CMF}}(r, \mu))\mathrm{d}x$$

Again, we find a contribution only for $x_{\rm CMF} \approx 0$, i.e., $x \approx \mu v'(r)$. Thus, we can replace r_0 by r and μ_0 by μ: only those frequencies/resonance zones contribute that are located at (or close to) the considered location (r, μ).

Realizing that $\phi(x_{\rm CMF}){\rm d}x = -{\rm d}\Phi$ with $\Phi(x_{\rm CMF} = x - \mu v') \rightarrow 1$ for $x \rightarrow -\infty$ and $\Phi(x_{\rm CMF} = x - \mu v') \rightarrow 0$ for $x \rightarrow \infty$, we find

$$\bar{I}(r,\mu) \approx \int_0^1 \left[I_{\rm core}(p){\rm e}^{-\tau_{\rm S}(r,\mu)\Phi(x_{\rm CMF})} + S(r)\left(1 - {\rm e}^{-\tau_{\rm S}(r,\mu)\Phi(x_{\rm CMF})}\right)\right] {\rm d}\Phi =$$

$$= I_{\rm core}(p)\frac{1-{\rm e}^{-\tau_{\rm S}(r,\mu)}}{\tau_{\rm S}(r,\mu)} + S(r)\left(1 - \frac{1-{\rm e}^{-\tau_{\rm S}(r,\mu)}}{\tau_{\rm S}(r,\mu)}\right),$$

which is thus *purely local.* Finally, by integrating over ${\rm d}\mu$, and accounting for the limits regarding the first term,

$$\bar{J}(r) = \beta_{\rm c}(r)I_{\rm core} + (1-\beta(r))\,S(r), \qquad \text{with} \tag{5.20}$$

$$\beta_{\rm c}(r)I_{\rm core} = \frac{1}{2}\int_{\mu_*}^1 I_{\rm core}(\mu,\bar{\nu})\frac{1-{\rm e}^{-\tau_{\rm S}(r,\mu)}}{\tau_{\rm S}(r,\mu)}{\rm d}\mu, \qquad \text{and}$$

$$\beta(r) = \frac{1}{2}\int_{-1}^1 \frac{1-{\rm e}^{-\tau_{\rm S}(r,\mu)}}{\tau_{\rm S}(r,\mu)}{\rm d}\mu \qquad \text{(escape probability)}.$$

We note the following:

(i) The angular integration does *not* require a highly resolved angular grid, since the interaction between x, μ and r has already been accounted for.

(ii) The core intensity has to be emitted (evaluated) at the core for an observer's (rest) frame frequency of $\bar{\nu} \approx \tilde{\nu}\,(1 + \mu v(r)/c)$, in order to display a local CMF-frequency of $\nu_{\rm CMF} \approx \tilde{\nu}$ at $[\mu v(r)]$, corresponding to a local $x_{\rm CMF} = 0$. This ensures that the resonance zone is illuminated by the full core intensity, and that there are no self-shadowing effects (at least if there are no line-overlap effects).

Sobolev optical depth for prototypical resonance lines. The Sobolev optical depth at (r, μ),

$$\tau_S(r,\mu) = \frac{\chi_{\rm l}(r)}{|Q(r,\mu)|} = \frac{\bar{\chi}_{\rm L}(r)}{\Delta\nu_{\rm D}\left|\mu^2\frac{{\rm d}v'}{{\rm d}r} + (1-\mu^2)\frac{v'}{r}\right|},$$

results for radial rays ($\mu = 1$) in

$$\tau_S(r) = \bar{\chi}_{\rm L}(r)\tilde{\lambda}/|{\rm d}v/{\rm d}r|.$$

For ground-state opacities of main ionization stages as present in many UV resonance lines, we have $\bar{\chi}_{\rm L}(r) \propto n_1(r) \propto \rho(r)$, and, exploiting the continuity equation, $\rho(r) = \dot{M}/(4\pi r^2 v(r))$, and the typical β-velocity law, we obtain

$$\tau_S(r) \propto \frac{1}{r^2\ v(r)\ \frac{{\rm d}v}{{\rm d}r}} = \frac{1}{\beta b R_* v_\infty^2}\left(\frac{v(r)}{v_\infty}\right)^{\frac{1}{\beta}-2},$$

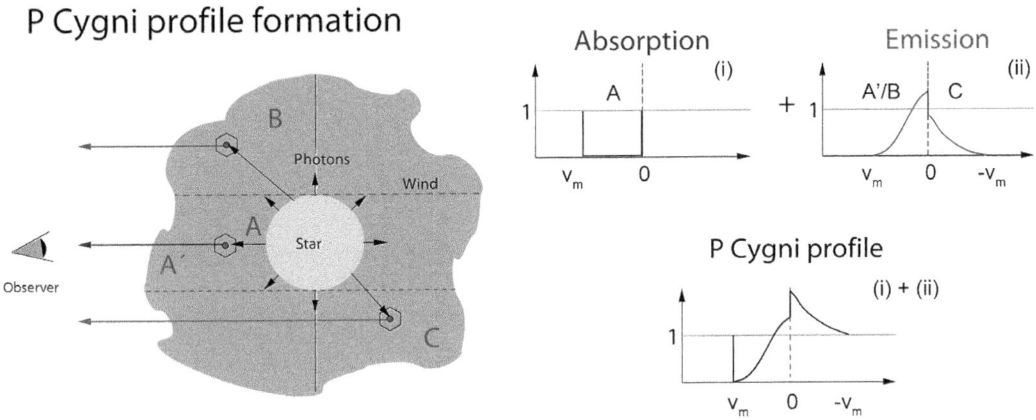

FIGURE 5.7. Principle of P Cygni-profile formation, for a strong resonance line, remaining optically thick until a maximum velocity, v_m. Due to Doppler shifts, all observer's frame frequencies corresponding to $[+v_m, -v_m]$ can contribute. (i) Absorption in region A in front of the stellar disk (approaching material $\rightarrow$ blue frequencies). (ii) Asymmetric emission from regions A/B in front hemisphere (blue emission due to approaching material), and region C (side lobes) in back hemisphere (red emission due to receding material). The emission itself is caused by line scattering; see the leftmost sketch.

with $b = 1 - (v_{\min}/v_\infty)^{1/\beta}$. For $\beta = 0.5$,[8] this implies $\tau_S(r) =$ const, while for a more typical situation with $\beta = 1$, $\tau_S(r) \propto v_\infty/v(r)$, and the optical depth decreases by roughly (and only) a factor of 100 from inside to outside. This explains why a typical UV P Cygni line, e.g., C IV 1548/1550 (Figure 5.1), remains optically thick throughout the *complete* wind, displaying a saturated absorption trough even for frequencies corresponding to v_∞ (see also Figure 5.7).

Limiting cases: source functions for purely scattering resonance lines. Very often, the source functions of the aforementioned resonance lines are dominated by line-scattering, and in such cases (see also Section 5.5),

$$S(r) = \bar{J}(r) = \frac{\beta_c(r) I_{\text{core}}}{\beta(r)}.$$

(a) In the optically thin limit, i.e., (locally) weak resonance lines, $\tau_S(r, \mu) \ll 1$,

$$\frac{1 - e^{-\tau_S(r,\mu)}}{\tau_S(r,\mu)} \rightarrow 1, \quad \text{and} \quad S(r) = \frac{\beta_c(r) I_{\text{core}}}{\beta(r)} \rightarrow W I_{\text{core}},$$

with dilution factor W (5.7). Thus, for large distances from the stellar core,

$$\left(\frac{r}{R_*}\right)^2 S(r) \rightarrow \frac{I_{\text{core}}}{4} = \text{const},$$

and the source function dilutes quadratically. This can be also understood in terms of an alternative argumentation: For an optically thin line, $\bar{J} \approx J_\nu$, and for an optically thin continuum as assumed here $J_\nu \rightarrow W I_{\text{core}}$ (see Section 5.3.1). Since $S = \bar{J}$, we thus have $S = W I_{\text{core}}$.

[8] This corresponds to a velocity field in line-driven winds when neglecting the so-called finite cone-angle correction factor, e.g., Castor et al. (1975).

(b) In the optically thick limit, $\tau_S(r,\mu) \gg 1$, which applies to strong UV resonance lines (discussed previously),

$$\frac{1-e^{-\tau_S(r,\mu)}}{\tau_S(r,\mu)} \rightarrow \frac{1}{\tau_S(r,\mu)},$$

and (after few calculations),

$$S(r) = \frac{\beta_c(r) I_{\rm core}}{\beta(r)} \rightarrow \left(\frac{R_*}{r}\right)^2 I_{\rm core} \frac{3}{4 + 8\left(\frac{{\rm d}\ln v}{{\rm d}\ln r}\right)^{-1}} \quad \text{for large radii.}$$

Since for large radii and a β velocity law ${\rm d}\ln v/{\rm d}\ln r \propto R_*/r$, the source function of an optically thick resonance line becomes proportional to $(R_*/r)^3$, and *decreases faster* than in the optically thin case. Also, this behaviour is important to understand the absorption troughs of UV P Cygni lines at high velocities.

Radiative line acceleration. In Sobolev approximation, the radiative line acceleration due to *one* line is provided by[9]

$$g_{\rm rad} = \frac{4\pi}{c}\frac{\bar{\chi}_{\rm L}}{\rho}\frac{1}{2}\int \bar{I}(\mu)\mu{\rm d}\mu \approx \frac{2\pi}{c}\frac{\bar{\chi}_{\rm L}}{\rho}\int_{\mu_*}^{1} I_{\rm core}(\mu,\bar{\nu})\frac{1-{\rm e}^{-\tau_S(r,\mu)}}{\tau_S(r,\mu)}\mu{\rm d}\mu,$$

since the contribution from the source term (even in μ) cancels when integrating over $\mu{\rm d}\mu$ with $\mu \in [-1,1]$. In particular,

$$g_{\rm rad} \propto \frac{\bar{\chi}_{\rm L}}{\rho}, \qquad \text{and NOT} \quad \propto \frac{\bar{\chi}_{\rm L}}{\rho\Delta\nu_D} \qquad \text{(see Appendix B.2).}$$

In the optically thick case, $\tau_{\rm S} = \bar{\chi}_{\rm L}/(\Delta\nu_D\,|Q(r,\mu)|) \gg 1$,

$$g_{\rm rad} \stackrel{\tau_{\rm S}\gg 1}{\rightarrow} \frac{2\pi}{c}\frac{\bar{\chi}_{\rm L}}{\rho}\frac{1}{\bar{\chi}_{\rm L}/\Delta\nu_D}\int_{\mu_*}^{1} I_{\rm core}(\mu,\bar{\nu})\,|Q(r,\mu)|\,\mu{\rm d}\mu,$$

and the line acceleration becomes independent of $\bar{\chi}_{\rm L}$

$$g_{\rm rad} \stackrel{\tau_{\rm S}\gg 1}{\rightarrow} \frac{2\pi\Delta\nu_D}{c\rho}\int_{\mu_*}^{1} I_{\rm core}(\mu,\bar{\nu})\,|Q(r,\mu)|\;\mu\,{\rm d}\mu,$$

with $|Q(r,\mu)| = \left|\mu^2{\rm d}v'/{\rm d}r + (1-\mu^2)v'/r\right|$.

Sobolev length. The Sobolev length is roughly the (half-)width of the resonance zone. More precisely, it is the length scale on which $v(r)$ changes by one $v_{\rm th}$ unit, accounting for the most decisive part of the line profile:

$$\Delta v = v_{\rm th} := \left|\frac{{\rm d}v}{{\rm d}r}\right| L_{\rm Sob} \quad \Rightarrow \quad L_{\rm Sob} = \frac{v_{\rm th}}{|{\rm d}v/{\rm d}r|} = \frac{1}{|{\rm d}v'/{\rm d}r|} \quad \text{for radial rays, and}$$

$$\Delta v = v_{\rm th} := \left|\frac{{\rm d}(\mu v)}{{\rm d}z}\right| L_{\rm Sob} \quad \Rightarrow \quad L_{\rm Sob} = \frac{v_{\rm th}}{\left|\mu^2\frac{{\rm d}v}{{\rm d}r} + (1-\mu^2)\frac{v}{r}\right|} = \frac{1}{\left|\mu^2\frac{{\rm d}v'}{{\rm d}r} + (1-\mu^2)\frac{v'}{r}\right|}$$

[9] Compare with (5.11), and account for the fact that *when integrated over frequency*, there is no force associated with emission, since there is no net momentum transfer due to an emission process presumed to be isotropic; see also Castor (1974).

for spherical symmetry. Most generally, the Sobolev length in direction $\mathbf{n}$ is

$$L_{\rm Sob} = \frac{v_{\rm th}}{|\mathbf{n} \cdot \nabla(\mathbf{n} \cdot \mathbf{v})|}$$

We note that for low microturbulent velocities, $L_{\rm Sob}$ depends on $m_{\rm ion}^{-1/2}$, i.e., decreases significantly from H to Fe.

Range of validity of the Sobolev approximation. Let's define a characteristic length scale, l_x, for a macrovariable x, defined via

$$\frac{{\rm d}x}{{\rm d}r} l_x = x, \quad \text{i.e.,} \quad l_x = \left(\frac{{\rm d}\ln x}{{\rm d}r}\right)^{-1}.$$

To warrant the validity of the Sobolev approximation (SA), $L_{\rm Sob}$ must be smaller than l_x,

$$\left|\frac{L_{\rm Sob}}{l_x}\right| = \left|\frac{{\rm d}\ln x}{{\rm d}v/v_{\rm th}}\right| < 1.$$

As an important example, we consider the (frequency-integrated) line opacity, assumed within the SA as being roughly constant over the resonance zone when evaluating the optical-depth integrals. For many (UV) resonance lines, $\bar{\chi}_{\rm L}(r) \propto \rho(r)$ (discussed previously), and a typical velocity field reads $v(r) = v_\infty(1 - \frac{R_*}{r})^\beta$, with $\beta \approx 1$, and neglecting the quantity $b \approx 1$ that plays no role here. Then,

$$\left|\frac{L_{\rm Sob}}{l_{\bar{\chi}_{\rm L}}}\right| = \frac{v_{\rm th}}{v} + \frac{2v_{\rm th}}{v_\infty}\frac{r}{R_*},$$

and the Sobolev approximation is valid (regarding an opacity $\propto \rho$) as long as the following are true:

(i) $v(r) > v_{\rm th}$ (sometimes, the SA is also called a supersonic approximation, though in view of this result it should be called superthermal).

(ii) $r/R_* < v_\infty/(2v_{\rm th}) = O(100)$, i.e., for all relevant radii.

As it turns out, a similar condition applies for the source function. The following are the only regions (in a smooth wind) where the SA inevitably fails:

- The subthermal region, where the density increases exponentially within a very extended resonance zone
- The transition zone between the quasihydrostatic photosphere and wind, where the resonance zone is still broad, but the velocity field has a significant curvature, and not a constant gradient[10].

Interestingly, the SA is almost perfectly valid in a supernova remnant, due to its homologous expansion, $v \propto r$, i.e., a constant gradient. However, when applied to line-driven winds, the SA fails in correctly describing the reaction of the line acceleration onto disturbances. Most important, the so-called line-driven instability (LDI) cannot be represented in the framework of the SA (e.g., Owocki and Rybicki, 1984).

5.4.3 *Extensions of the Sobolev Theory*

Particularly in the 1980s and 1990s, the 'standard' Sobolev theory (adopting an optically thin continuum and constant velocity gradients, line opacities and source functions over the resonance zone) has been extended towards more complex physical scenarios.

[10] Unfortunately, this zone is very important for the radiative line acceleration, and is badly described when using the SA (see Owocki and Puls, 1999).

Coupling with continuum. Hummer and Rybicki (1985) accounted for continua of arbitrary optical depth, and coupled the line with the continuum transfer. In this case,

$$\bar{J}(r) = \overline{\beta_{\rm c}(r) I_{\rm inc}} + (1-\beta(r))\, S_{\rm L}(r) + (S_{\rm c}(r) - S_{\rm L}(r))\, \bar{U}(\tau_{\rm S}, \beta_{\rm P}),$$

$$\text{with} \qquad \overline{\beta_{\rm c}(r) I_{\rm inc}} = \frac{1}{2}\int_{-1}^{1} I^{\rm inc}(r,\mu) \frac{1-{\rm e}^{-\tau_{\rm S}(r,\mu)}}{\tau_{\rm S}(r,\mu)} {\rm d}\mu, \tag{5.21}$$

and $I^{\rm inc}(r,\mu)$ the intensity incident to the considered location (resonance zone), usually the continuum intensity. In (5.21), $\beta(r)$ is the (conventional) escape probability,[11] $S_{\rm c}(r)$ the continuum source function and $\bar{U}(\tau_{\rm S}, \beta_{\rm P})$ a function describing the actual coupling of the opacities in the resonance zone, with $\beta_{\rm P} = \frac{\chi_{\rm c}}{\bar{\chi}_{\rm L}/\Delta\nu_D}$ the ratio between continuum and line opacity. The function $\bar{U}$ can be obtained, e.g., from precalculated tables (Taresch et al., 1997).

Often the last term in (5.21) can be neglected, but at least the first term (modified compared to the previous expressions) needs to be considered when the continuum is nonnegligible. To evaluate this term, one either uses the intensities from the continuum transfer, or one applies the following reasoning (unpublished thus far):

$$\begin{aligned}
\overline{\beta_{\rm c}(r) I_{\rm inc}} &= \frac{1}{2}\int_{-1}^{1} I^{\rm inc}(r,\mu) \frac{1-{\rm e}^{-\tau_{\rm S}(r,\mu)}}{\tau_{\rm S}(r,\mu)} {\rm d}\mu \\
&\overset{\tau_{\rm S}\gg 1, {\rm d}v/{\rm d}r>0}{\rightarrow} \frac{1}{2\chi_l}\int_{-1}^{1} I^{\rm inc}(r,\mu) \left[\mu^2 \frac{{\rm d}v'}{{\rm d}r} + (1-\mu^2)\frac{v'}{r}\right] {\rm d}\mu = \\
&= \frac{1}{\chi_l}\left[K_\nu(r)\left(\frac{{\rm d}v'}{{\rm d}r} - \frac{v'}{r}\right) + J_\nu(r)\frac{v'}{r}\right] = \\
&= \frac{1}{\chi_l} J_\nu(r)\left[f_\nu(r)\left(\frac{{\rm d}v'}{{\rm d}r} - \frac{v'}{r}\right) + \frac{v'}{r}\right] = J_\nu(r)\frac{1}{\tau_{\rm S}(r, \mu = \sqrt{f_\nu(r)})},
\end{aligned}$$

where all moments refer to continuum quantities (calculated in the spirit of Section 5.3.2), and $f_\nu(r) = \frac{K_\nu}{J_\nu}$ is the (conventional) Eddington factor. Including the optically thin case, one finds, to a good approximation

$$\overline{\beta_{\rm c}(r) I_{\rm inc}} \approx J_\nu(r) \frac{1-{\rm e}^{-\tau_{\rm S}(r,\mu=\sqrt{f_\nu(r)})}}{\tau_{\rm S}(r, \mu = \sqrt{f_\nu(r)})},$$

and avoids the angular integration by evaluating the integrand at $\mu = \sqrt{f_\nu(r)}$[12].

Inclusion of source-function gradients. As firstly shown by Sobolev (1957) and Castor (1974), the inclusion of source-function gradients is important when calculating the line acceleration. Though a constant source function does not contribute (due to cancellation effects when integrating over $\mu{\rm d}\mu$, discussed previously), corresponding gradients do contribute, and might become decisive, particularly in the inner wind regime (see also Owocki and Puls, 1999). Puls and Hummer (1988) improved on the previous

[11] Escape probabilities for static atmospheres are discussed in Section 1.10.

[12] A similar reasoning yields a fair approximation for the escape probability, $\beta(r) \approx \frac{1-{\rm e}^{-\tau_{\rm S}(r,\mu=\sqrt{1/3})}}{\tau_{\rm S}(r,\mu=\sqrt{1/3})}$.

works, and included corresponding continuum terms that turned out to be significant as well.

Inclusion of multiline effects. Multiline effects are essential when calculating the *total* line acceleration, $\sum_i g^i_{\rm rad}$, present in a wind (Puls, 1987). In addition to 'conventional' line-overlap effects (similar rest-wavelengths of different lines), lines can also interact with each other due to Doppler-induced frequency shifts. E.g., for the same $\nu_{\rm obs}$, there might be an interaction between a line with $\tilde{\nu}_1$ from the inner wind and a line with $\tilde{\nu}_2$ from the more outer part, if

$$\frac{\tilde{\nu}_1 - \tilde{\nu}_2}{\tilde{\nu}_1} \approx \frac{(\mu v)_2}{c} - \frac{(\mu v)_1}{c} > 0.$$

In other words, the radiation incident at $(\mu v)_2$ (determining the radiation field for the line with $\tilde{\nu}_2$) has already been processed before, by the bluewards line with $\tilde{\nu}_1$ at $(\mu v)_1$. See also Friend and Castor (1983).

Nonmonotonic velocity fields lead to more than one resonance zone, and need to be considered, e.g., when calculating the *approximate* line acceleration in time-dependent winds. Also in this case, the basic Sobolev theory can be adapted to include such effects (see Rybicki and Hummer, 1978, for the case of two coupled resonance zones, and Puls et al., 1993, for a generalization and application to instable-wind models).

SEI (Sobolev with exact integration). When calculating line profiles (specifically, UV P Cygni lines; for the profile formation principle, see Figure 5.7), and using the SA to determine *both* the source function *and* the emergent profile, the resulting accuracy is quite low, when compared to more 'exact' methods. A better approach is to calculate the scattering integral (and thus the source function, either in a complete NLTE or a two-level approach) using the SA, and then to derive the emergent line profile from an 'exact' formal solution[13] using such a source function.

To our knowledge, this had been first noted by Hamann (1981), and was explicitly suggested by Lamers et al. (1987), under the acronym SEI. Such an approach was also, and independently, used by Puls (1987), for the case of a large number of overlapping (UV) lines, in the context of NLTE wind modeling/spectrum synthesis.

5.4.4 *Comoving Frame (CMF) Transport*

Obviously, the calculation of the radiation field in an environment with significant (supersonic) velocity fields is either time consuming, if performed in the observer's frame (many grid points, frequencies and angles), or only approximate (but fast), when done using the SA. In the latter case, there are additional difficulties when considering not only one isolated line in an optically thin continuum, but more realistic situations as occurring in NLTE atmosphere modeling (many lines, various continua, multiline effects, etc.).

A quite simple and fast way out of this dilemma is possible when the velocity field is monotonic, after transforming to the comoving frame.[14]

[13] When calculating the formal solution via an integral method, it is advantageous to remap all quantities onto a microgrid of resolution $\approx v_{\rm th}/3$, to ensure a correct treatment of the resonance zone (e.g., Santolaya-Rey et al., 1997).

[14] A CMF solution is also possible for nonmonotonic velocity fields, at least in principle, but the algorithm becomes very complex.

The stationary RTE in the CMF: heuristic derivation. We start in the observer's frame, using the p-z geometry (now again for the front hemisphere only):

$$\pm\frac{\mathrm{d}I^{\pm}(z,p,\nu)}{\mathrm{d}z} = \eta_\nu\left(r,\nu\left(1-\frac{\mu v}{c}\right)\right) - \chi_\nu\left(r,\nu\left(1-\frac{\mu v}{c}\right)\right) I^{\pm}(z,p,\nu),$$

where in the following all CMF quantities are denoted by a sub- (or super-)script "0" (not to be confused with the denotation for the resonance zone), e.g., $\nu_0 \stackrel{\Delta}{=} \nu_{\rm CMF} = \nu\,(1-\mu v/c)$.

A velocity field produces Doppler shifts, aberration and advection terms (discussed later in this section); formally, all of these are $O(v/c)$ effects, but for lines the Doppler shifts become already significant if $v = O(v_{\rm th})$, due to the rapid change of the profile function. In the following heuristic approach, we concentrate on these Doppler shifts alone and neglect the rest (see also Lucy, 1971).

Since $\nu_0 = \nu_0(\nu, z) = \nu\,(1 - \mu v/c)$, the spatial derivative in the RTE needs to account for the change of ν_0 with z (see Figure 5.8):

$$\left.\frac{\mathrm{d}}{\mathrm{d}z}\right|_\nu = \left.\frac{\partial}{\partial z}\right|_{\nu_0} + \left.\frac{\partial}{\partial \nu_0}\right|_{z_0}\left.\frac{\partial \nu_0}{\partial z}\right|_\nu, \text{ with } \left.\frac{\partial \nu_0}{\partial z}\right|_\nu = -\frac{\nu}{c}\frac{\partial(\mu v)}{\partial z} \stackrel{O(v/c)}{\approx} \mp\frac{\nu_0}{c}\tilde{Q}(r,\mu). \quad (5.22)$$

Here, we have approximated $r \approx r_0$ and $\mu \approx \mu_0$, and accounted for the fact that when using $z > 0$ exclusively, $\partial(\mu v)/\partial z = \pm\,\tilde{Q}(r,\mu)$ for $\mu > 0$ and $\mu < 0$, respectively. $\tilde{Q}$ corresponds to our 'conventional' Q (e.g., (5.18)), but is evaluated using v instead of v'.

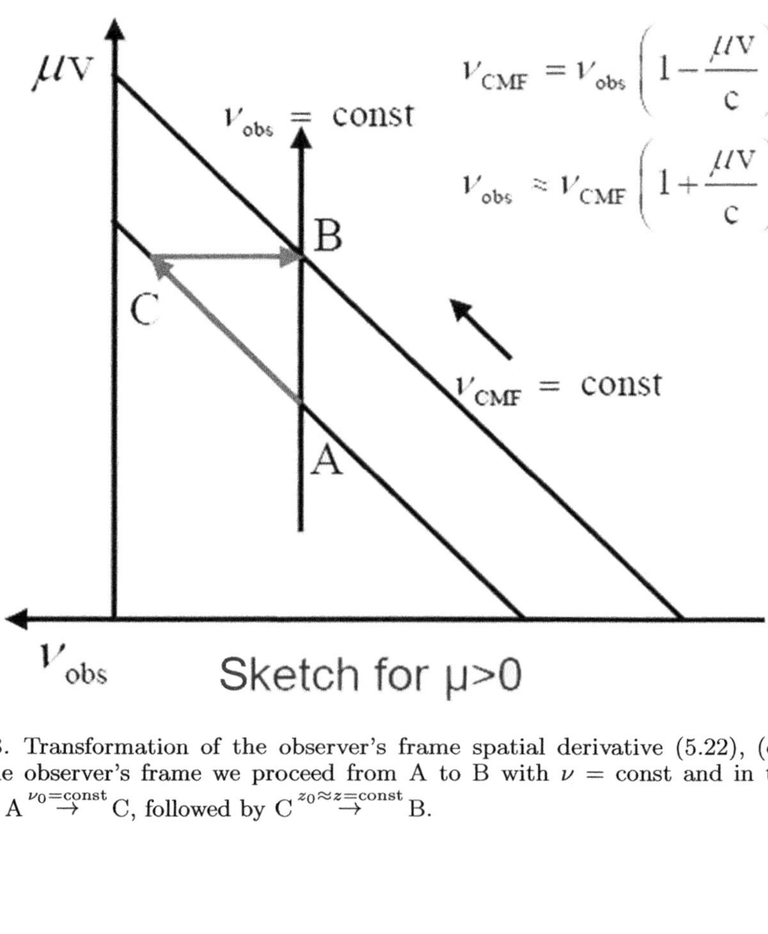

FIGURE 5.8. Transformation of the observer's frame spatial derivative (5.22), $(\mathrm{d}/\mathrm{d}z)_{\nu=\rm const}$: While in the observer's frame we proceed from A to B with ν = const and in the CMF we proceed via A $\stackrel{\nu_0=\rm const}{\rightarrow}$ C, followed by C $\stackrel{z_0\approx z=\rm const}{\rightarrow}$ B.

Thus, the RTE becomes

... in spherical symmetry, and using a p-z geometry, with $r_0 \approx r$, $z_0 \approx z$, $\mu_0 \approx \mu$,

$$\pm\frac{\partial I_0^{\pm}(z,p,\nu_0)}{\partial z} - \frac{\nu_0 \tilde{Q}(r,\mu)}{c}\frac{\partial I_0^{\pm}(z,p,\nu_0)}{\partial \nu_0} = \eta_0(r,\nu_0) - \chi_0(r,\nu_0)I_0^{\pm}(z,p,\nu_0). \quad (5.23)$$

While the spatial derivative enters with $\pm$ for outwards and inwards radiation, respectively, the frequency derivative has the same sign in both cases. This, again, is due to the fact that the gradient of (μv) is always positive in a spherically expanding medium (as long as $v(r)$ is monotonically increasing), irrespective of direction.

... in spherical symmetry with $r_0 \approx r$, $\mu_0 \approx \mu$

$$\mu\frac{\partial I_0(r,\mu,\nu_0)}{\partial r} + \frac{(1-\mu^2)}{r}\frac{\partial I_0(r,\mu,\nu_0)}{\partial \mu} - \frac{\nu_0\tilde{Q}(r,\mu)}{c}\frac{\partial I_0(r,\mu,\nu_0)}{\partial \nu_0}$$
$$= \eta_0(r,\nu_0) - \chi_0(r,\nu_0)I_0(r,\mu,\nu_0),$$

... and in plane-parallel symmetry with $z_0 \approx z$, $\mu_0 \approx \mu$

$$\mu\frac{\partial I_0(z,\mu,\nu_0)}{\partial z} - \frac{\nu_0\mu^2(\mathrm{d}v/\mathrm{d}z)}{c}\frac{\partial I_0(z,\mu,\nu_0)}{\partial \nu_0} = \eta_0(z,\nu_0) - \chi_0(z,\nu_0)I_0(z,\mu,\nu_0).$$

- The full transformation of the RTE (including time-dependent terms) can be found, e.g., in Castor (1972).
- Mihalas et al. (1976) showed that aberration terms (involving changes in direction μ) and advection terms (arising 'from gradients or from a "sweeping up" of radiation by the transformation' to the CMF) can be neglected when $v \ll c$ as considered here, while the frequency derivatives are most important. Thus far, the preceding equations are sufficient as long as $v \ll c$ (but: SN remnants with $v/c \lesssim 0.04 \ldots 0.15$(!)).
- In the preceding equations, particularly I_0, η_0, and χ_0 are comoving frame variables, and η_0 and χ_0 are *isotropic*.
- Consequently, for each line (if treated as a single one), only a small frequency range covering the variation of $\phi(\approx \pm 3v_{\mathrm{th}})$ needs to be considered.
- If only one line is considered, the RT is performed exclusively in the corresponding resonance zone.
- The CMF RTE is a partial differential equation (PDE) of hyperbolic type, and poses an initial boundary value problem, i.e., it requires boundary conditions in space and initial values in frequency.
- For larger frequency ranges, it might be useful to differentiate via

$$\frac{\nu_0\tilde{Q}(r,\mu)}{c}\frac{\partial}{\partial \nu_0} = \frac{\tilde{Q}(r,\mu)}{c}\frac{\partial}{\partial \ln \nu_0}.$$

Characteristics of the homogeneous equation. Often, the CMF equation of RT (e.g., (5.23)) is expressed in terms of Doppler units with respect to v_∞, $x_0 = (\nu_0 - \tilde{\nu})/\Delta\nu_\infty$, and $\Delta\nu_\infty = \nu_0\, v_\infty/c$, where $\tilde{\nu}$ is an arbitrary reference frequency close to ν_0 (e.g., the line-center frequency, if only one line is considered). Measuring v in units of v_∞ ($v'' = v/v_\infty$), and accounting for

$$\mathrm{d}x_0 = \frac{c}{v_\infty}\frac{\tilde{\nu}}{\nu_0}\frac{\mathrm{d}\nu_0}{\nu_0} \approx \frac{c}{v_\infty \nu_0}\mathrm{d}\nu_0 = \frac{\mathrm{d}\nu_0}{\Delta\nu_\infty},$$

we find

$$\pm\frac{\partial I_0^{\pm}(z,p,x_0)}{\partial z} - P(r,\mu)\frac{\partial I_0^{\pm}(z,p,x_0)}{\partial x_0} = \eta_0(r,x_0) - \chi_0(r,x_0)I_0^{\pm}(z,p,x_0) \tag{5.24}$$

with

$$P(r,\mu) = \frac{\mathrm{d}(\mu v'')}{\mathrm{d}z} = \left(\mu^2\frac{\mathrm{d}v''}{\mathrm{d}r} + (1-\mu^2)\frac{v''}{r}\right).$$

The characteristics of the homogeneous (RHS = 0) PDE are the curves (generally: hypersurfaces) along which $I_0^{\pm}$ remains constant if there is no absorption/emission, and need to be known, e.g., if we want to investigate the interaction (irradiation) of two lines (from red edge of first to blue edge of second line), in case of a negligible continuum. For the type of PDE considered here, these characteristics are given by (see standard textbooks)

$$\frac{\mathrm{d}x_0}{\mathrm{d}z} = \mp P(z),$$

and integration results in

$$0 < \Delta x_0 = x_{0,B} - x_0 = \mp\int_z^{z_B} P(z)\mathrm{d}z = \mp\left[\mu v''(z_B) - \mu v''(z)\right] = \mp\Delta\mu v''. \text{ Thus,}$$
$$I_0^{\pm}(\mu v''(z), x_0) = I_0^{\pm}(\mu v''(z_B), x_{0,B}) = I_0^{\pm}(\mu v''(z) \mp \Delta x_0, x_0 + \Delta x_0).$$

See Figure 5.9. Without absorption and emission, *all* photons are 'only' redshifted with regard to the CMF, from $x_0 + \Delta x_0$ to x_0, both when propagating outwards from $\mu v''(z) - \Delta x_0$ to $\mu v''(z)$, and when propagating inwards from $\mu v''(z) + \Delta x_0$ to $\mu v''(z)$. The corresponding *observer's frame* intensity at x, $I^{\pm}(z,x)$, remains constant, of course.

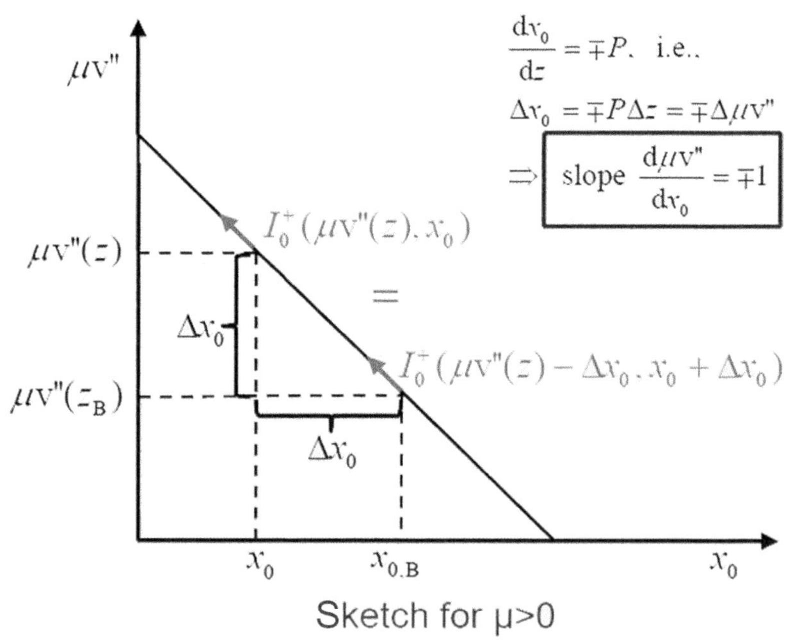

FIGURE 5.9. Characteristics of the homogeneous RTE in the CMF (5.24). See text.

Sobolev limit. From the comoving frame RTE, (5.24), one can also derive the Sobolev limit. Since we are in the CMF, this equation needs to be solved only in those regions of x_0 where the profile function is nonnegligible. This, however, corresponds to the resonance zone, where the SA assumes that all macrovariables (except v) are spatially constant. In this spirit, when neglecting the spatial derivative, the Sobolev limit can be easily obtained.

We will show this here for the case of one purely absorbing line with transition frequency $\tilde{\nu}$ and positive μ (no continuum), the generalization is left as an exercise for the reader (or see Lucy, 1971; Puls, 1991). From

$$-P(r,\mu)\frac{\partial I_0^+(z,p,x_0)}{\partial x_0} = -\chi_0(r,x_0)I_0^+(z,p,x_0),$$

where (z,r,μ) refer to the resonance zone, we obtain[15]

$$\ln\left[I_0^+(z,p,x_0)/I_0^{\rm inc}(z,p,x_{0,B})\right] = \frac{\bar{\chi}_{\rm L}(r)}{\Delta\nu_\infty P(r,\mu)}\int\limits_{x_{0,B}}^{x_0}\phi(x){\rm d}x$$

$$I_0^+(z,p,x_0) = I_0^{\rm inc}(z,p,x_{0,B})\,\exp\left[-\tau_S(r,\mu)\,\Phi(x_0)\right],$$

q.e.d. (One might compare with (5.19), and note that the preceding solution is already evaluated in the resonance zone.)

Solution methods. The basic approach to numerically solve the CMF RTE in spherically expanding atmospheres is similar to the treatment of the (quasi-)isotropic continuum (Section 5.3.2).

Method 1 (formal solution): Here, 'only' the discretized CMF RTE for the Feautrier variables u^0 and v^0 is solved, with $u^0 = \frac{1}{2}(I_0^+ + I_0^-)$ and $v^0 = \frac{1}{2}(I_0^+ - I_0^-)$. In p-z geometry, we thus have (cf. (5.24))

$$\begin{aligned}\frac{\partial u^0}{\partial z} - P\frac{\partial v^0}{\partial {\rm x}_0} &= -\chi_0 v^0\\ \frac{\partial v^0}{\partial z} - P\frac{\partial u^0}{\partial {\rm x}_0} &= \chi_0(S_0 - u^0),\end{aligned} \tag{5.25}$$

where x_0 is the CMF frequency in Doppler units with respect to v_∞, and $P(r,\mu) = ({\rm d}\mu v/v_\infty)/{\rm d}z = \left(\mu^2\frac{{\rm d}\mu v/v_\infty}{dr} + (1-\mu^2)\frac{v/v_\infty}{r}\right)$.

(5.25) is a system of two coupled, first-order PDEs, where (almost) all variables are to be evaluated in the CMF and depend on z (as a function of impact parameter p), r and x_0.

Spatial boundary values are specified as before (see Section 5.3.2), plus a 'blue-wing' boundary condition at the bluemost frequency, from the solution of a pure continuum transport. Special attention needs to be paid if the integration extends over a larger frequency range (more than one line), by a careful formulation of the outer boundary condition for optically thick conditions[16] (e.g., bluewards from the He II Lyman edge); otherwise, numerical artefacts might be created and transported through the spatial and frequency grid.

[15] Since x_0 is in units of v_∞, the corresponding profile function is normalized with regard to $\Delta\nu_\infty$

[16] Unfortunately, such formulations are usually not published.

The preceding PDEs are discretized using either of the following:

- A fully implicit scheme[17] (second-order in space, first-order in frequency), that is unconditionally stable (Mihalas et al., 1975).
- A semi-implicit (Crank–Nicholson) scheme, which is of higher accuracy, since it is of second order in frequency. If used in the formulation by Hamann (1981) (and *not* in the formulation by Mihalas et al., 1975), this is unconditionally stable as well.

Method 2 (variable Eddington factors): Here one uses the CMF-moments equations to obtain the moments of the radiation field (in the CMF). Contrasted with the corresponding observer's frame equations for isotropic opacities/emissivities (Section 5.3.1), also the third moment of the specific intensity, N^0_ν, enters the equations:

$$\frac{1}{r^2}\frac{\partial\left(r^2 H^0_\nu\right)}{\partial r} - \frac{\nu_0}{c}\left[\frac{v}{r}\frac{\partial(J^0_\nu - K^0_\nu)}{\partial \nu_0} + \frac{\mathrm{d}v}{\mathrm{d}r}\frac{\partial K^0_\nu}{\partial \nu_0}\right] = \eta_0(\nu_0) - \chi_0(\nu_0)J^0_\nu$$

$$\frac{\partial K^0_\nu}{\partial r} + \frac{3K^0_\nu - J^0_\nu}{r} - \frac{\nu_0}{c}\left[\frac{v}{r}\frac{\partial(H^0_\nu - N^0_\nu)}{\partial \nu_0} + \frac{\mathrm{d}v}{\mathrm{d}r}\frac{\partial N^0_\nu}{\partial \nu_0}\right] = -\chi_0(\nu_0)H^0_\nu$$

By means of the sphericality factor q_ν (5.13) and the Eddington factors $f^0_\nu = K^0_\nu/J^0_\nu$ and $g^0_\nu = N^0_\nu/H^0_\nu$ (calculated from the formal solution), we obtain again a coupled system of first-order PDEs for $r^2 J^0_\nu$ and $r^2 H^0_\nu$, that can be solved by discretization:

$$-\frac{\partial\left(r^2 H^0_\nu\right)}{\partial r} + \frac{\nu_0}{c}\left[\frac{\mathrm{d}v}{\mathrm{d}r} - \frac{v}{r}\right]\frac{\partial\left(f^0_\nu r^2 J^0_\nu\right)}{\partial \nu_0} + \frac{\nu_0}{c}\frac{v}{r}\frac{\partial\left(r^2 J^0_\nu\right)}{\partial \nu_0} = \chi_0(\nu_0)\left(r^2 J^0_\nu - r^2 S^0_\nu\right)$$

$$-\frac{\partial\left(q_\nu f^0_\nu r^2 J^0_\nu\right)}{q_\nu \partial r} + \frac{\nu_0}{c}\left[\frac{\mathrm{d}v}{\mathrm{d}r} - \frac{v}{r}\right]\frac{\partial\left(g^0_\nu r^2 H^0_\nu\right)}{\partial \nu_0} + \frac{\nu_0}{c}\frac{v}{r}\frac{\partial\left(r^2 H^0_\nu\right)}{\partial \nu_0} = \chi_0(\nu_0) r^2 H^0_\nu$$

In this case, a Rybicki scheme might be used if the source function can be separated into scattering and true absorption/emission components[18].

Radiative acceleration. To calculate the radiative acceleration, in the observer's frame we would need to evaluate (see (5.11))

$$g_{\mathrm{rad}} = \frac{1}{c\rho}\int \mathrm{d}\nu \oint \left(\chi\left(\nu(1-\mu v/c)\right) I_\nu(\mu) - \eta\left(\nu(1-\mu v/c)\right)\right)\mathbf{n}\mathrm{d}\Omega,$$

since the (line-)opacities and emissivities are angle dependent when a velocity field is present.

Because of the isotropy of χ^0_ν and η^0_ν in the comoving frame, however, this expression becomes considerably simplified when evaluated in the CMF,

$$\mathbf{g}^0_{\mathrm{rad}} = \frac{4\pi}{c\rho}\int \chi^0_\nu H^0_\nu \mathrm{d}\nu, \quad \text{since} \oint \chi^0_\nu I_0(\mu,\nu_0)\mathbf{n}\mathrm{d}\Omega = 4\pi\chi^0_\nu H^0_\nu, \quad \text{and} \oint \eta^0_\nu \mathbf{n}\mathrm{d}\Omega = 0.$$

Interestingly (and fortunately), one can show (e.g., Mihalas, 1978, chapter 15.3), that this expression is not only valid when used within the fluid frame (=CMF) equations of motion, but also, to order (v/c), in the corresponding inertial frame formulation. Namely, when the moments of the radiation field contained in the coupled matter–radiation equation of motion are expressed in terms of their CMF counterparts, and if the CMF moments equations (which we have just shown) are used, a delicate cancellation of

[17] For this scheme, an almost optimum local approximate lambda operator (ALO) can be calculated in parallel (see the next section and Puls, 1991).

[18] See equations (1.46) and (1.47) in Section 1.9.4.

terms ensures that also in the inertial frame the preceding expression for $\mathbf{g}^0_{\rm rad} \overset{O(v/c)}{\rightarrow} \mathbf{g}_{\rm rad}$ can be used for the radiative acceleration.

5.5 Accelerated Lambda Iteration (ALI) and 'Preconditioning'

The content of this section is not *directly* related to radiative transfer, but important if an NLTE treatment[19] of the plasma is required. This is particularly true for the atmospheres of hot stars, where the radiative rates dominate over the collisional ones in the line-forming region, due to a strong radiation field (and low densities in the stellar wind). Part of this section overlaps with the contents of Sections 3.7 and 3.8, but most issues are discussed here with special emphasis on the conditions in rapidly expanding atmospheres, providing an additional perspective.

Basically, there are two methods to obtain a consistent solution for the radiation field and the occupation numbers: (i) the complete-linearization method (Auer and Mihalas, 1969), used, e.g., in CMFGEN (Appendix A); and (ii) the ALI (Werner and Husfeld, 1985), used, e.g., in PoWR, WM-basic and FASTWIND (also Appendix A). Generally, the ALI method is easier to programme and has a faster performance per iteration step, but often requires more iterations than complete linearization.

The basic idea of the original (not accelerated) lambda iteration is as follows: One (a) starts with guessing values for the occupation numbers (e.g., from LTE or simplified NLTE conditions); (b) calculates corresponding opacities and source functions; (c) solves the RTE and calculates the mean intensities and scattering integrals; and (d) solves the rate equations involving J_ν and $\bar{J}$, i.e., calculates new occupation numbers. Subsequently, steps (b) to (d) are carried out again, and the process is iterated until (at least in terms of wishful thinking) convergence is obtained.

In practice, however, the convergence of this iteration (if at all achieved) is particularly slow for optically thick, scattering dominated processes, and it is rather difficult or even impossible to define an appropriate convergence criterion. These difficulties base on the fact that during each iteration, information is propagated only over $\Delta\tau_\nu \approx 1$. The *Accelerated* Lambda Iteration has been developed to get rid of these problems.

5.5.1 *A Simple Example*

To obtain more insight into the difficulties outlined earlier, we first concentrate on a simple showcase, namely a purely scattering line (e.g., a UV-resonance line) in Sobolev approximation. Then, we have the following:

(i) $S = \bar{J}$	Most simple 'rate equation' (e.g., from two-level atom without collisions)
(ii) $\bar{J} = (1-\beta)S + \beta_{\rm c} I_{\rm core}$	'Formal solution' (Sobolev solution for line transfer in optically thin continua, see (5.20))

Let's assume that the opacities are known and remain constant over the iteration, which is not too wrong for resonance lines.

Method A: Using (i) and (ii) in parallel, it is possible to obtain a consistent analytic solution,

$$S = (1-\beta)S + \beta_{\rm c} I_{\rm core} \Rightarrow S = \frac{\beta_{\rm c} I_{\rm core}}{\beta} \text{ (balance between irradiation and escape)}$$

[19] This accounts for the coupling between radiation field and occupation numbers via rate equations.

Method B: Alternatively, we apply the lambda iteration. We start with a guess value for the source function, S^0, and calculate the scattering integral, $\bar{J}^0$, using (ii). Then we determine a new iterate for the source function, S^1, using (i):

$$S^1 = (1-\beta)S^0 + \beta_c I_{\rm core}.$$

Generally,

$$\left.\begin{aligned} S^n &= (1-\beta)S^{n-1} + \beta_c I_{\rm core} \\ S^{n-1} &= (1-\beta)S^{n-2} + \beta_c I_{\rm core} \end{aligned}\right\} S^n - S^{n-1} := \Delta S^n = (1-\beta)\Delta S^{n-1}, \qquad (5.26)$$

and for optically thick lines where $\beta \rightarrow 1/\tau_S$ and thus $\beta \ll 1$, it turns out that $\Delta S^n \approx \Delta S^{n-1}$, i.e., no reasonable convergence criterion can be defined.

From this example, two questions are obvious: When do we consider the solution as converged? And how does the direct solution (*Method A*) and the Lambda-iterated solution (*Method B*) compare? To answer these questions, we investigate the limiting value of S^n for $n \rightarrow \infty$.

$$\begin{aligned} S^n &= (1-\beta)S^{n-1} + \beta_c I_{\rm core} = (1-\beta)\left[(1-\beta)S^{n-2} + \beta_c I_{\rm core}\right] + \beta_c I_{\rm core} = \\ &= \ldots = (1-\beta)^n S^o + \beta_c I_{core}\left[(1-\beta)^{n-1} + (1-\beta)^{n-2} + \cdots + 1\right]. \end{aligned}$$

Accounting for $\sum\limits_{\rm i=0}^{\rm n-1} q^i = \frac{1-q^n}{1-q}$, we obtain

$$S^n = (1-\beta)^n S^0 + \beta_c I_{\rm core}\frac{1-(1-\beta)^n}{\beta} \overset{n\rightarrow\infty}{\rightarrow} \frac{\beta_c I_{\rm core}}{\beta},$$

i.e., indeed the Lambda-iterated solution (from *Method B*) converges (very slowly) to the correct one from *Method A* (and becomes independent from the start value).

How many iteration steps are required? For $\beta \ll 1$, we can approximate $(1-\beta)^n \approx (1-n\beta)$, and to ensure convergence, we must have $(1-\beta)^n \rightarrow 0$, i.e., $n \gtrsim 1/\beta \rightarrow \tau_S$. Thus, we would need roughly the same number of iterations as the size of τ_S, which (i) can be very large for resonance lines, $n \approx \tau_S$ up to $O(10^5 \ldots 10^6)$, and (ii) shows that indeed, per iteration step, information corresponding to only $\Delta\tau = 1$ is propagated.

5.5.2 *Accelerated Lambda Iteration*

Generalizing the aforementioned simplified problem, we need to fulfil the following requirement for a consistent solution of the coupled problem (RT and rate equations)

$$S^n \overbrace{=}^{\text{via rate equations}} f(J^n) = f\left(\overbrace{\Lambda\left[S^n\right]}^{\text{formal solution}}\right),$$

which is a nonlinear and, except for the Sobolev case, nonlocal problem. In contrast, the lambda iteration provides us with

$$S^n = f(J^{n-1}) = f(\Lambda\left[S^{n-1}\right]),$$

which displays the well-known convergence problems. We stress that Λ is an affine operator,[20] due to the boundary conditions.

[20] Linear transformation plus displacement.

In the following, we consider continuum (J) and line problems ($\bar{J}$) in parallel. A generalization of results for continuum quantities to line conditions is straightforward, by solving for all line frequencies and integrating over the profile function.

For values on a 1-D spatial grid (with N grid points), we may rewrite the formal solution of RT in form of an affine relation,

$$\mathbf{J} = \mathbf{\Lambda} \cdot \mathbf{S} + \mathbf{\Phi},$$

where $\mathbf{J}$, $\mathbf{S}$ and $\mathbf{\Phi}$ are vectors of length N, and $\mathbf{\Lambda}$ is a matrix with $N \times N$ elements. $\mathbf{\Phi}$ corresponds to the boundary conditions, $\mathbf{J}(\mathbf{S} = \mathbf{0})$.

If required, the elements Φ_i and Λ_{ij} might be derived (in 1-D) from $N + 1$ formal solutions for $\mathbf{S} = \mathbf{0}$, $\mathbf{S} = \mathbf{e}_1, \ldots, \mathbf{S} = \mathbf{e}_N$, respectively, where

$$\mathbf{e}_j = (0, \ldots, 0, 1, 0, \ldots, 0)^T$$

is the unit vector in direction j.

ALI is based on the idea of operator splitting (e.g., Cannon, 1973), namely to split[21]

$$\mathbf{\Lambda} = \mathbf{\Lambda}^A + \left(\mathbf{\Lambda} - \mathbf{\Lambda}^{\mathbf{A}}\right),$$

the lambda operator into an approximate operator (with a linear component that should be easily invertible) and a rest part. Then we can approximate

$$\mathbf{J}^n \approx \mathbf{\Lambda}^A \left[\mathbf{S}^n\right] + \left(\mathbf{\Lambda} - \mathbf{\Lambda}^A\right) \left[\mathbf{S}^{n-1}\right],$$

where equality is obtained for $n \to \infty$, when $\mathbf{S}^{n-1} \to \mathbf{S}^n$.

This is the essential clue, since now we have a relation (at step n) between $\mathbf{J}^n$ and $\mathbf{S}^n$, and not only between $\mathbf{J}^{n-1}$ and $\mathbf{S}^{n-1}$.

Also the ALO, $\mathbf{\Lambda}^A$, needs to be of affine type, i.e., $\mathbf{\Lambda}^A \left[\mathbf{S}\right] = \mathbf{\Lambda}^* \cdot \mathbf{S} + \mathbf{\Phi}^*$, but even then

$$\begin{aligned} \mathbf{J}^n &= \left[\mathbf{\Lambda}^* \cdot \mathbf{S}^n + \mathbf{\Phi}^*\right] + \mathbf{J}^{n-1} - \left[\mathbf{\Lambda}^* \cdot \mathbf{S}^{n-1} + \mathbf{\Phi}^*\right], \text{ i.e.,} \\ \mathbf{J}^n &= \mathbf{\Lambda}^* \cdot \mathbf{S}^n + \mathbf{\Delta J}^{n-1}, \text{ with } \mathbf{\Delta J}^{n-1} = \mathbf{J}^{n-1} - \mathbf{\Lambda}^* \cdot \mathbf{S}^{n-1}, \end{aligned} \tag{5.27}$$

only the linear part of the ALO, $\mathbf{\Lambda}^*$, needs to be specified, assuming that $\mathbf{\Phi}^*$ remains constant over the iteration.

We note that $\mathbf{\Delta J}^{n-1}$ depends only on $\mathbf{S}^{n-1}$, and can be calculated from the formal solution for $\mathbf{J}^{n-1}$ (and specified $\mathbf{\Lambda}^*$).

ALI in practice. To illustrate how ALI works in practice, we consider a continuum problem with scattering, or – again – a two-level atom,

$$\mathbf{S} = \boldsymbol{\xi} \mathbf{J} + \boldsymbol{\psi},$$

where $\boldsymbol{\xi}$ is a diagonal matrix (containing the scattering fractions, $0 \le \xi_{ii} \le 1$), and $\boldsymbol{\psi}$ is a vector (containing the Planck functions). Then,

$$\mathbf{S}^n = \boldsymbol{\xi} \left(\mathbf{\Lambda}^* \mathbf{S}^n + \mathbf{\Delta} \boldsymbol{J}^{n-1}\right) + \boldsymbol{\psi},$$

[21] Similar to the Jacobi iteration in boundary value problems.

and we obtain an explicit expression for $\mathbf{S}^n$,

$$\mathbf{S}^n = (\mathbf{1} - \xi\mathbf{\Lambda}^*)^{-1}\left(\xi\mathbf{\Delta J}^{n-1} + \psi\right) = (\mathbf{1} - \xi\mathbf{\Lambda}^*)^{-1}\left(\xi(\mathbf{\Lambda} - \mathbf{\Lambda}^*)\mathbf{S}^{n-1} + \psi\right), \qquad (5.28)$$

which constitutes the ALI scheme for simple source functions (those that can be analytically separated into a scattering and thermal part).

With $\mathbf{\Delta S}^n := \mathbf{S}^n - \mathbf{S}^\infty$ (deviation from the 'true' source function $\mathbf{S}^\infty$, in contrast to the definition in (5.26)), we finally find (after a few algebraic manipulations)

$$\mathbf{\Delta S}^n = \mathbf{A}\mathbf{\Delta S}^{n-1}, \qquad \text{with amplification matrix} \quad \mathbf{A} = (\mathbf{1} - \xi\mathbf{\Lambda}^*)^{-1}\left(\xi(\mathbf{\Lambda} - \mathbf{\Lambda}^*)\right).$$

One can show that under typical conditions $\mathbf{A}$ has a complete set of real and orthogonal eigenvectors, and real eigenvalues λ (e.g., Puls and Herrero, 1988). Expanding $\mathbf{\Delta S}$ in terms of these eigenvectors, for large n we obtain $\mathbf{\Delta S}^n \approx \lambda_{\max}^n \mathbf{\Delta S}^0$, where $\lambda_{\max} = \pm\max(|\lambda_i|)$, and the minus sign applies when the corresponding eigenvalue is negative. Thus, the ALI scheme converges if $|\lambda_{\max}| < 1$. For *static* problems, Olson et al. (1986) showed that in particular

$$|\lambda_{\max}| < 1, \qquad \text{if} \quad \mathbf{\Lambda}^* = \text{diag(linear part of}\,\mathbf{\Lambda}).$$

A very elegant method to calculate the corresponding $\mathbf{\Lambda}^*$ has been provided by Rybicki and Hummer (1991, appendix). For the case of CMF line transfer, on the other hand, Puls (1991) developed an almost optimum, purely local ALO (see Figure 5.10).

- Both of these ALOs can be calculated in parallel with the corresponding RT, on very fast timescales.
- Since the CMF line transfer has an essentially local character in rapidly expanding atmospheres (taking place only in the narrow resonance zone), a local ALO is sufficient when solving the rate equations under such conditions.
- For local ALOs, an overestimation of the exact diagonal leads to divergence in most cases.

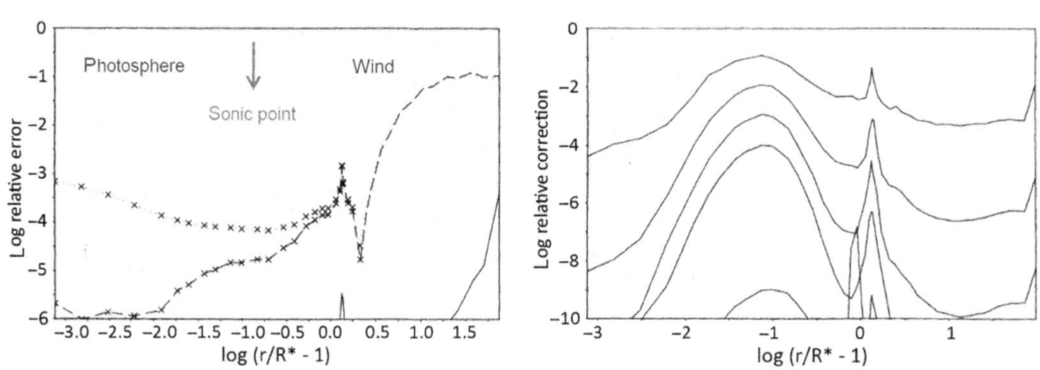

FIGURE 5.10. Local ALO, $\mathbf{\Lambda}^*$, from Puls (1991), and corresponding ALI cycle, for a CMF line source function in an expanding atmosphere. The displayed example refers to a strong, purely scattering line. *Left*: relative deviation (absolute value, logarithmic) among (i) solid – the exact diagonal of the Lambda operator and $\mathbf{\Lambda}^*$ (relative differences mostly below 10^{-6}); (ii) dotted – the exact diagonal and $(1-\beta)$ (cf. 5.29); and (iii) dashed – the exact diagonal and $(1-\beta-\bar{U})$, when using the SA with continuum. Since $(1-\beta)$ overestimates the exact diagonal in most regions (not visible since absolute values are displayed), it cannot be used as an ALO. *Right*: Relative corrections $\Delta S^n/S^n$ for subsequent iterations, as a function of radius. Adapted from Puls (1991). Reproduced with permission © ESO.

For non-local ALOs and more sophisticated iteration schemes (e.g., required in multi-D calculations), we refer to Trujillo Bueno and Fabiani Bendicho (1995) and references therein. See also Hennicker et al. (2018).

Comparison between ALI scheme and Sobolev approach (line case). Assuming a local ALO, for each depth point we have the correspondence

$$\left.\begin{array}{ll}\text{ALI:} & \bar{J}^n = \Lambda^* S^n + \underbrace{\Delta \bar{J}^{n-1}}_{\bar{J}^{n-1}-\Lambda^* S^{n-1}} \\ \text{Sobolev:} & \bar{J}^n = (1-\beta) S^n + \beta_{\rm c} I_{\rm core}\end{array}\right\} \Lambda^* \triangleq (1-\beta), \text{ and } \Delta \bar{J}^{n-1} \triangleq \beta_{\rm c} I_{\rm core} \quad (5.29)$$

We note that when comparing with the SA with continuum, this correspondence would read $\Lambda^* \triangleq (1-\beta-\bar{U})$. See Figure 5.10 and (5.21).

5.5.3 *Implementation into Rate Equations – 'Preconditioning'*

In the last section of this overview, we discuss how the ALI approach is implemented into the rate equations. We start with the definition of the so-called net line rate, Z_{ul}, quantifying the net transition rate for a line transition with upper and lower levels u, l, and corresponding occupation numbers, n_u, n_l, respectively:

$$Z_{ul} = n_u A_{ul} \left(1 - \frac{\bar{J}}{S}\right),$$

where A_{ul} is the Einstein coefficient for spontaneous emission, and $Z_{ul} > 0$ if the downward transitions dominate. The corresponding line source function is given by

$$S = \frac{n_u A_{ul}}{n_l B_{lu} - n_u B_{ul}},$$

with Einstein coefficients for absorption, B_{lu}, and induced emission, B_{ul}.

Without ALI, and applying the conventional (not accelerated) lambda iteration, S^n would be calculated using $\bar{J}^{n-1}$ in the rate equations,

$$\frac{\bar{J}^{n-1}}{S^n} = \bar{J}^{n-1}\left(\frac{n_l B_{lu} - n_u B_{ul}}{n_u A_{ul}}\right)$$

$$\Rightarrow Z_{ul} = n_u \cdot \underbrace{\left(A_{ul} + B_{ul}\bar{J}^{n-1}\right)}_{\text{downward line rate}} - n_l \cdot \underbrace{\left(B_{lu}\bar{J}^{n-1}\right)}_{\text{upward line rate}},$$

where the second equation denotes the downward and upward rates for the considered line transition within the rate matrix.

With ALI and local ALO, S^n is calculated using $\bar{J}^n = \Lambda^* S^n + \Delta \bar{J}^{n-1}$:

$$\frac{\bar{J}^n}{S^n} = \Lambda^* + \frac{\Delta \bar{J}^{n-1}}{S^n}$$

$$\Rightarrow Z_{ul} = n_u A_{ul}\left(1 - \Lambda^* - \frac{\Delta \bar{J}^{n-1}}{S^n}\right) =$$

$$= n_u \cdot \underbrace{\left(A_{ul}(1-\Lambda^*) + B_{ul}\Delta \bar{J}^{n-1}\right)}_{\text{downward line rate}} - n_l \cdot \underbrace{\left(B_{lu}\Delta \bar{J}^{n-1}\right)}_{\text{upward line rate}}$$

A comparison of the line rates,

$$A_{ul} \to A_{ul}(1-\Lambda^*)$$
$$B_{ul}\bar{J}^{n-1} \to B_{ul}\Delta\bar{J}^{n-1}$$
$$B_{lu}\bar{J}^{n-1} \to B_{lu}\Delta\bar{J}^{n-1},$$

shows that all rates become smaller in the ALI formulation, since the inefficient part (the optically thick line core, where upward and downward rates are equal) is analytically cancelled, and only the efficient part (the optically thin wings) survives. This modification of the line rates when using the ALI scheme has been named 'preconditioning' by Rybicki and Hummer (1991), and the corresponding rates are sometimes called 'effective' or 'reduced' rates.

Reduced rates for Sobolev transport. Similar to the preceding reasoning, we now investigate the consequence of using the SA scattering integrals in the rate equations,

$$\begin{aligned} Z_{ul} &= n_u\left(A_{ul}+B_{ul}\bar{J}\right) - n_l B_{lu}\bar{J} = \\ &= n_u\left(A_{ul}+B_{ul}\left[(1-\beta)S+\beta_{\rm c}I_{\rm core}\right]\right) - n_l B_{lu}\left[(1-\beta)S+\beta_{\rm c}I_{\rm core}\right] = \cdots = \\ &= n_u\left(A_{ul}\beta+B_{ul}\beta_{\rm c}I_{\rm core}\right) - n_l B_{lu}\beta_{\rm c}I_{\rm core}. \end{aligned}$$

Also here, the contribution from the optically thick core cancels analytically. By comparing this contribution with the analogous result from the ALI approach, we again find the correspondence (see (5.29))

$$\Lambda^* \triangleq (1-\beta), \qquad \text{and} \quad \Delta\bar{J}^{n-1} \triangleq \beta_{\rm c}I_{\rm core}.$$

If one would use the Sobolev approximation with continuum (5.21), this correpondence would read

$$\Lambda^* \triangleq (1-\beta-\bar{U}), \qquad \text{and} \quad \Delta\bar{J}^{n-1} \triangleq \beta_{\rm c}(r)I_{\rm inc}+\bar{U}S_c.$$

5.6 Further Issues and Applications

Due to space (and time) limitations, a variety of additional issues could not be treated in this overview. In the following, we will provide important keywords in this context (certainly not a complete list), marked in italics if *directly* related to specific RT problems in early-type stellar atmospheres.

- Temperature structure: radiative equilibrium vs. thermal electron balance
- Energy equation, adiabatic expansion and cooling in the outermost wind
- *LDI and impact of a diffuse radiation field*
- Inhomogeneous winds, shocks and X-ray emission
- Examples/applications
  - *Supersonic 'microturbulence' vs. nonmonotonic v-fields*
  - Supersonic macro turbulence
  - (Quasi-)recombination lines
  - Optical-depth invariants and scaling relations
  - H_α in O-stars and AB-supergiants
  - Impact of wind on weaker lines, and specifically N III λ4640
  - *IR/radio excess*
  - *IR lines: inverted levels (or close to inversion)*

  - X-rays: impact on resonance lines/'superionization'
  - Emission lines in WRs
- Wind inhomogeneities/clumping
  - *Micro- and macroclumping, porosity*
  - *Clumping in RTE*
  - H_α vs. He II λ4686
  - *Velocity-porosity*
  - Clumping – coupling with rate equations
- Outlook
  - *Multi-D problems/formulation*
  - *Time dependence, relativistic treatment*
  - *Nonradial line forces (e.g., in rotating winds)*
  - *Polarization (linear, circular → B-fields)*

5.7 Appendix A: NLTE Model Atmosphere Codes for Hot Stars

Table 5.1 compares presently available atmospheric codes that can be used for the spectroscopic analysis of hot stars. Since the codes Detail/Surface and TLUSTY calculate occupation numbers/spectra within hydrostatic, plane-parallel atmospheres, they are "only" suited for the analysis of stars with negligible winds (see also end of Section 5.2). The different computation times are majorly caused by the different approaches to deal with line-blocking/blanketing. The overall agreement between the various codes (within their domain of application) is quite satisfactory, though certain discrepancies are found in specific parameter ranges, particularly regarding EUV ionizing fluxes (Puls et al., 2005; Simón-Díaz and Stasińska, 2008).

5.8 Appendix B: Further Comments on the Line Profile Function

5.8.1 *Appendix B.1: Depth Dependent Thermal Speeds*

To avoid a depth dependence of the frequency grid when measuring frequencies in depth dependent Doppler units, one uses a *fiducial* thermal speed, $v^*_{\rm th}$, such that

$$x = \frac{\nu - \tilde{\nu}}{\Delta\nu^*_{\rm D}} \quad \text{with} \quad \Delta\nu^*_{\rm D} = \frac{\tilde{\nu} v^*_{\rm th}}{c}.$$

Let

$$\delta(r) = \frac{\Delta\nu_{\rm D}(r)}{\Delta\nu^*_{\rm D}} = \frac{v_{\rm th}(r)}{v^*_{\rm th}}, \qquad \text{then} \quad \frac{\nu - \tilde{\nu} - \mu v(r)\tilde{\nu}/c}{\Delta\nu_{\rm D}(r)} = \frac{x - \mu v'(r)}{\delta(r)},$$

now with $v'(r) = v(r)/v^*_{\rm th}$ (cf. Section 5.4.1). In this notation,

$$\phi_\nu(x_{\rm CMF}, r) = \phi_\nu(x - \mu v', r) = \frac{1}{\Delta\nu^*_{\rm D}\delta(r)\sqrt{\pi}} \exp\left[-\left(\frac{x - \mu v'(r)}{\delta(r)}\right)^2\right],$$

with units "per frequency" (s), or alternatively,

$$\chi_\nu(x_{\rm CMF}, r) = \frac{\bar{\chi}_{\rm L}(r)}{\Delta\nu^*_{\rm D}} \phi(x_{\rm CMF}, r),$$

with dimensionless

$$\phi(x_{\rm CMF}, r) = \frac{1}{\delta(r)\sqrt{\pi}} \exp\left[-\left(\frac{x - \mu v'(r)}{\delta(r)}\right)^2\right], \qquad \text{and} \quad \frac{\bar{\chi}_{\rm L}(r)}{\Delta\nu^*_{\rm D}} = \frac{\bar{\chi}_{\rm L}(r)\tilde{\lambda}}{v^*_{\rm th}}.$$

TABLE 5.1. Comparison of state-of-the-art, NLTE, line-blanketed model atmosphere codes

Code	DETAIL/ SURFACE[1]	TLUSTY[2]	CMFGEN[3]	FASTWIND[4]	PHOENIX[5]	PoWR[6]	WM-basic[7]
Geometry	Plane-parallel	Plane-parallel	Spherical	Spherical	Spherical/ pl.-para./3-D	Spherical	Spherical
Blanketing	LTE	Yes	Yes	Approx.	Yes	Yes	Yes
Radiative line transfer	Observer's frame	Observer's frame	CMF	CMF/ Sobolev	CMF/ obs.frame	CMF	Sobolev
Temperature structure	Radiative equilibrium	Radiative equilibrium	Radiative equilibrium	e^- therm. balance	Radiative equilibrium	Radiative equilibrium	e^- therm. balance
Diagnostic range	No limitations	No limitations	No limitations	Optical/IR	No limitations	No limitations	UV
Major application	Hot stars with negligible winds	Hot stars with negligible winds	OB(A)-stars WRs, SNe	OB-stars, early A-sgs	Stars below 10 kK,SNe	WRs, O-stars	Hot stars w. dense winds, ion. fluxes, SNe
Comments	No wind	No wind	Start model required	User-specified atomic models	Molecules included	–	No clumping
Execution time	Few minutes	Hour(s)	Hours	Few minutes to 0.5 h	Hours	Hours	1–2 h

[1] Giddings (1981), Butler and Giddings (1985); [2] Hubený (1998); [3] Hillier and Miller (1998); [4] Puls et al. (2005); [5] Hauschildt (1992); [6] Gräfener et al. (2002); [7] Pauldrach et al. (2001).

5.8.2 *Appendix B.2: Integrals Involving the Profile Function: Which Normalization to Use?*

- Spatial integrals of type $\int \chi^{\text{line}}(\nu_{\text{CMF}}, r) f_\nu(r) \mathrm{d}r$

$$\to \int \frac{\bar{\chi}_{\text{L}}(r)}{\Delta\nu_{\text{D}}} \phi(x_{\text{CMF}}, r) f_\nu(r) \mathrm{d}r,$$

  e.g., optical depth if $f_\nu(r) = 1$.
- Frequency integrals of type $\int f_\nu(r)\, \phi(\nu_{\text{CMF}}, r) \mathrm{d}\nu$

$$\to \int f(\nu(x), r)\, \phi(x_{\text{CMF}}, r) \mathrm{d}x,$$

  e.g., scattering integrals, if $f_\nu(r) = J_\nu(r)$.
- Frequency integrals of type $\int \chi^{\text{line}}(\nu_{\text{CMF}}, r) f_\nu(r) \mathrm{d}\nu$

$$\to \bar{\chi}_{\text{L}}(r) \int f(\nu(x), r)\, \phi(x_{\text{CMF}}, r) \mathrm{d}x,$$

  e.g., in the context of line acceleration, $g_{\text{rad}}(r)$, see Section 5.4.3

If applicable, one needs to use $\Delta\nu_{\text{D}}^*$ instead of $\Delta\nu_{\text{D}}$. Moreover,

$$\phi(\nu_{\text{CMF}}, r) = \frac{\phi(x_{\text{CMF}}, r)}{\Delta\nu_{\text{D}}}, \quad \text{i.e.,} \quad \phi(\nu_{\text{CMF}}, r) \mathrm{d}\nu = \phi(x_{\text{CMF}}, r) \mathrm{d}x,$$

and $\phi(\nu_{\text{CMF}}, r) = \phi_\nu(x_{\text{CMF}}, r)$ normalized with respect to frequency, whereas $\phi(x_{\text{CMF}}, r)$ normalized with respect to x.

REFERENCES

Auer, L. H. and Mihalas, D. 1969. Non-LTE Model Atmospheres. III. A Complete-Linearization Method. *ApJ*, **158**(Nov.), 641.

Butler, K. and Giddings, J. R. 1985. *Newsl. Anal. Astron. Spectra*, No. 9

Cannon, C. J. 1973. Frequency-Quadrature Perturbations in Radiative-Transfer Theory. *ApJ*, **185**(October), 621–630

Castor, J. I. 1972. Radiative Transfer in Spherically Symmetric Flows. *ApJ*, **178**(December), 779–792

Castor, J. I. 1974. On the Force Associated with Absorption of Spectral Line Radiation. *MNRAS*, **169**(November), 279–306

Castor, J. I., Abbott, D. C. and Klein, R. I. 1975. Radiation-Driven Winds in of Stars. *ApJ*, **195**(January), 157–174

Friend, D. B. and Abbott, D. C. 1986. The Theory of Radiatively Driven Stellar Winds. III Wind Models with Finite Disk Correction and Rotation. *ApJ*, **311**(December), 701–707

Friend, D. B. and Castor, J. I. 1983. Stellar Winds Driven by Multiline Scattering. *ApJ*, **272**(September), 259–272

Gabler, R., Gabler, A., Kudritzki, R. P., Puls, J. and Pauldrach, A. 1989. Unified NLTE Model Atmospheres Including Spherical Extension and Stellar Winds: Method and First Results. *A&A*, **226**(December), 162–182

Giddings, J. R. 1981. PhD Thesis, University of London

Gräfener, G., Koesterke, L. and Hamann, W. R. 2002. Line-Blanketed Model Atmospheres for WR Stars. *A&A*, **387**(May), 244–257

Hamann, W.-R. 1980. The Expanding Envelope of Zeta Puppis: A Detailed UV-Line Fit. *A&A*, **84**(April), 342–349

Hamann, W.-R. 1981. Line Formation in Expanding Atmospheres: On the Validity of the Sobolev Approximation. *A&A*, **93**(January), 353–361

Hauschildt, P. H. 1992. A Fast Operator Perturbation Method for the Solution of the Special Relativistic Equation of Radiative Transfer in Spherical Symmetry. *J. Quant. Spec. Radiat. Transf.*, **47**(6), 433–453

Hennicker, L., Puls, J., Kee, N. D. and Sundqvist, J. O. 2018. 3D Radiative Transfer: Continuum and Line Scattering in Non-Spherical Winds from OB Stars. *A&A*, **616**(September), A140

Hillier, D. J. and Miller, D. L. 1998. The Treatment of Non-LTE Line Blanketing in Spherically Expanding Outflows. *ApJ*, **496**(1), 407–427

Hubený, I. 1998. Non-LTE Line-Blanketed Model Atmospheres of Hot Stars. Page 139 of: Chan, Kwing Lam, Cheng, K. S. and Singh, H. P. (eds.), *1997 Pacific Rim Conference on Stellar Astrophysics*. Astronomical Society of the Pacific Conference Series, vol. 138

Hubený, I. and Mihalas, D. 2014. *Theory of Stellar Atmospheres*. Princeton University Press.

Hummer, D. G. and Rybicki, G. B. 1971. Radiative Transfer in Spherically Symmetric Systems: The Conservative Grey Case. *MNRAS*, **152**, 1

Hummer, D. G. and Rybicki, G. B. 1985. The Sobolev Approximation for Line Formation with Continuous Opacity. *ApJ*, **293**(June), 258–267

Kudritzki, R.-P. and Puls, J. 2000. Winds from Hot Stars. *ARA&A*, **38**, 613–666

Lamers, H. J. G. L. M., Cerruti-Sola, M. and Perinotto, M. 1987. The "SEI" Method for Accurate and Efficient Calculations of Line Profiles in Spherically Symmetric Stellar Winds. *ApJ*, **314**(March), 726–738

Lucy, L. B. 1971. The Formation of Resonance Lines in Extended and Expanding Atmospheres. *ApJ*, **163**(January), 95

Lucy, L. B. and Solomon, P. M. 1970. Mass Loss by Hot Stars. *ApJ*, **159**(March), 879–893

Mihalas, D. 1978. *Stellar Atmospheres*, 2nd edition W. H. Freeman and Co.

Mihalas, D., Kunasz, P. B. and Hummer, D. G. 1975. Solution of the Comoving Frame Equation of Transfer in Spherically Symmetric Flows. I Computational Method for Equivalent-Two-Level-Atom Source Functions. *ApJ*, **202**(December), 465–489

Mihalas, D., Kunasz, P. B. and Hummer, D. G. 1976. Solution of the Comoving-Frame Equation of Transfer in Spherically Symmetric Flows. III. Effect of Aberration and Advection Terms. *ApJ*, **206**(June), 515–524

Olson, G. L., Auer, L. H. and Buchler, J. R. 1986. A Rapidly Convergent Iterative Solution of the Non-LTE Line Radiation Transfer Problem. *J. Quant. Spec. Radiat. Transf.*, **35**(June), 431–442

Owocki, S. P. and Puls, J. 1996. Nonlocal Escape-Integral Approximations for the Line Force in Structured Line-driven Stellar Winds. *ApJ*, **462**(May), 894

Owocki, S. P. and Puls, J. 1999. Line-Driven Stellar Winds: The Dynamical Role of Diffuse Radiation Gradients and Limitations to the Sobolev Approach. *ApJ*, **510**(January), 355–368

Owocki, S. P. and Rybicki, G. B. 1984. Instabilities in Line-Driven Stellar Winds. I Dependence on Perturbation Wavelength. *ApJ*, **284**(September), 337–350

Pauldrach, A., Puls, J. and Kudritzki, R. P. 1986. Radiation-Driven Winds of Hot Luminous Stars: Improvements of the Theory and First Results. *A&A*, **164**(August), 86–100

Pauldrach, A. W. A., Hoffmann, T. L. and Lennon, M. 2001. Radiation-Driven Winds of Hot Luminous Stars. XIII. A Description of NLTE Line Blocking and Blanketing towards Realistic Models for Expanding Atmospheres. *A&A*, **375**(August), 161–195

Puls, J. 1987. Radiation-Driven Winds of Hot Luminous Stars. IV The Influence of Multi-Line Effects. *A&A*, **184**(October), 227–248

Puls, J. 1991. Approximate Lambda-Operators Working at Optimum Convergence Rate. II Line Transfer in Expanding Atmospheres. *A&A*, **248**(August), 581–589

Puls, J. and Herrero, A. 1988. Approximate Lambda-Operators Working at Optimum Convergence Rate. I Theory and Application to Hydrostatic Atmospheres. *A&A*, **204**(October), 219–228

Puls, J. and Hummer, D. G. 1988. The Sobolev Approximation for the Line Force and Line Source Function in a Spherically-Symmetrical Stellar Wind with Continuum Opacity. *A&A*, **191**(February), 87–98

Puls, J., Owocki, S. P. and Fullerton, A. W. 1993. On the Synthesis of Resonance Lines in Dynamical Models of Structured Hot-Star Winds. *A&A*, **279**(November), 457–476

Puls, J., Urbaneja, M. A., Venero, R., Repolust, T., Springmann, U., Jokuthy, A. and Mokiem, M. R. 2005. Atmospheric NLTE-Models for the Spectroscopic Analysis of Blue Stars with Winds. II. Line-Blanketed Models. *A&A*, **435**(May), 669–698

Puls, J., Vink, J. S. and Najarro, F. 2008. Mass loss from hot massive stars. *A&A Rev.*, **16**(Dec.), 209–325.

Rybicki, G. B. 1991 (Jan). Recent Advances in Computational Methods. Page 1 of: Crivellari, L., Hubený, I. and Hummer, D. G. (eds.), *NATO Advanced Science Institutes (ASI)* Series C, vol. 341.

Rybicki, G. B., and Hummer, D. G. 1978. A generalization of the Sobolev method for flows with nonlocal radiative coupling. *ApJ*, **219**(Jan.), 654–675.

Rybicki, G. B., and Hummer, D. G. 1991. An accelerated lambda iteration method for multilevel radiative transfer. I - Non-overlapping lines with background continuum. *A&A*, **245**(May), 171–181.

Santolaya-Rey, A. E., Puls, J., and Herrero, A. 1997. Atmospheric NLTE-models for the spectroscopic analysis of luminous blue Stars with Winds. *A&A*, **323**(July), 488–512.

Simón-Díaz, S. and Stasińska, G. 2008. The Ionizing Radiation from Massive Stars and Its Impact on HII Regions: Results from Modern Model Atmospheres. *MNRAS*, **389**(September), 1009–1021

Sobolev, V. V. 1957. The Diffusion of Lα Radiation in Nebulae and Stellar Envelopes. *Soviet Ast.*, **1**(October), 678

Sobolev, V. V. 1960. *Moving Envelopes of Stars.* Harvard University Press

Taresch, G., Kudritzki, R. P., Hurwitz, M., et al. 1997. Quantitative Analysis of the FUV, UV and Optical Spectrum of the O3 Star HD 93129A. *A&A*, **321**(May), 531–548

Trujillo Bueno, J. and Fabiani Bendicho, P. 1995. A Novel Iterative Scheme for the Very Fast and Accurate Solution of Non-LTE Radiative Transfer Problems. *ApJ*, **455**(December), 646

Werner, K. and Husfeld, D. 1985. Multi-Level Non-LTE Line Formation Calculations Using Approximate Lambda-Operators. *A&A*, **148**(July), 417–422

6. Phenomenology and Physics of Late-Type Stars

MARIA BERGEMANN, CAMILLA JUUL HANSEN AND TIMOTHY C. BEERS

Abstract

In this chapter, we present the basics of the physics and phenomenology of FGKM-type stars. This review is based on recent developments in the observational and theoretical domains of stellar physics, including a variety of techniques – spectroscopy, interferometry, photometry – and large-scale stellar surveys. We focus on the advances in radiative transfer modelling and spectroscopy of stars across the full metallicity range. To provide the reader with the essential supplementary information, we also give a brief qualitative account of the structure and evolution of low- and intermediate-mass stars and of stellar nucleosynthesis. We also provide a brief overview of new models of stellar atmospheres and stellar spectra, with emphasis on non local thermodynamic equilibrium (NLTE) and hydrodynamics. Lastly, we discuss some of the relevant observational studies of stellar abundances in the context of stellar populations, evolution of metal-poor stars and Galactic archeology.

6.1 Introduction

Stellar astrophysics has evidenced revolutionary advances during the past decades. The advent of large-scale stellar surveys on telescopes equipped with highly sensitive detectors has transformed our understanding of the physics of stellar atmospheres and their interiors. Detailed abundances, stellar parameters and kinematics, which are now available for more than a million stars, are also prerequisites for understanding the evolution of the Milky Way and its stellar populations. The main reason for the unique constraining power of stars is that they form an exceptionally diverse class of cosmic objects, with physical conditions spanning many orders of magnitude. In the Sun alone, density changes from $\rho_{\rm core} \sim 10^2$ g/cm^3 in the core to $\rho_{\rm surface} \sim 10^{-7}$ g/cm^3 in the photosphere (and five orders of magnitude less in the chromosphere), and the local kinetic temperature drops from $T_{\rm core} \sim 10^7$ K to $T_{\rm surface}$[1] $\sim 5,800$ K in the outer layers. For a red supergiant with a mass of 10 $M_{\odot}$, this change is even more dramatic, from $\rho_{\rm core} \sim 10^5$ g/cm^3 to $\sim 10^{-15}$ g/cm^3 at base of the wind above the photosphere.

The immense variation of the internal stellar conditions reflects the complex nature of phenomena occurring in their interiors and surfaces. Physical theory allows the existence of 'living'[2] stars as small as only 8% of the mass and radius of our own star, the ultracool M dwarfs, but also stars that exceed the radius of the Solar System in size. The extreme contrast between the macrophysics, that is, the global phenomena acting on the surfaces and in interiors of stars, such as convection, pulsation, winds and mass loss, and also the microphysics, which is needed to model the production and transport of energy, implies an

[1] Here we refer to the *effective* temperature of the Sun, i.e., the temperature of a black body radiator that emits the same integrated flux as the Sun.

[2] Here we do not discuss systems that are affected by interactions, such as subdwarfs in binary systems, and stellar remnants, i.e., white dwarfs, neutron stars and black holes, which may have much smaller radii.

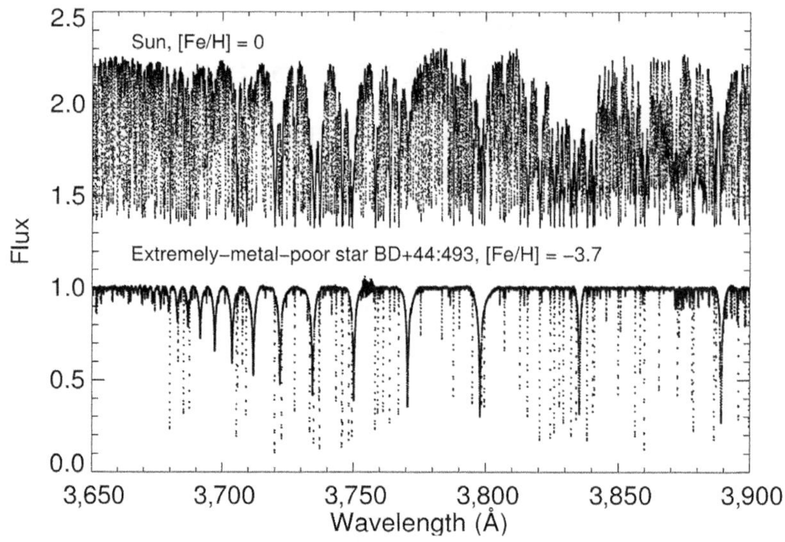

FIGURE 6.1. Blue regions of high-resolution spectra observed for the Sun (KPNO) and the HDS/Subaru spectrum of a carbon-enhanced metal-poor (CEMP) star, BD+44°493 (Ito et al., 2013). Note the dramatic contrast in the number of spectral lines seen in these stars, which differ in metallicity by about four orders of magnitude.

extraordinary complexity of the theoretical models required to make robust predictions of the properties of stars. Thus, stars are indeed the ideal plasma physics laboratories, offering a window into conditions that cannot be achieved elsewhere.

In this review, we focus on the so-called 'cool', late-type stars, which represent spectral classes F, G, K and M (Figure 6.1). The effective temperature, $T_{\rm eff}$, at the surface of these objects does not change greatly, at most by a factor of two, from a typical red supergiant ($T_{\rm eff} \sim$ 3,400 K) to a typical hot main-sequence turnoff star ($T_{\rm eff} \sim 6,500$ K). However, cool stars possess other remarkable characteristics, which make them the most robust and useful tracers of the evolutionary history of their parent populations. First, the unevolved stars on the main sequence preserve the natal composition of the gas cloud from which they formed. We note, though, the effects such as atomic diffusion and mixing, which may modify the surface composition. Secondly, the lifetimes of cool stars range from a few million years for the most massive stars to over a Hubble time for the least massive stars. Thirdly, cool stars span an enormous range in luminosity (L $= 4\pi R^2 \sigma_{\rm SB} T_{\rm eff}^4$), which reflects the variation in the radius R and in the total emitted flux ($F_{tot} = \sigma_{\rm SB} T_{\rm eff}^4$). Indeed, the most luminous stars, with L $\sim 10^5$ $L_\odot$, can nowadays be observed with 10 m telescopes to distances of millions of parsecs (Mpc; Evans et al., 2011). Also, ongoing large stellar surveys are discovering low-metallicity stars with ages greater than 13 Gyr (Frebel and Norris, 2015). These are the best candidates for probing the physical state of the very early Galaxy and the conditions in the early Universe, at a level of detail superior to any study of ultrahigh-redshift galaxies.

The physics of cool stars is extremely complex, therefore in this chapter we are not aiming to provide a comprehensive overview of all the major developments that have occurred over the past decade. Neither is this meant to be an in-depth study of any particular subject. The main goal of this article is to set the scene, and to give the flavour of the current research in modelling and observations of late-type stars, with a focus on their atmospheres. We will also mention the applications of these data in studies of other astronomical domains. Models of stellar spectra are used in most fields in contemporary astrophysics, from studies of stellar evolution to exoplanet spectra, populations synthesis and galaxy formation. Observations of stellar abundances even have implications for cosmology. Therefore, it seems very timely to review the present state of these studies, and draw an outlook for future studies.

This chapter is structured as follows. We summarize the main observational constraints on the physical properties of cool stars and outline briefly their evolution in Section 6.2. Model atmosphere and synthetic spectra, which are needed for the quantitative diagnostic of the observations, are reviewed in Section 6.3. Section 6.4 presents a brief recap of stellar nucleosynthesis. Section 6.5 introduces the reader to the fields of stellar populations and Galactic chemical evolution, and mentions some outstanding science cases, which have been resolved and/or triggered by new observations or by application of the new models. We close with some conclusions and an outlook for the future in Section 6.6.

6.2 Observations of Cool Stars

6.2.1 *Terminology*

Astronomers like acronyms, and in this chapter the reader will encounter many. In this section, a short overview of various types of stars and terminology will be given. The fundamental parameters of stars, which set their evolutionary paths, are their mass and chemical composition. However, from an observer's perspective, it is common to characterize stars by their effective temperature, surface gravity and metallicity. The latter classification is also approximate, as it stems from the historical effort to employ one-dimensional hydrostatic plane-parallel model atmospheres, which are semi-infinite (and thus have no size); the gravitational acceleration at the surface, log (g), as well as the bolometric flux, are assumed to be constant as well. Throughout the discussion in the next sections, it should be kept in mind that these are just a set of approximations, and the standard quantities, $T_{\rm eff}$, log(g), and [Fe/H], are not the fundamental parameters[3] of stars, per se.

Metallicity[4] is usually denoted as [Fe/H], i.e., the logarithmic ratio of the number of absorbing atoms of iron to hydrogen per unit volume in the star scaled to the solar ratio (*odot*). The abundances are defined in the following way:

$$\log(A_{\rm Fe}) = \log(N_{\rm Fe}/N_{\rm H}) + 12 \tag{6.1}$$

$$[{\rm Fe/H}] = \log(A_{\rm Fe,star}) - \log(A_{\rm Fe,\odot}) \tag{6.2}$$

$$[{\rm Mg/Fe}] = m[Mg/H] - [{\rm Fe/H}] = \log(A_{Mg,star}) - \log(A_{Mg,\odot}) - [{\rm Fe/H}]. \tag{6.3}$$

In this notation, the Sun has [Fe/H] $= 0$ and *zero* abundance of any other element relative to metallicity, i.e., for example [Mg/Fe] $= 0$. Based on this value, the star will either be subsolar ([Fe/H] < 0) or supersolar ([Fe/H] > 0). It is sometimes convenient to have a finer division of the subsolar metallicities, because chemical traces of many nuclear processes are most easily found at various levels below solar metallicity (see Table 6.1), and in some specialized research fields, only individual metallicity bins are important.

[3] There is an ambiguity as to how to define the fundamental parameters. Some refer to mass (M), luminosity (L) and radius (R); see Wittkowski (2004). From the perspective of stellar evolution, the fundamental parameters are mass and detailed chemical abundances, and possibly the rate of mass loss, as these define how a star will evolve and what luminosity, mass and radius it will have at any point in time.

[4] In astronomy, every element beyond He is referred to as a 'metal'.

TABLE 6.1. A common classification of cool stars by metallicity (Beers and Christlieb, 2005).

Term	Acronym	[Fe/H]
Metal-rich	MR	[Fe/H] > 0
Metal-poor	MP	[Fe/H] < -1
Very metal-poor	VMP	[Fe/H] < -2
Extremely metal-poor	EMP	[Fe/H] < -3
Ultra metal-poor	UMP	[Fe/H] < -4
Hyper metal-poor	HMP	[Fe/H] < -5
Mega metal-poor	MMP	[Fe/H] < -6

These notations are not binding, and it is common in astronomical literature to refer to all cool stars with [Fe/H] < -1 simply as metal-poor stars.

The effect of metallicity on the blue regions of stellar spectra is shown in Figure 6.1. This wavelength regime is most sensitive to the metallicity of a star, because it hosts the largest number of metallic lines.

The classification based on mass is derived from stellar evolution theory and represents qualitatively different paths taken by stars in a given mass range. Stars less than 2 $M_\odot$ are usually referred to as low-mass stars. Stars with mass from 2 to $\sim$ 7 to 10 $M_\odot$ are intermediate-mass objects, and those with even higher masses are 'massive' stars. The evolutionary paths of these objects are described briefly in the next section.

6.2.2 *A Brief Sketch of the Evolution of Low- and Intermediate-Mass Stars*

Stars live through one of two ways – contraction or nuclear burning. For the star to end up on the main sequence, where nuclear burning in the core sets in, the star must first undergo a phase of contraction[5] to achieve a temperature in the core sufficient to initiate nuclear reactions. Stellar evolution then proceeds as follows. Nuclei with the lowest nuclear charge, starting with hydrogen, burn first, until the fuel is exhausted in the core and the core contracts, increasing its temperature, until the conditions allow ignition of nuclei with the next lowest Coulomb barrier, and so forth. But, before the next fusion stage begins in the core, fusion of the lighter elements occurs in the layers outside the core, forming the well-known 'onion-shell' structure in massive stars. In this way, the chemical composition inside the star changes. Internal mixing processes during later evolutionary stages can bring a portion of this material to the surface, which is the region we can observe and derive abundances for.

Figure 6.2 shows a typical diagram representing the evolution of cool stars. The definition of a cool star is very broad, and includes both low- and high-mass stars in different evolutionary stages, or even in different systems, for example, single stars and binaries. In particular, we find main-sequence stars, subgiants, red giants and asymptotic giant branch stars, which are simply different stages of evolution for low-mass stars.

- Main sequence (MS)

  Inspecting Figure 6.2 more closely, we find that the lower part of the diagram is occupied by stars on the MS, those that fuse hydrogen into helium in the core. Stars spend the majority of their lifetimes on the MS. For a typical 1 $M_{\rm Sun}$ star such as

[5] The Kelvin–Helmholtz contraction, caused by energy loss at the surface of a star and by virtue of the virial theorem, leads to compression and heating in the core.

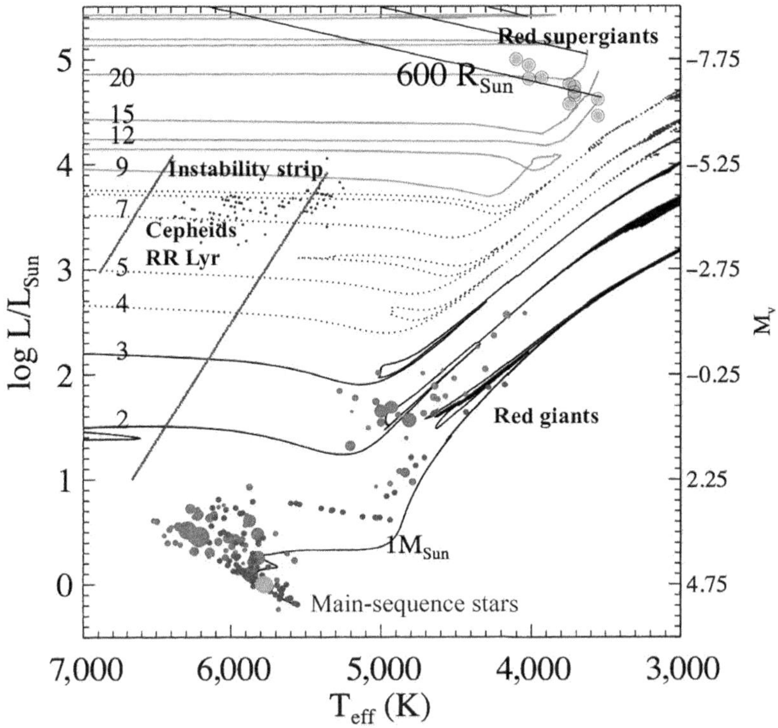

FIGURE 6.2. The evolutionary $T_{\rm eff}$ luminosity diagram for cool stars. Observed data for Cepheids are from Genovali et al. (2014). Other observed datasets were taken from Ruchti et al. (2013), Gazak et al. (2014) and Pace et al. (2012).

the Sun, the MS lifetime is $\sim$ 9 Gyr. Hydrogen burning proceeds predominantly by pp chain or CNO cycle for stars with masses below or above roughly 1.5 $M_{\rm Sun}$, respectively. Once the core H reservoir is depleted, the star contracts, and the layers surrounding the core heat up, so H burning is ignited in a shell. This is when the star turns off the main sequence. Turnoff stars are extremely important for studies of stellar populations, because their ages can be accurately determined through isochrone fitting (Soderblom, 2010).

- Turnoff (TO) and subgiant (SG)
  After the TO, the star becomes more luminous, but its surface temperature drops, and it moves onto the subgiant branch, where it undergoes hydrogen shell burning. Hydrogen shell burning is always dominated by the CNO cycle. The SG branch, although representing a short evolutionary stage, is also important as an age diagnostic for individual stars, because this is the locus where the tracks of different ages and metallicities do not overlap. The mass of the He core grows through the ashes of H burning in the shell, and, at a certain point, the core becomes degenerate[6]. This is when the star moves onto the red giant branch. Due to the expansion and the surface cooling of a star, the envelope becomes convective, allowing more efficient outward energy transport.

[6] This only happens for lower-mass stars, where electron degeneracy pressure due to the exclusion principle becomes the dominant source of pressure, supporting the structure of the star against gravity.

- Red giant branch (RGB)
  Subsequent H burning in the shell goes on via the CNO cycle. The star moves almost vertically in the $T_{\rm eff}$-log(g) diagram, maintaining its total radiative flux ($F_{\rm Bol} = \sigma_{\rm SB}\ T_{\rm eff}^4$), but rapidly increasing the luminosity and radius. RGB stars are more luminous and cooler than the MS stars, typically with $T_{\rm eff} < 5,200$ K and surface gravities of log(g) < 3, yet luminosities up to $10^3\ L_\odot$. The first dredge-up takes place on the ascending part of the RGB, when the products of H burning are mixed up to the surface through large-scale convection. The surface is then enriched in ^{13}C, He and N, but depleted in Be, Li and ^{12}C.

  The tip of the RGB is determined by the luminosity at which low-mass stars with cores supported by electron degeneracy experience the He flash, in which He ignition occurs as a runaway process, due to electron degeneracy decoupling pressure from temperature. When enough thermal energy has been released, degeneracy is lifted, and the runaway process halts. All stars at the tip of the RGB possess He cores of $\approx$0.5 $M_\odot$ and similar luminosities, which makes them a powerful diagnostic of extragalactic distances, as they are 'standard candles' (Salaris, 2012). The most recent studies indicate that the impact of physics on stellar evolutionary models is significant, and may have an effect on the determination of the Hubble constant, H_0, using the tip of the RGB method, on the order of few percent (Serenelli et al., 2017).
- Horizontal branch (HB)
  After the tip of the RGB stage, the star moves into a new equilibrium configuration, in which He burning occurs in the core under nondegenerate conditions, with H burning in a shell. This phase lasts about 10^8 years, and the star radiates with a luminosity of $\sim$ 50–100 $L_\odot$. HB stars are easier to unambiguously identify in globular clusters (GCs) than in the field, and they offer an independent way of determining the age of a GC, because the HB morphology depends on both age and metallicity (Catelan, 2009). The coolest end of the HB is also termed the 'red clump' (RC). The stars of solarlike metallicity pile up there, because the masses of He degenerate cores are nearly equal. However, the mass of the H-burning envelope also matters for the location of a core He-burning star in the $T_{\rm eff}$-log(g) plane, with larger envelopes corresponding to cooler $T_{\rm eff}$. The location of a star on the HB also depends on its metallicity. In metal-poor GCs, the HB is extended to higher temperatures (or bluer colours), due to the presence of old low-metallicity stars with thin H envelopes.

  It is barely possible to spectroscopically distinguish between ascending and descending (towards the RC and core He-burning phase) stage, although attempts have been made recently using indirect methods coupled to the data from asteroseismology (Ness et al., 2016).
- Asymptotic giant branch (AGB)
  After exhaustion of He in the core, the star consists of an electron degenerate C/O-core, a shell that burns He to C and O (O is the most abundant ash of He burning), an intershell consisting of He, followed by a thin shell where H burns to He. The outer H-rich envelope is convective. During this phase, the star experiences multiple He shell flashes. Slow neutron-capture (*s*-process) nucleosynthesis occurs in ^{13}C-rich pockets, and the products are mixed up to the surface during the thermal pulses (Karakas and Lattanzio, 2014).

The final stages of stellar evolution will depend critically on the initial mass and metallicity of a star, but also on the mass loss. These stages are summarized in Table 6.2.

Other members of the cool star group include objects on the instability strip (RR Lyrae, Cepheids), as well as red supergiants (RSGs), which are massive OB stars that have evolved from the main sequence. RSGs have enormous radii, up to 1500 $R_\odot$, but

TABLE 6.2. Evolutionary paths of stars with different ZAMS masses (courtesy A. Serenelli). This rough classification is based on the type of He ignition. One may also distinguish low- from intermediate-mass stars by the physical properties of the core. Low-mass stars $M \lessapprox 1.5\ M_{\odot}$ have a convective core on the main sequence.

Class	Mass range ($M_{\odot}$) approximate	Main nuclear energy source Evolutionary endpoint
Brown dwarfs	0.01–0.08	Deuterium burning Lithium burning (massive ones) **H-degenerate cores (> Hubble time)**
Very low-mass stars	0.08–0.45	Hydrogen burning **Helium core white dwarfs (> Hubble time)**
Low-mass stars	0.45–2	Hydrogen burning Helium burning (degenerate ignition) **Carbon/oxygen white dwarfs**
Intermediate mass stars	2–7/10	Hydrogen burning Helium burning (nondegenerate ignition) **Carbon/oxygen white dwarfs**
Massive stars	7/8–10/12	H-burning, He-burning Carbon burning (degenerate ignition) **Oxygen/neon white dwarfs** **E-capture supernovae (neutron stars)**
Very massive stars	> 10/12	H-burning, He-burning, C-burning Ne-burning, O-burning, Si-burning – – > Fe-cores **Core collapse supernovae (neutron stars, black holes)** **No remnant for pair-instability SNe (?)**

Note: The text in bold indicates the final evolutionary stage or the remnant

temperatures rarely greater than $\sim 4,300$ K (Davies et al., 2013). The atmospheres of RSGs are extremely diffuse, with gravitational acceleration at the surface of $\log(g) \lesssim 1$.

It is common to distinguish pulsating stars, those on the instability strip, from nonpulsating stars. One has to keep in mind the complexity behind this picture, as all stars oscillate, although the physical mechanisms driving pulsations are different (Gautschy and Saio, 1995). Oscillation periods range from minutes to years, and oscillation amplitudes – from parts per million (ppm) to magnitudes. Oscillations of the Sun are stochastically excited by mass motions in the outer convective zone, and are particularly strong at frequencies around 3 mHz – the so-called five-minute oscillations. This is also the mechanism that drives low-amplitude oscillations of MS and RGB stars (solarlike oscillations; Chaplin and Miglio, 2013), but with periods that scale roughly inversely with the square root of the mean density of the star.

Cepheids and RR Lyrae are the most prominent members of the radially pulsating group, which change their radii by up to 20% during pulsations. Cepheids occur in two types: Type I, young intermediate-to-massive stars with 5-10 $M_{\odot}$, and Type II, old metal-poor stars with very low mass ($M \sim 0.5\ M_{Sun}$). Type I Cepheids have short periods, from less than a day to a few weeks, whereas Type II Cepheids pulsate with longer periods, of a few to ~ 50 days (Bono et al., 1997; Wallerstein, 2002). Cepheids are classical fundamental mode pulsators, which experience self-excitation, driven mostly by the κ effect, the increase in opacity with heating and compression of singly ionized helium. These conditions are met in the regions of partial ionization of He or H. Physically, Type II cepheids are thought to be objects that experience shell flashes, either on the evolution towards the AGB or on the AGB, and are temporarily moved into the instability strip (Wallerstein, 2002). Cepheids are very important as extragalactic distance indicators, and can be used to calibrate SN Ia for the determination of the Hubble constant (Madore and Freedman, 1998; Riess et al., 2011; Zhang et al., 2017).

RR Lyrae are another class of pulsating (or variable) stars. If the H envelope mass is low, the star ends up on the blue end of the horizontal branch, and is dynamically unstable, which results in cyclic pulsation. These stars have short pulsation periods, typically between 0.5 and 0.8 days (Catelan, 2009). RR Lyrae are low-mass stars, which means that they are long-lived, and the ones found in the Milky Way today are typically $\sim$ 10–12 Gyr old. These stars are bright and identifiable via their distinctive light curves, and can be used to probe the earliest epochs of the formation of the Galaxy, in particular its halo and the bulge (Lee, 1992; Hansen et al., 2011, 2016b; Soszyński et al., 2014; Marconi and Minniti, 2018).

Many stars occur in multiple systems, which affects not only the evolutionary path of the system, but also their surface composition. One prominent example are the blue stragglers (BS), which are thought to form through two channels, referred to as 'collisional BS' or 'mass-transfer BS'. The latter refers to mass transfer from the primary massive star, which has experienced Roche lobe overflow onto the less-massive secondary. This hypothesis is supported by observations of BS in GCs, and in particular, depletion of C and O in their atmospheres (Ferraro et al., 2006).

6.2.3 *Multimessenger Diagnostics*

Modern instruments offer a great variety of tools that can be used to perform multiwavelength observations of stars. Photometric observations, i.e., observations of the integrated light in a given bandpass, are most useful to globally map stellar populations or to perform deep imaging in smaller parts of the sky, centred on smaller objects (globular or open star clusters). Observations in different photometric bands offer potential diagnostics of the stellar atmosphere. The bluer bands are, in particular, quite important as $T_{\rm eff}$ and metallicity indicators. Filters in the near-UV that cover the Balmer jump can also be used as an indicator of the surface gravity (Casagrande et al., 2011). Infrared photometry is a particularly powerful means to probe the high-extinction regions of the Milky Way, such as the disc and the bulge. Imaging from space facilities, such as the Hubble Space Telescope (HST) or Gaia, offer much cleaner information, devoid of the blurring and absorption/emission signatures of the Earth's atmosphere. This is particularly useful for studies of the content of distant compact stellar populations, such as globular clusters, where high angular resolution is essential.

Spectroscopy is a very powerful method to explore the surface properties of stars. Modern instruments, such as the optical spectrographs Ultraviolet and Visual Echelle Spectrograph (UVES) at the Very Large Telescope (VLT), High Resolution Echelle Spectrometer (HIRES) at Keck, High Dispersion Spectrograph (HDS) at Subaru, or Magellan Inamori Kyocera Echelle (MIKE) at Magellan, provide high resolving power, $R > 30{,}000$, with full wavelength coverage across the optical region of several thousand angstroms from blue to red. For a bright star with magnitude $V < 12$, one may reach very high signal-to-noise ratios, thus resolving the weakest spectral lines, representing low-abundance exotic elements such as silver (Hansen et al., 2012). Spectroscopic information typically derives from a number of diagnostic features, which can be combined, e.g., by exploiting the requirement of excitation–ionization balance to constrain the total flux ($T_{\rm eff}$) or pressure (gravitational acceleration) at the surface (Bergemann et al., 2012; Ruchti et al., 2013). Metallicity, or iron abundance, can be derived through the analysis

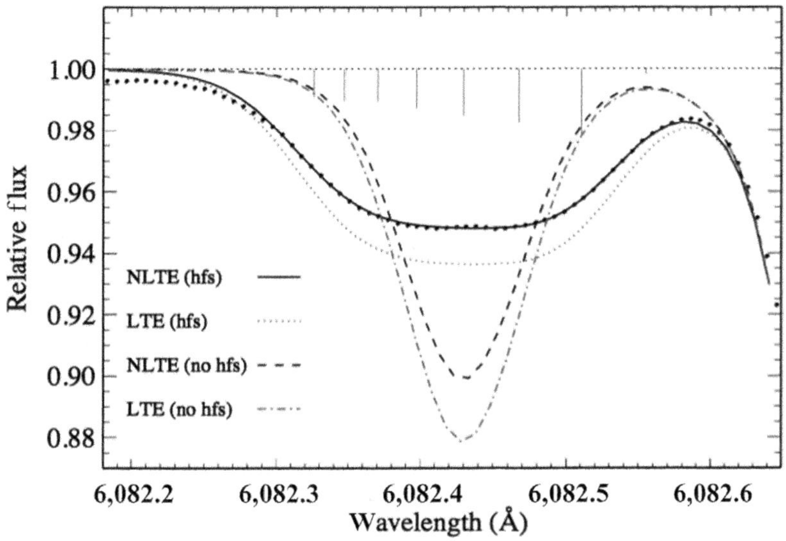

FIGURE 6.3. A line of neutral cobalt in the solar spectrum, computed using local thermodynamic equilibrium (LTE) and NLTE, with and without HFS, respectively.

of numerous iron lines. In a typical high-resolution optical spectrum of a late-type star, there are > 1 million detectable lines of Fe I. The abundances of other elements can be derived from individual spectral lines, which are found all over, from the near-UV to the near-IR.

While the positions of spectral lines appear to be completely random, they are of course not, as the wavelength (or energy) simply reflects the arrangement of the energy levels in the electronic configuration system of a given atom. This, in turn, depends on the atomic properties, such as the mass of the nucleus, its magnetic moment and the number and degree of filling of the electronic shells. Some elements with nonzero magnetic moments, particularly those with odd numbers of neutrons and odd numbers of protons in the nucleus, experience splitting of energy levels due to the interaction of nuclear magnetic moments with magnetic fields produced by the outer electrons. The larger the magnetic moment, such as 5/2 for Mn and 7/2 for Co, the more prominent is the hyperfine splitting (HFS) of spectral lines, which is clearly distinguishable in a high-resolution stellar spectrum (Bergemann and Gehren, 2008; Bergemann et al., 2010). For some Co I lines, for example, the splitting exceeds 0.5 Å, far larger than the typical Doppler broadening caused by the thermal motions of atoms in the stellar atmosphere (Figure 6.3). Furthermore, most elements are represented by a number of stable and unstable isotopes in different abundance proportions. Barium is one of the species that is strongly affected by isotopic splitting, in addition to HFS.

The shapes and positions of spectral lines also convey an enormous amount of information about the stellar surface: From the overall wavelength shift of the lines, one may measure the radial velocity of the star, from the line shape, the stellar rotation velocity, and, with sufficiently high-resolution spectra, the direction and velocity of mass flows at the surface (Dravins, 1999; Allende Prieto et al., 2002; Ramírez et al., 2008). At very high spatial and temporal resolution, like that available from observations of the Sun (Asplund et al., 2009; Lind et al., 2017; Bergemann et al., 2019), one may even determine the properties of convective motions from the blue- and redshifts of individual spectral lines, which react sensitively to the atmospheric dynamics. Some spectral lines, like the Mn I 5395 Å (Danilovic et al., 2016), Ni I 6767.8 Å and H_α 6562.8 Å lines, are also sensitive to magnetic field and velocities in stellar chromospheres (Leenaarts et al., 2012).

Stellar spectra are generally a remarkable source of information. However, not all required information can be derived from a spectrum. In particular, among the most important determinations is the evolutionary state of stars, which constrain the mass, the age or even the position of a star on the RGB. Recent purely empirical studies have shown that mass might be correlated with the shape of the hydrogen α line (Bergemann et al., 2016); the strength of the Li I line at 6707 Å is sometimes also used as a proxy for age (Soderblom, 2010). Additional chemical clocks, such as [C/Fe] and [N/Fe] (Martig et al., 2016), [Y/Mg] (Spina et al., 2016) or chronometer pairs based on the radioactive species Th and U (Hill et al., 2017; Placco et al., 2017; Hansen et al., 2018a), have also been used.

One particularly important development during the past decade is related to the advent of large-scale spectroscopic surveys. Many surveys have been carried out on 2-, 4-, or even 8-m class telescopes, like the Apache Point Observatory Galactic Evolution Experiment (APOGEE) and Sloan Digital Sky Survey (SDSS) Sloan Extension for Galactic Understanding and Exploration (SEGUE) surveys (at the 2.5-m telescope at Apache Point Observatory), Galactic Archaeology with HERMES (at the 3.8-m Anglo-Australian Telescope [AAT]), and Gaia-European Organisation for Astronomical Research in the Southern Hemisphere (ESO) (at the 8-m VLT). The targets identified in low-resolution programs are often followed up at higher resolution with larger facilities, such as Keck and Subaru (Beers and Christlieb, 2005). Future large-scale spectroscopic efforts include Dark Energy Spectroscopic Instrument (DESI), 4-Metre Multi-Object Spectroscopic Telescope (4MOST), Multi Object Optical and Near-infrared Spectrograph (MOONS), William Herschel Telescope Enhanced Area Velocity Explore (WEAVE) and Prime Focus Spectography (PFS). The IR wavelength regime will be particularly important for the future studies. The next generation of extremely large telescopes, such as the Giant Magellan Telescope (GMT), Thirty Meter Telescope (TMT) and Extremely Large Telescope (ELT), will be diffraction limited telescopes at IR wavelengths allowing for adaptive optics (AO) supported multiobject spectroscopy.

6.3 Modelling Atmospheres and Spectra of Cool Stars

Spectroscopy of cool stars greatly relies on models, in particular, on models of stellar atmospheres and model (synthetic) spectra. The models are usually computed in an ab initio fashion, by consistently solving a set of equations to describe the energy balance and transfer in the atmospheres. The predicted observables include the stellar energy distribution (SED), photometric magnitudes, limb-darkening, spatially resolved intensities and detailed line profiles, which can be compared to the observations.

The classical one-dimensional (1-D) models have traditionally assumed several simplifying approximations, the assumption of hydrostatic equilibrium and LTE. Hydrostatic equilibrium describes the balance between gas pressure and gravitational force, $dP/dz = -g\rho(z)$, which can be regarded as the simplest form of the Navier–Stokes equations, ignoring any mass motions.

To make the models more realistic, it is common to factor in other terms, such as radiative and turbulent pressures. The latter, $dP/dz \sim -\rho\nu^2$, where ν is the characteristic velocity. For energy conservation, the assumptions of flux constancy, i.e., no sources or sinks of energy, and radiative equilibrium (for the radiative flux calculations) are typically used. The convective energy transfer is treated using approximate recipes, such as the 'mixing-length theory'. The convective flux is $F_{\text{conv}} \sim a_{\text{MLT}}^2/H_p^{3/2}$, where a_{MLT} is the mixing length and H_p is the scale height, $H_p = -dr/dlnP$. In the hydrostatic approximation, H_p is proportional to $\sim P/\rho g$, the typical distance over which density drops by a factor of e. The mixing length describes the characteristic distance, travelled by a fluid element adiabatically (no exchange of heat energy with the environment), after which the fluid element dissolves into the environment. In stellar atmosphere applications, this motion is nonadiabatic, because of radiative losses that are included in the models.

LTE is a simple approximation, based on the Boltzmann and Saha formulae that describe the distribution of atoms among their internal energy states (excitation and ionization). These approximate formulae are very useful. They allow one to compute the excitation and ionization distributions with just a few parameters of state. The Saha equation has the following form $N_{r+1}/N_r \approx 1/n_e U_{r+1}/U_r e^{-E(Z)/k_B T}$, where U is the partition function, E ionization potential of the ionization stage r of an atom with a charge Z, n_e the concentration (number density) of free electrons and N_{r+1} and N_r the total population densities of two ionization stages. This form of the Saha equation only holds for low-density plasma, when Coulomb interactions between ions and electrons have no impact on the atomic structure, thus the 'screening' of the ionization potential can be neglected.

These simplified 1-D LTE models are broadly used, mainly because of their computational tractability. Huge grids of 1-D LTE plane-parallel stellar atmospheres models have been computed already (Kurucz, 1993; Gustafsson et al., 2008). These models are also useful to explore the effect of individual parameters, such as opacities, on the energy balance.

Recent advances in the field focus on two major directions. New three-dimensional (3-D) models (Collet et al., 2007; Nordlund et al., 2009; Freytag et al., 2012) deal with some major problems of the classical models, through the inclusion of full 3D hydrodynamics, at the expense of some simplifications related to radiative transfer (see Figure 6.4). These models solve the compressible Euler equations for the conservation of

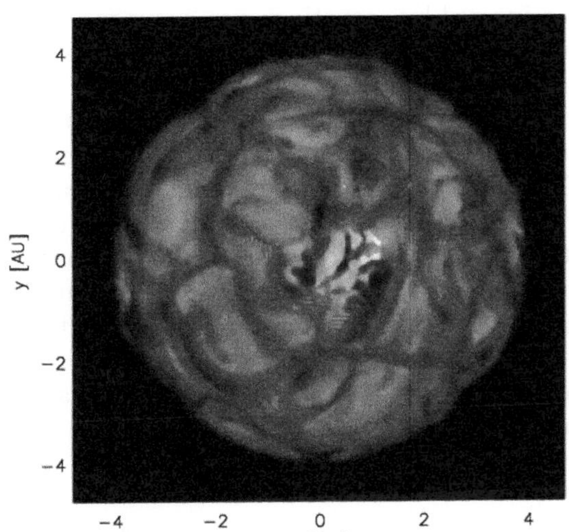

FIGURE 6.4. A 3-D hydrodynamical simulation of a red supergiant star (adapted from Chiavassa et al., 2011).

mass, momentum and energy, and exist in two variations: (a) the box-in-a-star model, which assume that the gravitation potential is constant ($dF/dz = -g$, and applied to solarlike stars); and (b) the star-in-a-box models, which include the entire star in the simulation box, and adopt a spherical gravitational potential with a softened $1/r$ profile (Freytag et al., 2002).[7] The latter variant applies to pulsating stars and stars with very extended envelopes.

Recent studies of stellar convection have shown that the characteristic size of the granules at the surface scales with the pressure scale height (Schwarzschild, 1975; Freytag et al., 1997; Chiavassa et al., 2009; Trampedach et al., 2013; Tremblay et al., 2013). For red supergiants, H_p is huge, because of their very low densities at the surface and large radii, thus in the models of red supergiants enormous convective cells emerge (Chiavassa et al., 2011). The apparent size of granules depends on wavelength that reflects the depth dependence of the granular properties. In the deeper layers, close to the depths where the 1.6 μm (opacity minimum) continuum forms, the cells have a size of 1.8–2.5 AU in diameter. These cells evolve on the timescales of several years and penetrate deep into the star. In the outermost layers, at small optical depths, granules are of only a fraction of an AU in size, and vary on shorter timescales of less than a few months. For giants and supergiants, the photocentre also changes because of variability, which has an impact on the accuracy and estimate of the parallaxes (Ludwig, 2006; Chiavassa et al., 2011).

Solarlike stars are covered by thousands of cells, with smaller diametres of only 10^3 km. The granulation pattern in theoretical models shows remarkable similarity when contrasted with spatially resolved observations of the Sun, not only in terms of the size of granules, but also intensity contrast and variability, suggesting that the simulations offer a realistic description of the physics of these stars (Nordlund et al., 2009).

The 2-D and 3-D convective simulations have also been used to explore the mechanism of pulsations consistently with convection (Freytag and Höfner, 2008; Freytag et al., 2017). Early 1D simulations assumed a sinusoidally moving piston, i.e., arbitrarily injected kinetic energy at the bottom to simulate the stellar pulsations (numerically, variable inner boundary conditions). In contrast, 3-D radiative-hydrodynamics models show pulsations on the timescales of years, i.e., global periodic variations of the physical surface properties, which are usually referred to as 'pulsations'.[8] These models also show that shocks, which are triggered by acoustic waves, move material from the inner regions to the outer regions, where dust forms and the radiative pressure on dust leads to stellar winds, consistent with observations of AGB stars.

Recently, it has become possible to perform detailed spectrum-synthesis calculations using 2-D radiation-hydrodynamics (RHD) simulations of short-period Cepheids (Freytag et al., 2012). The RHD simulations are still limited in their treatment of radiative transfer by including grey opacities, but they are capable of producing self-excited oscillations due to the κ mechanism starting from initial hydrostatic conditions.

This enables a test of the validity of the quasistatic approximation, which has been used to study pulsating stars for decades. Intriguingly, new results employing RHD models show that the hydrostatic 1-D model atmospheres may be used to provide unbiased estimates of stellar parameters, but only for the phases of maximum expansion and the beginning of the compression phase, corresponding to the photometric phases of 0.3–0.65

[7] The potential in the core region is softened to $1/\sqrt{(}r^2 + a^2)$, where a is the softening parameter.

[8] Note that in these simulations it is difficult to distinguish radial from nonradial modes, since the change in volume is accompanied by a change in shape.

(where 0 is set to be the maximum light phase) (Vasilyev et al., 2017). It has also been shown that the long-standing problem of the 'K-term', i.e., the putative residual negative line-of-sight velocity of the Galactic Cepheids (Parenago, 1945; Stibbs, 1956; Pont et al., 1994; Nardetto et al., 2009) is mainly the result of a convective blueshift of spectral lines (Vasilyev et al., 2017).

Another important ingredient of stellar atmosphere models is NLTE. This term involves a large number of physical processes that describe the various channels through which gas particles interact with the radiation field, including excitation, ionization, recombination and charge transfer (Barklem, 2016). NLTE calculations are now possible for very large atomic models, such as those of neutral iron or cobalt (Bergemann et al., 2010). The available theoretical evidence indicates that the effects of deviations from LTE are typically not negligible, and vary both in sign and in magnitude. Whereas the spectral lines of the lighter elements, such as Li and O, usually exhibit negative NLTE effects (in the sense that NLTE abundances are lower than LTE ones), the optical lines of Fe-peak elements show positive effects. The situation is then reversed for the lines of even heavier species, like Cu, which again exhibit increasingly negative departures from LTE. These are mainly related to the abundance of the element, the ionization potential of the ion in a given ionization stage and the excitation potential of the energy levels involved in the transition. Fe-peak elements are mostly affected by overionization, whereas lighter species and the lines of singly ionized (majority species) are mainly sensitive to resonance-line scattering. The analysis of NLTE diagrams, such as in Figure 6.5, indicate that the NLTE effects are not completely random, but follow certain regularities, permitting predictions of the magnitude of the NLTE corrections even for the elements for which detailed calculations are not yet available. In particular, this behaviour suggests that the NLTE corrections for the optical Ni lines should be minor and positive, of ~ 0.05 dex. For heavier species with higher ionization potentials, slightly negative NLTE effects are expected.

It is now also possible to combine 3-D models and NLTE, either in the form of NLTE calculations with mean 3-D ($<$ 3-D$>$) models (Osorio and Barklem, 2016; Bergemann et al., 2017a) or full 3-D calculations, where radiative transfer is performed through 3-D cubes of physical parameters (Amarsi and Asplund, 2017). The available evidence is still sparse, however, as comparison of mean 3-D NLTE with 1-D LTE, as well as with full 3-D NLTE abundances, has only been performed for a limited number of chemical elements. The results indicate that, for the solarlike stars, full 3-D NLTE, <3-D> NLTE and 1-D NLTE results are consistent within the standard error of abundance determination

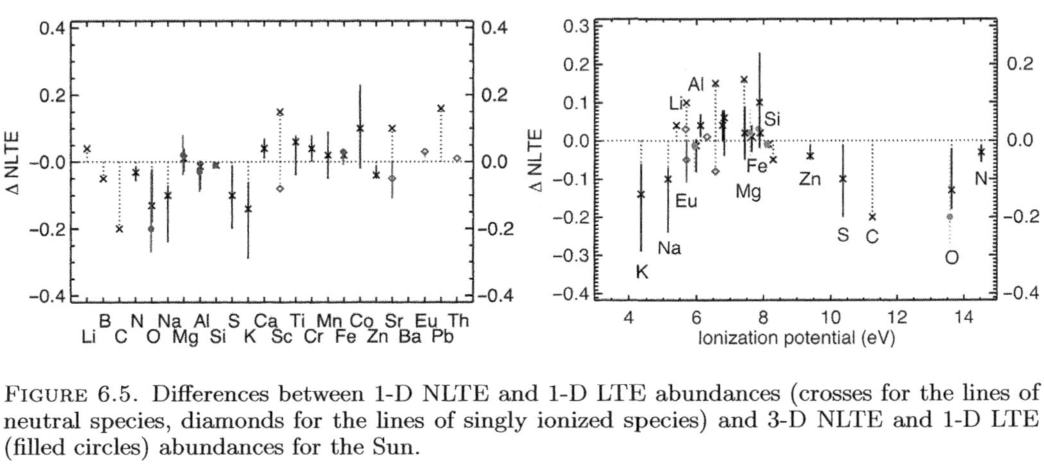

FIGURE 6.5. Differences between 1-D NLTE and 1-D LTE abundances (crosses for the lines of neutral species, diamonds for the lines of singly ionized species) and 3-D NLTE and 1-D LTE (filled circles) abundances for the Sun.

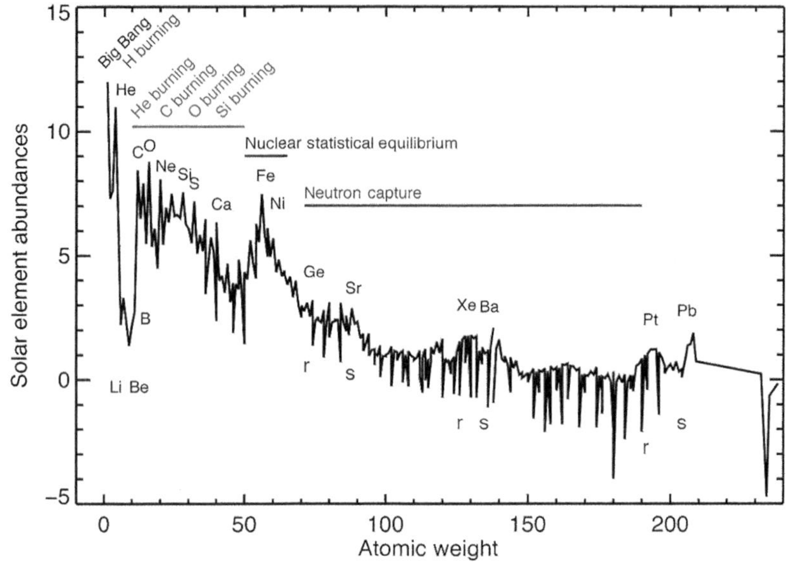

FIGURE 6.6. The abundances in the solar photosphere in the logarithmic scale (based on Lodders et al., 2009). The abundance of hydrogen is set to 12. The main element-production channels are indicated.

(i.e., within 0.03 dex or even less), for Li, O, Na, Mg, Al and Fe (Figure 6.5). Comparing this to the typical differences between 1-D LTE and 1-D NLTE estimates, which may differ by an order of magnitude, the available evidence suggests that full 3-D NLTE calculations and NLTE calculations with averaged 3-D models, as well the NLTE analysis with the standard 1-D hydrostatic models, provide a solid basis for high-precision abundance diagnostics of late-type stellar spectra.

Finally, it is important to assess the accuracy of the models when applied to different regimes of the electromagnetic spectrum. For the Sun, there is sufficient evidence that chromospheric radiation transport plays an increasingly important role in the UV (Hauschildt et al., 1999; Hall, 2008; Linsky, 2017). The chromosphere is also visible in the cores of the strong lines, which is a particularly limiting factor in the infrared, where the number of spectral lines is far lower than at optical wavelengths. Whether the UV region is appreciably affected by the chromospheric radiation transport in stars other than the Sun is not yet well understood. Observational evidence suggests that the chromosphere may affect even very metal-poor stars such as HD 84937 ([Fe/H]= −2.0; Spite et al. 2017). Because of the absence of ab initio chromospheric models for such stars, no conclusions on the accuracy of the UV diagnostics can yet be drawn.

6.4 A Recap on Stellar Nucleosynthesis

Stars are the factories for all chemical elements beyond H and He in the Universe. Production of chemical elements occurs in stars of different types and different evolutionary phases, as different conditions are needed to reach the necessary temperatures and pressures to ignite specific nuclear reactions. Theoretical evidence, corroborated by meticulous observations of element abundances in supernovae (SNe), AGB and novae ejecta (José and Iliadis, 2011), indicates that there are several major groups of chemical elements that are coproduced under similarconditions and thus can be used to trace specific production sites. These sites are described in the following, and are shown schematically on the cosmic abundance chart (Figure 6.6).

- Hydrogen (H, and deuterium, D), He, a tiny amount of Li and minuscule fractions of B and Be are produced in Big Bang nucleosynthesis (BBN), during the first few minutes, when the temperature drops from 10^{32} K to 10^{9} K. Li production is still not entirely understood, and it is very sensitive to the photon-to-baryon number in the BBN (Steigman, 2007), which offers an interesting way to constrain the Big Bang models through observations of Li abundances in the oldest stars (Sbordone et al., 2010). Stellar models suggest that ^{7}Li can be destroyed by hot-bottom-burning (Lattanzio et al., 1996). Also, recent observations provide evidence for the production of Li in novae (Tajitsu et al., 2015). An important astrophysical puzzle is the depletion of Li in the oldest Galactic stars; the abundance of Li in their atmospheres is a factor of three to four less than what is expected from primordial nucleosynthesis (Fields, 2012). B and Be are produced mainly by cosmic ray spallation reactions. Some solarlike young stars show extreme underabundance of Li and Be, which has been one of the prime puzzles in stellar evolution, but recent models with episodic bursty accretion and rotational mixing appear to resolve the puzzle (Viallet and Baraffe, 2012).
- Carbon, nitrogen and oxygen are formed in intermediate-mass stars (AGB phase) and in massive stars. Carbon is produced via triple-α reactions during He burning, which has a low probability and only works because of a fundamentally crucial Hoyle resonance, an excited energy state of ^{12}C, which matches exactly the interaction energy of ^{8}Be and an α-particle. Nitrogen can be formed in two ways; if it is formed directly from He, it is called primary nitrogen. This can take place in AGB or massive stars and requires a large degree of stellar rotation and mixing (Chiappini et al., 2005; Spite et al., 2005). If, on the other hand, N is formed at the expense of C and O in the CNO cycle, it is referred to as secondary N (Spite et al., 2005).
- The production of alpha elements goes on via a sequence of ^{4}He captures on nuclei. The α-capture on ^{12}C makes stable ^{16}O, releasing some energy. This is then followed by the α-ladder, producing stable isotopes of Mg, Si, S, Ca and unstable isotopes ^{44}Ti (decays to ^{44}Ca), ^{48}Cr (decays to ^{48}Ti), ^{52}Fe (decays to ^{52}Cr) and ^{56}Ni. The latter has the maximum binding energy per nucleon and decays to the dominant isotope of iron, ^{56}Fe.
- The odd-Z elements ^{23}Na and ^{27}Al are produced during the carbon- and neon-burning phases in intermediate-mass and massive stars, $M > 3\ M_{\odot}$. They are thought to be so-called secondary nuclei, requiring preexisting neutron-rich seed nuclei (Gehren et al., 2006).
- Fe-peak elements require very high temperatures and pressures for their production. Furthermore, neutron-rich (odd-Z) elements, such as Mn and Co, also depend on the neutron enrichment. These conditions are usually satisfied during the late stages of the evolution of massive stars (which are then composed of multiple shells, in which nuclear burning takes place) and during explosions. Recent models of SN Ia explosions (Seitenzahl et al., 2013; Seitenzahl and Townsley, 2017) indicate that the production of Mn is an extremely sensitive tracer of Type Ia SNe. A fraction of Ti and Cr is produced during hydrostatic O burning. Heavier elements are produced in explosive conditions in supernovae, when the outward propagating shock wave heats the matter to sufficiently high temperatures necessary for the production of these high-Z species. Some species are produced by incomplete Si burning, such as Cr and Mn, while others are produced in the region of complete Si burning (e.g., Co) and α-rich freeze-out from nuclear statistical equilibrium (Thielemann et al., 2007). The

difference between the two lies essentially in the peak temperature in the burning layer. Incomplete Si burning typically occurs at $4 \cdot 10^9$ K $<$ T $<$ $5 \cdot 10^9$ K, whereas compete Si burning occurs at T $> 5 \cdot 10^9$ K (Nomoto et al., 2005; Thielemann et al., 2007).

- Thermonuclear fusion does not produce elements beyond Fe. The reason is that charged-particle reactions past ^{56}Ni are endoergic, i.e., they no longer produce energy, but consume it. Thus, most elements heavier than iron form by capturing neutrons onto heavy existing isotopes and subsequent β-decays. Only very few proton-rich nuclei are produced by a *p*-process,[9] which may be associated with core-collapse supernovae[10] or SN Ia.

There are two main channels through which the elements beyond the Fe peak can be formed, either through a slow neutron-capture (*s*-) process, where the capture rate is much longer than the subsequent beta-decay rate, or the rapid neutron-capture (*r*-) process, where the captures happen much faster than the decay, driving the process far from the so-called valley of stability.

- The *s*-process is a secondary process, requiring preexisting seed nuclei, and it typically takes place in AGB stars, which burn H and He in shells. The He-rich layer, which separates the two shells, hosts ^{13}C-rich pockets, which form through captures of protons from the convective envelope onto ^{12}C nuclei. Then, the reaction ^{13}C(α,n)^{16}O produces free neutrons, which are captured onto existing seeds, typically Fe-peak nuclei, creating nuclei with atomic numbers $A \sim 90 - 205$. This process is thought to be responsible for the main *s*-process component. In more massive stars, M$>$ 15 $M_{\odot}$, *s*-process nucleosynthesis can take place through the ^{22}Ne(α,n) ^{25}Mg reaction, at higher temperatures and densities (Frischknecht et al., 2016). This process is thought to produce the weak *s*-process component ($A < 90$). None of these processes can be observed directly, and to date there remain many unknowns associated with them, where the size of the ^{13}C-pockets remain one of the largest uncertainties. Recent studies detected the radioactive Tc, which is only produced by the *s*-process in pulsating AGB stars (Merrill, 1952; Neyskens et al., 2015).
- Approximately half of the heavy elements are, however, not formed by the *s*-process, but are attributed to the *r*-process, a primary process that can form the heaviest isotopes starting out with just protons and many neutrons. This means that no preexisting heavy isotope is needed in this process. The *r*-process requires extreme conditions, such as very high neutron densities ($> 10^{23}$cm^{-3}; Kratz et al., 2007), high temperatures, high entropies and fast expansion of the explosion. Previously, the *r*-process was thought to be primarily associated with core-collapse SNe explosions. However, recent 2-D SNe models have shown that the 9-11M$_{\odot}$ SNe are not sufficiently neutron rich to produce elements heavier than Sn, i.e., $Z \geq 50$ (Wanajo et al., 2011).

Recent observations of the 'kilonova' SSS17a (Drout et al., 2017; Kasen et al., 2017; Pian et al., 2017; Shappee et al., 2017; Smartt et al., 2017) provide strong indications that the *r*-process is likely to be hosted by the merger of two neutron stars (NSM), an event that has recently received a great deal of attention because of its association with the gravitational wave detection GW170817 by the advanced LIGO–VIRGO collaboration. NSMs are extremely rich in neutrons and can easily host the *r*-process. A merger of two neutron stars emits gravitational waves, produces a kilonova event

[9] Only 35 stable p-nuclei exist (Travaglio et al., 2018).

[10] This could be a νp-process taking place in core-collapse supernovae as proposed by, e.g., Fröhlich et al. (2006).

(Metzger et al., 2010) and powers a short gamma ray burst. The afterglow of a kilonova is powered by the radioactive decay of unstable heavy isotopes, and a boost in the abundance of lanthanides detectable in the near-IR is expected (Tanvir et al., 2013; Martin et al., 2015). However, one should keep in mind that neutron stars are also the endpoints of the evolution of massive stars, and so far it has not been possible to separate the contribution of heavy element material from rare core-collapse SNe explosions or ejecta of a NSM based on observations (Thielemann et al., 2017). Massive supernovae can also form the r-process elements, provided they have strong magnetic fields and explode asymmetrically with jets (Nishimura et al., 2017; Mösta et al., 2018). This needs to be probed in more detail with refined models and observations.

6.5 Stellar Populations and Galactic Chemical Evolution (GCE)

The chemical abundance measurements in stars, as well as theoretical studies of element production, have an important application in astrophysics – studies of the chemical evolution of stellar populations and galaxies.

As massive stars evolve faster, their ejecta will pollute the interstellar medium (ISM) earlier (on shorter timescales) than is the case for lower-mass stars. The latter take longer to evolve and to eject their processed material. Therefore, lower-mass stars will experience a time delay with respect to more massive objects, which end their lives as SNe, leaving behind neutron stars, black holes or nothing at all. Thus, by observing stars with different masses and ages, we can study the chemical evolution of galaxies, and, hence, the associated processes that influenced their formation, such as mergers, infall and outflows, galactic winds, the initial mass function and star-formation activity (Tinsley, 1979; Lacey and Fall, 1983; Goetz and Koeppen, 1992). On the other hand, old stars, which are also expected to be the most metal-poor ones, probe the first heavy element pollution of the ISM after the Big Bang, and they are excellent testbeds of stellar yields, the first initial mass function and early stellar evolution (Bromm and Larson, 2004; Matteucci and Calura, 2005).

A few observational trends stand out clearly when we study Galactic Chemical Evolution. Stars with metallicity below [Fe/H]~ -2 show a large star-to-star abundance scatter in neutron-capture elements, such as Ba, compared to elements such as Mg, which exhibit little scatter (Figure 6.7). Since most of these very metal-poor stars are halo stars, when identified by their space motions (kinematics),[11] this observation provides constraints on the Galactic halo formation. In particular, the heavy-element scatter may indicate that the Galaxy was inhomogeneous in the early stages, and the halo could have hosted transient star-forming regions that were dense and sufficiently massive to sustain prolonged self-enrichment by Type II and Type Ia SNe (Gilmore and Wyse, 1998; Karlsson, 2005; Karlsson and Gustafsson, 2005).

The scatter in s-process elements in stars with [Fe/H] < -2 may also indicate that the massive objects might have ejected different amounts of elements, i.e., having yields that are dependent on mass and metallicity (Maeder, 1990). The fact that the scatter decreases at metallicities above [Fe/H]≈ -2 is consistent with the general enrichment dominated by several distinct nucleosynthesis channels, and reflects the background metallicity level. The relative pollution by an individual object becomes less significant as

[11] Halo stars exhibit a wide range of orbital eccentricities, including low and high eccentricity. Compared to the disc stars, which have exclusively low-eccentricity orbits, halo stars also have much larger velocity dispersions.

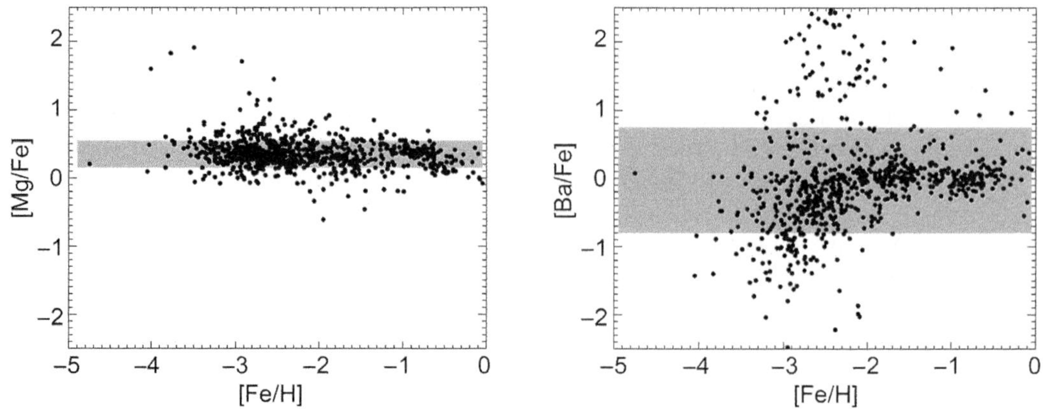

FIGURE 6.7. Left panel: [Mg/Fe] vs. [Fe/H], right panel: [Ba/Fe] vs. [Fe/H]. Note the difference in star-to-star scatter as indicated by the standard deviation around the mean (grey-shaded region). Data from the Frebel (2010) compilation. Figure adapted from Hansen et al. (2014).

the base metallicity level increases. This is best seen in the distributions of, e.g., elements produced by the *s*- and *r*-process. The change in scatter could also be related to the onset of AGB pollution, as these stars will have had sufficient time to evolve and eject material rich in C, N and *s*-process elements (Busso et al., 1999).

Under closer inspection, observations suggest that the Galactic halo population is even more complex. It has now been firmly established that galaxy halos form through mergers and accretion of subhalos (Searle and Zinn, 1978; White and Frenk, 1991; Bullock and Johnston, 2005). This implies not only that some scatter in chemical abundances of the halo stars is expected, but also that chemical-abundance diagrams may contain specific 'features', indicating stellar populations which were not formed in situ in the Milky Way, but were added to our Galaxy over time. The known dwarf galaxies exhibit a very large range of star formation histories (Tolstoy et al., 2009; Kirby et al., 2011; Leaman, 2012). Recent observations indicate that the halo indeed has a metal-rich ([Fe/H] ~ -1) stellar component that hosts low-[α/Fe] stars (Nissen and Schuster, 1997, 2010), suggesting their origin in dwarf galaxies. The metal-rich part of the halo and the thick-disc stars overlap in metallicity, with the thick disc extending to [Fe/H] ~ -1.8. However, in contrast to the thick-disc population, the halo stars show an inverse correlation between [Fe/H] and [α/Fe] (Bergemann et al., 2017b); the slope of the trend line is important for chemical-evolution models.

The structure of the Galactic halo is still a matter of a debate. Carollo et al. (2007, 2010) and Beers et al. (2012) have argued that 'the halo' comprises both inner-halo and outer-halo populations of stars, based on their observed spatial distribution, kinematics and chemical abundances (the inner-halo population peaking at metallicity [Fe/H] ~ -1.6; the outer-halo population peaking at [Fe/H] ~ -2.2). This evidence may suggest that the populations mark different eras of formation, and differences in the mass distributions of the subgalactic fragments (minihalos) that merged or were accreted during their formation. Some numerical galaxy-formation simulations lend support to this view (Font et al., 2011; McCarthy et al., 2012; Tissera et al., 2013, 2014, 2018). However, others have questioned this interpretation (Schönrich et al., 2011, 2014; Deason et al., 2013), suggesting that the secondary (retrograde rotating) halo component should be much weaker, if present.

Generally, the interpretation of chemical abundance-age-kinematics diagrams in the context of Galaxy formation scenarios is nontrivial. Although the overall abundances of heavy elements generally increases with time, detailed observations show that the metallicity of stars does not uniquely correlate with age, neither during the Galactic halo nor during disc formation (Holmberg et al., 2009; Bergemann et al., 2014). Galaxy-formation simulations suggest that the oldest stars are expected to be found in the Galactic bulge, the most metal-rich component of the Milky Way (White and Springel, 2000; Brook et al., 2007; Starkenburg et al., 2017; Griffen et al., 2018).

The apparent distribution of stars in the chemical-abundance plane is, however, highly sensitive to the selection function of the observational program (Bergemann et al., 2014; Thompson et al., 2018). 'Monometallic' or 'monoabundance' stellar populations, i.e., groups of stars exhibiting similar abundances, do not necessarily imply that they are coeval, or even formed within the same Galactic component. Conversely, stars formed in the halo, in the disc or in the bulge do not possess a well-defined characteristic signature in the chemical-abundance plane, and even less so in kinematics. In particular, and in contrast to the surface chemical composition of unevolved stars, stellar kinematics does change after stellar birth (Binney and Lacey, 1988). Stars experience dynamical disturbances in the Galaxy that change their orbits, and this kinematical rearrangement makes it difficult to tag stars and identify their parent population. Stars formed in the Galactic centre can be relocated into the outer bulge or inner halo (El-Badry et al., 2018), those formed in the disc plane acquire large altitudes above the plane (Laporte et al., 2017; Bergemann et al., 2018), and disc stars themselves experience radial migration via scattering of giant molecular clouds and transient spiral arms (Sellwood and Binney, 2002; Schönrich and Binney, 2009a,b).

More metal-rich stars, mostly members of the disc, show another interesting correlation: The relative abundances of their α-elements, $[\alpha/\mathrm{Fe}]$, indicate a break at $[\mathrm{Fe/H}] \sim -1$. This is commonly referred as a *knee*, and is thought to arise because of the onset of SN Ia pollution (Matteucci and Greggio, 1986). SN Type Ia events result from mass transfer in a binary system of low-mass stars. They have a typical time delay of ~ 1 Gyr, although there is extragalactic evidence for a fast SN Ia channel (Mannucci et al., 2006). SN Ia produce substantial amounts of Fe-peak elements, but very little α-elements (Figure 6.8), resulting in the downward trend. The proposed dual structure of the Galactic disc, which is seen in the $[\alpha/\mathrm{Fe}]$ versus [Fe/H] in some elements, but not in the others, is still a matter of debate (Lee et al., 2011; Bergemann et al., 2014; Mikolaitis et al., 2014). Bensby et al. (2014) found the bimodal distribution in [Ti/Fe], but not in [Mg/Fe], whereas Hawkins et al. (2016) observe bimodality in [Mg/Fe] but not in [Ti/Fe]. The old unsettled problem of whether the thick disc is a stand-alone population with a formation history different from that of the thin disc (Chiappini et al., 1997) is, in turn, complicated by the apparent complex structure of the former being not a monolithic population (Spagna et al., 2010; Curir et al., 2012; Schönrich and McMillan, 2017). Volume complete, statistically significant samples of stars need to be analysed in high-quality spectra and with updated models, to control for systematic errors and settle this debate.

6.5.1 *Old and Unique Chemical Tracers*

Among the low-metallicity stars in the halo there exist a number of stars with particularly important abundance patterns that are shaping our view of the origin of the elements.

The star CS 22982-052, one of the most strongly r-process-enhanced stars (an r-II star, following Beers and Christlieb, 2005, with $[\mathrm{Eu/Fe}] > +1.0$), was analysed most recently by Sneden et al., 2003. Up until 2013, this was the star for which we had

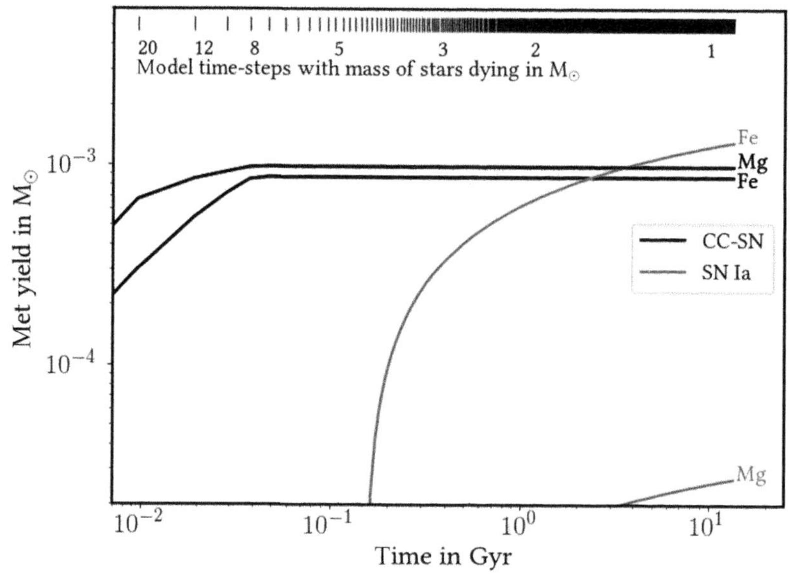

FIGURE 6.8. The evolution of the net yields of Mg and Fe in a simple stellar population (adapted from Rybizki et al., 2017). Reproduced with permission © ESO.

the most complete abundance information after the Sun. However, Siqueira Mello et al. (2013), based on both HST- and ground-based high-resolution spectra, analysed the r-II star CS 31082-001, and obtained measurements for a total of almost 70 stable elements in the periodic table (including the 'jewelry store' elements silver, gold and platinum, as well as the radioactive species thorium and uranium). Although these stars exhibit essentially identical elemental-abundance patterns among the heavy elements from Ba to Hf (matching a scaled-solar r-process distribution), they differ dramatically in their [Th/Eu] abundance ratios, a phenomenon referred to as the 'actinide boost' (see, e.g., Mashonkina et al. 2014). Roughly 30% of r-process-enhanced stars exhibit the actinide boost, which may provide a fundamental clue to the origin of the r-process. Over the past few decades, a total of $\sim 25-30$ r-II stars have been identified. This number is expected to dramatically increase in the near future due to dedicated survey efforts now under way, e.g., the R-Process Alliance effort (Hansen et al., 2018b), whose goal is to bring the number of known r-II stars up to on the order of 100 or more. Such large, 'statistical' samples are required to search for the subtle elemental-abundance signatures that may be used to differentiate different astrophysical sites of the r-process.

By way of contrast, there also exist stars that appear highly deficient in their light and heavy r-process elements, relative to a scaled-solar pattern. HD 88609 and HD 122563 are two such stars (Honda et al., 2007) that have been used to support the operation of a 'weak' or 'limited' r-process, decoupled from the 'main' r-process (Hansen et al., 2012, 2014; Frebel and Beers, 2018).

To date, the most Fe-poor star yet found was discovered by Keller et al. (2014). This CEMP star is so Fe-poor that it only has an upper limit of $[Fe/H] < -7.5$. Comparison to yield predictions from supernovae (e.g., Heger and Woosley, 2002) showed that the chemistry matched that of the ejecta from a 60 $M_\odot$ Type II SN. However, later 3-D and NLTE corrections to both molecular and atomic line abundances indicated a slightly higher metallicity, as well as a less massive progenitor (Bessell et al., 2015; Nordlander

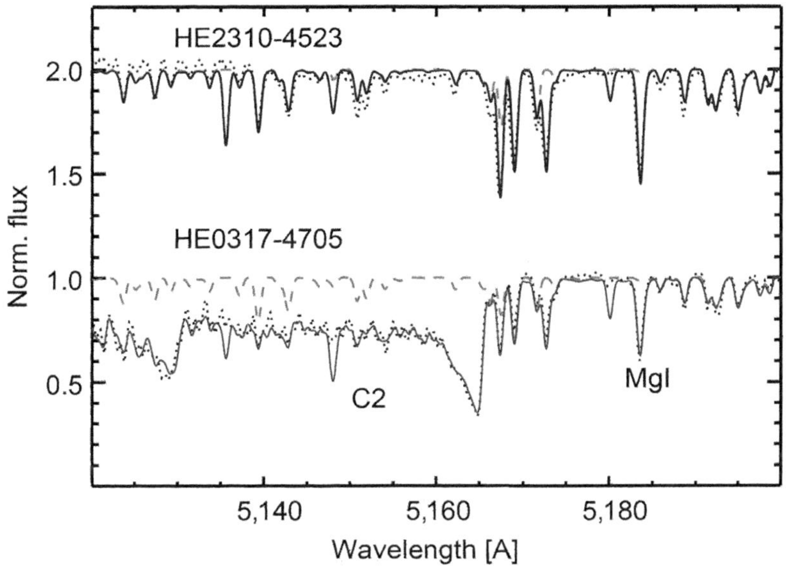

FIGURE 6.9. Observations of two stars (dots) with similar stellar parameters (and [Fe/H]= –2.5) but very differing chemical composition. Top: HE 2310-4523 compared to synthetic spectra (solid line) with [C/Fe] = +0.2 and [Mg/Fe] = +0.25 (Hansen et al. 2016), bottom: HE 0317-4705 and synthetic spectra (solid line) with [C/Fe] = +1.4, [Mg/Fe] = +0.5 Hansen et al. (2019). The dashed line represents no C and no Mg.

et al., 2017). Another interesting, hyper-metal-poor star, SDSS J102915+172927, was recognized a few years earlier by Caffau et al. (2011), from high-resolution spectroscopic follow-up of extremely metal-poor turnoff stars identified in the Sloan Digital Sky Survey. This star is unique in the sense that it has [Fe/H] ~ -5, but unlike almost all other such extreme stars, the star appears to lack a signature of carbon enrichment[12] (only weak upper limit on [C/Fe] has been reported). Recently, additional ultra, hyper and mega ([Fe/H] < -6) metal-poor TO stars from SDSS have been reported by Aguado et al. (2018), and references therein, all of which exhibit carbon enhancement (or have upper limits on [C/Fe]), indicating that they are CEMP stars.

Significant fractions of very metal-poor stars are in fact known to be CEMP stars. These stars can be further divided into subclasses based on their neutron-capture elemental-abundance patterns. CEMP-*s* stars are the most populated subclass, where the elements representing the slow neutron-capture process are prominent. At least 80% of the CEMP-*s* stars are binaries (Starkenburg et al., 2013; Hansen et al., 2016c), and almost certainly are associated with mass-transfer events from a former AGB companion. However, several stars in this subclass are suspected to be single stars, which might be explained by massive-star pollution (Choplin et al., 2017) of the natal cloud from which the low-mass star observed today formed. The fraction of single CEMP-*s* stars is still under debate, as these may suffer from observational biases and reside in wide binary systems with periods longer than 10^4 days (Abate et al., 2018). In contrast, the CEMP-no stars are characterized by a lack of neutron-capture overabundances. A number of lines of evidence point to their formation very early in the history of the Universe, and indeed some CEMP-no stars may well represent bonafide second-generation stars that have locked up

[12] Starkenburg et al. (2018) found a similar star in their Pristine survey.

the nucleosynthesis products of the very first (massive) stars (see Hansen et al., 2016a, and references therein). To date, a handful of stars have been identified as CEMP-r stars, which, in addition to their carbon enhancement, exhibit strong overabundances of r-process elements (interestingly, this class includes the canonical r-II star CS 22892-052). Their origin is still very much under discussion. Finally, a hybrid class, known as CEMP-r/s stars, includes carbon-enhanced stars with both r- and s-process elemental-abundance signatures present in their spectra. At first, it proved difficult to explain how both the r- and s-process enriched material got mixed into these stars; however, recent models have shown that a nucleosynthetic process, the so-called intermediate neutron-capture process (i-process), which may operate in AGB or massive stars, might explain this abundance pattern (Dardelet et al., 2014; Hampel et al., 2016; Clarkson et al., 2018).

Although the signature of increasing fractions of CEMP stars with decreasing [Fe/H] has been recognized for many years (Beers and Christlieb, 2005, and references therein), the most recent evaluation of their frequencies have been revised upwards by as much as a factor of two, compared with the previous result (Lee et al., 2013).[13] Two CEMP classes can be distinguished in a relatively simple way. Spite et al. (2013) and Bonifacio et al. (2015) demonstrated that the two dominant classes of CEMP stars have different enhancements of absolute C abundances ($A(\mathrm{C})$ = logϵ(C)=log (N_C/N_H) + 12). The CEMP-s stars exhibit a high level of $A(\mathrm{C})$, mainly owing to C-rich material being transferred in a binary system, while the CEMP-no stars (typically not binaries) have lower A(C).

Yoon et al. (2016) demonstrated that the morphology of the $A(\mathrm{C})$-[Fe/H] space is richer still, separating out three groups of CEMP stars, the Group I stars comprising CEMP-s stars, and the Group II and Group III stars comprising CEMP-no stars with apparently different progenitors ors and/or enrichment histories. Yoon et al. further showed that CEMP-s and CEMP-no stars can be distinguished based on medium-resolution spectroscopic measurement of $A(\mathrm{C})$ alone, with a success rate commensurate with that obtained from high-resolution spectroscopic measurement of Ba. From application of this approach, one can clearly see the very different behaviours of CEMP-no and CEMP-s stars in the revised frequency diagram (Figure 6.10). A recent study (Hansen et al., 2019) showed that Sr and Ba can be used, even in low-resolution spectra, to separate all CEMP subgroups. Comparison of these derived frequencies with predictions from population synthesis and numerical galaxy assembly models should prove illuminating.

6.6 Conclusions and Future Outlook

We have endeavoured to provide a brief overview of an extremely broad set of topics, and in truth have only brushed the surface. Many of the fields discussed in the present chapter are advancing rapidly, and will remain vibrant for a long time. It is easy to forget that no more than a century has passed since humans came to understand that the Milky Way is but one of perhaps 100 billion galaxies in the Universe, and that nuclear fusion powers individual stars over their lifetimes, no matter how brief or how long. As of this writing, the Gaia mission is making its second public data release, with new, precise information on distances, proper motions and spectral energy distributions for more than

[13] This revision arises for two primary reasons. First, CEMP stars near the TO were undercounted in the Lee et al. (2013) study, due to the diminished strength of the CH G-band for stars with $T_{\mathrm{eff}} > 5750\,\mathrm{K}$, and corrections needed to be applied to cooler RGB stars to account for the effect of carbon dilution from first dredge-up. The Yoon et al. (2018) study includes only SG and RGB stars, with the correction of Placco et al. (2014) applied.

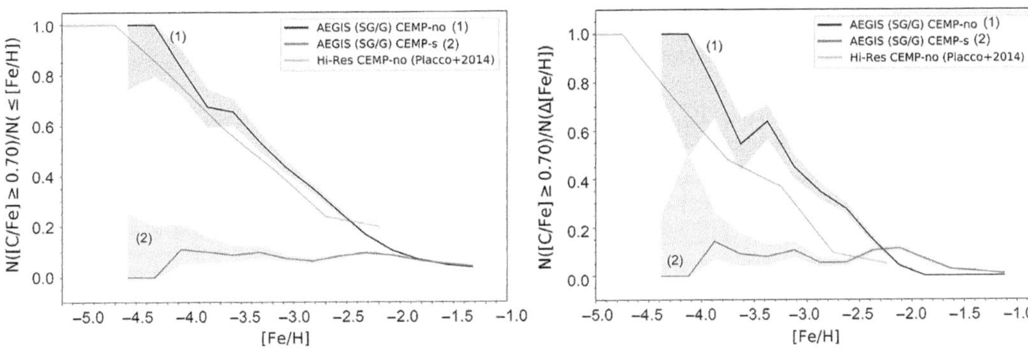

FIGURE 6.10. The cumulative (left) and differential (right) CEMP frequencies, as a function of [Fe/H], plotted separately for CEMP-no and CEMP-*s* stars, classified from medium-resolution spectroscopy. The light grey line indicates the frequencies of CEMP-no stars for a sample of stars with available high-resolution spectroscopic information, from Placco et al. (2014). Figure adapted from Yoon et al. (2018). © American Astronomical Society (AAS). Reproduced with permission.

a billion stars in the Galaxy, and radial velocities for many millions of the brighter stars. Significantly more detailed information will follow once the full dataset is acquired and distributed. This represents the beginning of a new era, where fundamental information for the stars of the Milky Way will enable researchers to confront outstanding issues with a level of detail that was not imaginable a few decades ago.

Our ability to model many aspects of the evolution of stars, in particular those that can be used to infer the nature of the earliest generations born in the Universe, is still in relative infancy. Present and future massive photometric and spectroscopic surveys of resolved stellar populations in the Milky Way will no doubt spur these modelling efforts to take on even greater challenges. The success of these efforts depends crucially on the contributions from astronomers and physicists who are only now beginning their careers, so they must be encouraged and nurtured in order to carry this quest for understanding to the next level.

REFERENCES

Abate, C., Pols, O. R. and Stancliffe, R. J. 2018. Understanding the Orbital Periods of CEMP-s Stars. *A&A*, **620**(November), A63

Aguado, D. S., Allende Prieto, C., González Hernández, J. I. and Rebolo, R. 2018. J0023+0307: A Mega Metal-Poor Dwarf Star from SDSS/BOSS. *ApJ*, **854**(February), L34

Allende Prieto, C., Lambert, D. L., Tull, R. G. and MacQueen, P. J. 2002. Convective Wavelength Shifts in the Spectra of Late-Type Stars. *ApJ*, **566**(February), L93–L96

Amarsi, A. M. and Asplund, M. 2017. The Solar Silicon Abundance Based on 3D Non-LTE Calculations. *MNRAS*, **464**(January), 264–273

Asplund, M., Grevesse, N., Sauval, A. J. and Scott, P. 2009. The Chemical Composition of the Sun. *ARA&A*, **47**(September), 481–522

Barklem, P. S. 2016. Accurate Abundance Analysis of Late-Type stars: Advances in Atomic Physics. *A&A Rev.*, **24**(May), 9

Beers, T. C. and Christlieb, N. 2005. The Discovery and Analysis of Very Metal-Poor Stars in the Galaxy. *ARA&A*, **43**(September), 531–580

Beers, T. C., Carollo, D., Ivezić, Ž., et al. 2012. The Case for the Dual Halo of the Milky Way. *ApJ*, **746**(February), 34

Bensby, T., Feltzing, S. and Oey, M. S. 2014. Exploring the Milky Way Stellar Disk: A Detailed Elemental Abundance Study of 714 F and G Dwarf Stars in the Solar Neighbourhood. *A&A*, **562**(February), A71

Bergemann, M. and Gehren, T. 2008. NLTE Abundances of Mn in a Sample of Metal-Poor Stars. *A&A*, **492**(December), 823–831

Bergemann, M., Collet, R., Amarsi, A. M., Kovalev, M., Ruchti, G. and Magic, Z. 2017a. Non-local Thermodynamic Equilibrium Stellar Spectroscopy with 1D and <3D> Models. I. Methods and Application to Magnesium Abundances in Standard Stars. *ApJ*, **847**(September), 15

Bergemann, M., Collet, R., Schönrich, R., et al. 2017b. Non-Local Thermodynamic Equilibrium Stellar Spectroscopy with 1D and <3D> Models. II. Chemical Properties of the Galactic Metal-Poor Disk and the Halo. *ApJ*, **847**(Sept.), 16

Bergemann, M., Gallagher, A. J., Eitner, P., et al. 2019. Observational Constraints on the Origin of the Elements. I. 3D NLTE Formation of Mn Lines in Late-Type stars. *arXiv e-prints arXiv*:1905.05200

Bergemann, M., Lind, K., Collet, R., Magic, Z. and Asplund, M. 2012. Non-LTE Line Formation of Fe in Late-Type Stars I. Standard Stars with 1D and <3D> Model Atmospheres. *MNRAS*, **427**(November), 27–49

Bergemann, M., Pickering, J. C. and Gehren, T. 2010. NLTE Analysis of CoI/CoII Lines in Spectra of Cool Stars with New Laboratory Hyperfine Splitting Constants. *MNRAS*, **401**(January), 1334–1346

Bergemann, M., Ruchti, G. R., Serenelli, A., et al. 2014. The Gaia-ESO Survey: Radial Metallicity Gradients and Age-Metallicity Relation of Stars in the Milky Way Disk. *A&A*, **565**(May), A89

Bergemann, M., Serenelli, A., Schönrich, R., et al. 2016. The Gaia-ESO Survey: Hydrogen Lines in Red Giants Directly Trace Stellar Mass. *A&A*, **594**(October), A120

Bergemann, M., Sesar, B., Cohen, J. G., et al. 2018. Two Chemically Similar Stellar Overdensities on Opposite Sides of the Plane of the Galactic Disk. *Nature*, **555**(March), 334–337

Bessell, M. S., Collet, R., Keller, S. C., et al. 2015. Nucleosynthesis in a Primordial Supernova: Carbon and Oxygen Abundances in SMSS J031300.36-670839.3. *ApJ*, **806**(June), L16

Binney, J. and Lacey, C. 1988. The Diffusion of Stars through Phase Space. *MNRAS*, **230**(February), 597–627

Bonifacio, P., Caffau, E., Spite, M., et al. 2015. TOPoS . II. On the Bimodality of Carbon Abundance in CEMP Stars Implications on the Early Chemical Evolution of Galaxies. *A&A*, **579**(July), A28

Bono, G., Caputo, F. and Santolamazza, P. 1997. Evolutionary Scenario for Metal-Poor Pulsating Stars. I. Type II Cepheids. *A&A*, **317**(January), 171–177

Bromm, V. and Larson, R. B. 2004. The First Stars. *ARA&A*, **42**(September), 79–118

Brook, C. B., Kawata, D., Scannapieco, E., Martel, H. and Gibson, B. K. 2007. The Spatial Distribution of the Galactic First Stars. II. Smoothed Particle Hydrodynamics Approach. *ApJ*, **661**(May), 10–18

Bullock, J. S. and Johnston, K. V. 2005. Tracing Galaxy Formation with Stellar Halos. I. Methods. *ApJ*, **635**(December), 931–949

Busso, M., Gallino, R. and Wasserburg, G. J. 1999. Nucleosynthesis in Asymptotic Giant Branch Stars: Relevance for Galactic Enrichment and Solar System Formation. *Annual Review of Astronomy and Astrophysics*, **37**(January), 239–309

Caffau, E., Bonifacio, P., François, P., et al. 2011. An Extremely Primitive Star in the Galactic Halo. *Nature*, **477**(September), 67–69

Carollo, D., Beers, T. C., Chiba, M., et al. 2010. Structure and Kinematics of the Stellar Halos and Thick Disks of the Milky Way Based on Calibration Stars from Sloan Digital Sky Survey DR7. *ApJ*, **712**(March), 692–727

Carollo, D., Beers, T. C., Lee, Y. S., et al. 2007. Two Stellar Components in the Halo of the Milky Way. *Nature*, **450**(December), 1020–1025

Casagrande, L., Schönrich, R., Asplund, M., et al. 2011. New Constraints on the Chemical Evolution of the Solar Neighbourhood and Galactic Disc(s): Improved Astrophysical Parameters for the Geneva–Copenhagen Survey. *A&A*, **530**(June), A138

Catelan, M. 2009. Horizontal Branch Stars: The Interplay between Observations and Theory, and Insights into the Formation of the Galaxy. *Ap&SS*, **320**(April), 261–309

Chaplin, W. J. and Miglio, A. 2013. Asteroseismology of Solar-Type and Red-Giant Stars. *ARA&A*, **51**(August), 353–392

Chiappini, C., Matteucci, F. and Ballero, S. K. 2005. The Origin of Nitrogen: Implications of Recent Measurements of N/O in Galactic Metal-Poor Halo Stars. *A&A*, **437**(July), 429–436

Chiappini, C., Matteucci, F. and Gratton, R. 1997. The Chemical Evolution of the Galaxy: The Two-Infall Model. *ApJ*, **477**(March), 765–780

Chiavassa, A., Plez, B., Josselin, E. and Freytag, B. 2009. Radiative Hydrodynamics Simulations of Red Supergiant Stars. I. Interpretation of Interferometric Observations. *A&A*, **506**(November), 1351–1365

Chiavassa, A., Pasquato, E., Jorissen, A., et al. 2011. Radiative Hydrodynamic Simulations of Red Supergiant Stars. III. Spectro-Photocentric Variability, Photometric Variability, and Consequences on Gaia Measurements. *A&A*, **528**(April), A120

Choplin, A., Hirschi, R., Meynet, G. and Ekström, S. 2017. Are Some CEMP-s Stars the Daughters of Spinstars? *A&A*, **607**(November), L3

Clarkson, O., Herwig, F. and Pignatari, M. 2018. Pop III i-Process Nucleosynthesis and the Elemental Abundances of SMSS J0313–6708 and the Most Iron-Poor Stars. *MNRAS*, **474**(February), L37–L41

Collet, R., Asplund, M. and Trampedach, R. 2007. Three-Dimensional Hydrodynamical Simulations of Surface Convection in Red Giant Stars: Impact on Spectral Line Formation and Abundance Analysis. *A&A*, **469**(July), 687–706

Curir, A., Lattanzi, M. G., Spagna, A., et al. 2012. The Thick Disk Rotation-Metallicity Correlation as a Fossil of an 'Inverse Chemical Gradient' in the Early Galaxy. *A&A*, **545**(September), A133

Danilovic, S., Solanki, S. K., Livingston, W., Krivova, N. and Vince, I. 2016. Variation of the Mn I 539.4 nm Line with the Solar Cycle. *A&A*, **587**(March), A33

Dardelet, L., Ritter, C., Prado, P., et al. 2014. i Process and CEMP-s+r Stars. Page 145 of: *XIII Nuclei in the Cosmos (NIC XIII).*

Davies, B., Kudritzki, R.-P., Plez, B., et al. 2013. The Temperatures of Red Supergiants. *ApJ*, **767**(April), 3

Deason, A. J., Van der Marel, R. P., Guhathakurta, P., Sohn, S. T. and Brown, T. M. 2013. The Velocity Anisotropy of Distant Milky Way Halo Stars from Hubble Space Telescope Proper Motions. *ApJ*, **766**(March), *arXiv:*1302.5111 [astro-ph.GA]

Dravins, D. 1999. Stellar Surface Convection, Line Asymmetries, and Wavelength Shifts. Page 268 of: Hearnshaw, J. B., and Scarfe, C. D. (eds.), *IAU Colloq. 170: Precise Stellar Radial Velocities.* Astronomical Society of the Pacific Conference Series, vol. 185

Drout, M. R., Piro, A. L., Shappee, B. J., et al. 2017. Light Curves of the Neutron Star Merger GW170817/SSS17a: Implications for r-Process Nucleosynthesis. *Science*, **358**(December), 1570–1574

El-Badry, K., Bland-Hawthorn, J., Wetzel, A., et al. 2018. Where Are the Most Ancient Stars in the Milky Way? *ArXiv e-prints*, April

Evans, C. J., Davies, B., Kudritzki, R.-P., et al. 2011. Stellar Metallicities beyond the Local Group: The Potential of J-Band Spectroscopy with Extremely Large Telescopes. *A&A*, **527**(March), A50

Ferraro, F. R., Sabbi, E., Gratton, R., et al. 2006. Discovery of Carbon/Oxygen-Depleted Blue Straggler Stars in 47 Tucanae: The Chemical Signature of a Mass Transfer Formation Process. *ApJ*, **647**(August), L53–L56

Fields, B. 2012 (October). The Universe as an Accelerator: The Primordial Lithium Problem and Dark Matter. Page PB.001 of: *APS Division of Nuclear Physics Meeting Abstracts*

Font, A. S., McCarthy, I. G., Crain, R. A., et al. 2011. Cosmological Simulations of the Formation of the Stellar Haloes around Disc Galaxies. *MNRAS*, **416**(October), 2802–2820

Frebel, A. 2010. Stellar Archaeology: Exploring the Universe with Metal-Poor Stars. *Astronomische Nachrichten*, **331**(May), 474–488

Frebel, A. and Beers, T. C. 2018. The Formation of the Heaviest Elements. *Physics Today*, **71**(1), 30–37

Frebel, A. and Norris, J. E. 2015. Near-Field Cosmology with Extremely Metal-Poor Stars. *ARA&A*, **53**(August), 631–688

Freytag, B. and Höfner, S. 2008. Three-Dimensional Simulations of the Atmosphere of an AGB Star. *A&A*, **483**(May), 571–583

Freytag, B., Holweger, H., Steffen, M. and Ludwig, H.-G. 1997. On the Scale of Photospheric Convection. Page 316 of: Paresce, F. (ed.), *Science with the VLT Interferometer*. Springer

Freytag, B., Liljegren, S. and Höfner, S. 2017. Global 3D Radiation-Hydrodynamics Models of AGB Stars: Effects of Convection and Radial Pulsations on Atmospheric Structures. *A&A*, **600**(April), A137

Freytag, B., Steffen, M. and Dorch, B. 2002. Spots on the Surface of Betelgeuse – Results from New 3D Stellar Convection Models. *Astronomische Nachrichten*, **323**(July), 213–219

Freytag, B., Steffen, M., Ludwig, H.-G., Wedemeyer-Böhm, S., Schaffenberger, W., and Steiner, O. 2012. Simulations of stellar convection with CO5BOLD. *Journal of Computational Physics*, **231**(Feb.), 919–959.

Frischknecht, U., Hirschi, R., Pignatari, M., et al. 2016. s-Process Production in Rotating Massive Stars at Solar and Low Metallicities. *MNRAS*, **456**(February), 1803–1825

Fröhlich, C., Martínez-Pinedo, G., Liebendörfer, M., et al. 2006. Neutrino-Induced Nucleosynthesis of A $>$ 64 Nuclei: The νp Process. *Physical Review Letters*, **96**(14), 142502

Gautschy, A. and Saio, H. 1995. Stellar Pulsations across the HR Diagram: Part 1. *ARA&A*, **33**, 75–114

Gazak, J.-Z., Davies, B., Kudritzki, R., Bergemann, M. and Plez, B. 2014. Quantitative Spectroscopic J-Band Study of Red Supergiants in Perseus OB-1.*ApJ*, **788**, 58

Gehren, T., Shi, J. R., Zhang, H. W., Zhao, G. and Korn, A. J. 2006. Na, Mg and Al Abundances as a Population Discriminant for Nearby Metal-Poor Stars. *A&A*, **451**(June), 1065–1079

Genovali, K., Lemasle, B., Bono, G., et al. 2014. On the Fine Structure of the Cepheid Metallicity Gradient in the Galactic Thin Disk. *A&A*, **566**(June), A37

Gilmore, G. and Wyse, R. F. G. 1998. Element Ratios and the Formation of the Stellar Halo. *AJ*, **116**(August), 748–753

Goetz, M. and Koeppen, J. 1992. On Abundance Gradients in Spiral Galaxies. *A&A*, **262**(September), 455–467

Griffen, B. F., Dooley, G. A., Ji, A. P., O'Shea, B. W., Gómez, F. A. and Frebel, A. 2018. Tracing the First Stars and Galaxies of the Milky Way. *MNRAS*, **474**(February), 443–459

Gustafsson, B., Edvardsson, B., Eriksson, K., Jørgensen, U. G., Nordlund, Å. and Plez, B. 2008. A Grid of MARCS Model Atmospheres for Late-Type Stars. I. Methods and General Properties. *A&A*, **486**(August), 951–970

Hall, J. C. 2008. Stellar Chromospheric Activity. *Living Reviews in Solar Physics*, **5**(March), 2

Hampel, M., Stancliffe, R. J., Lugaro, M. and Meyer, B. S. 2016. The Intermediate Neutron-Capture Process and Carbon-Enhanced Metal-Poor Stars. *ApJ*, **831**(November), 171

Hansen, C. J., El-Souri, M., Monaco, L., et al. 2018a. Ages and Heavy Element Abundances from Very Metal-Poor Stars in the Sagittarius Dwarf Galaxy. *ApJ*, **855**(March), 83

Hansen, C. J., Hansen, T. T., Koch, A., et al. 2019a. Abundances and Kinematics of Carbon-Enhanced Metal-Poor Stars in the Galactic Halo*: A New Classification Scheme Based on Sr and Ba. *Astronomy and Astrophysics*, **623**, A128

Hansen, C. J., Montes, F. and Arcones, A. 2014. How Many Nucleosynthesis Processes Exist at Low Metallicity? *ApJ*, **797**(December), 123

Hansen, C. J., Nordström, B., Bonifacio, P., et al. 2011. First Stars. XIII. Two Extremely Metal-Poor RR Lyrae Stars. *A&A*, **527**(March), A65

Hansen, C. J., Nordström, B., Hansen, T. T., et al. 2016a. Abundances of Carbon-Enhanced Metal-Poor Stars as Constraints on Their Formation. *A&A*, **588**(April), A37

Hansen, C. J., Primas, F., Hartman, H., et al. 2012. Silver and Palladium Help Unveil the Nature of a Second r-Process. *A&A*, **545**(September), A31

Hansen, C. J., Rich, R. M., Koch, A., Xu, S., Kunder, A. and Ludwig, H.-G. 2016b. Chemical Abundances in a High-Velocity RR Lyrae Star near the Bulge. *A&A*, **590**(May), A39

Hansen, T. T., Andersen, J., Nordström, B., et al. 2016c. The Role of Binaries in the Enrichment of the Early Galactic Halo. III. Carbon-Enhanced Metal-Poor Stars – CEMP-s Stars. *A&A*, **588**(April), A3

Hansen, T. T., Holmbeck, E. M., Beers, T. C., et al. 2018b. The *R*-Process Alliance: First Release from the Southern Search for *r*-Process-Enhanced Stars in the Galactic Halo. *ArXiv e-prints*, April

Hauschildt, P. H., Allard, F. and Baron, E. 1999. The NextGen Model Atmosphere Grid for $3{,}000 \leq T_{\rm eff} \leq 10{,}000$ K. *ApJ*, **512**(February), 377–385

Hawkins, K., Masseron, T., Jofré, P., Gilmore, G., Elsworth, Y. and Hekker, S. 2016. An Accurate and Self-Consistent Chemical Abundance Catalogue for the APOGEE/Kepler Sample. *A&A*, **594**(October), A43

Heger, A. and Woosley, S. E. 2002. The Nucleosynthetic Signature of Population III. *ApJ*, **567**(March), 532–543

Hill, V., Christlieb, N., Beers, T. C., et al. 2017. The Hamburg/ESO r-Process Enhanced Star Survey (HERES). XI. The Highly r-Process-Enhanced Star CS 29497-004. *A&A*, **607**(November), A91

Holmberg, J., Nordström, B. and Andersen, J. 2009. The Geneva–Copenhagen survey of the Solar Neighbourhood. III. Improved Distances, Ages, and Kinematics. *A&A*, **501**(July), 941–947

Honda, S., Aoki, W., Ishimaru, Y. and Wanajo, S. 2007. Neutron-Capture Elements in the Very Metal-Poor Star HD 88609: Another Star with Excesses of Light Neutron-Capture Elements. *ApJ*, **666**(September), 1189–1197

Ito, H., Aoki, W., Beers, T. C., Tominaga, N., Honda, S. and Carollo, D. 2013. Chemical Analysis of the Ninth Magnitude Carbon-Enhanced Metal-Poor Star BD+44 493. *ApJ*, **773**(August), 33

José, J. and Iliadis, C. 2011. Nuclear Astrophysics: The Unfinished Quest for the Origin of the Elements. *Reports on Progress in Physics*, **74**(9), 096901

Karakas, A. I. and Lattanzio, J. C. 2014. The Dawes Review 2: Nucleosynthesis and Stellar Yields of Low- and Intermediate-Mass Single Stars. *Proc. Astron. Soc. Aust.*, **31**(July), e030

Karlsson, T. 2005. Stochastic Chemical Enrichment in Metal-Poor Systems. I. Theory. *A&A*, **439**(August), 93–106

Karlsson, T. and Gustafsson, B. 2005. Stochastic Chemical Enrichment in Metal-Poor Systems. II. Abundance Ratios and Scatter. *A&A*, **436**(June), 879–894

Kasen, D., Metzger, B., Barnes, J., Quataert, E. and Ramirez-Ruiz, E. 2017. Origin of the Heavy Elements in Binary Neutron-Star Mergers from a Gravitational-Wave Event. *Nature*, **551**(November), 80–84

Keller, S. C., Bessell, M. S., Frebel, A., et al. 2014. A Single Low-Energy, Iron-Poor Supernova as the Source of Metals in the Star SMSS J031300.36-670839.3. *Nature*, **506**(February), 463–466

Kirby, E. N., Cohen, J. G., Smith, G. H., Majewski, S. R., Sohn, S. T. and Guhathakurta, P. 2011. Multi-Element Abundance Measurements from Medium-Resolution Spectra. IV. Alpha Element Distributions in Milky Way Satellite Galaxies. *ApJ*, **727**(February), 79

Kratz, K.-L., Farouqi, K., Pfeiffer, B., Truran, J. W., Sneden, C. and Cowan, J. J. 2007. Explorations of the r-Processes: Comparisons between Calculations and Observations of

Low-Metallicity Stars. *ApJ*, **662**(June), 39–52

Kurucz, R. L. 1993. A New Opacity-Sampling Model Atmosphere Program for Arbitrary Abundances. Page 87 of: Dworetsky, M. M., Castelli, F. and Faraggiana, R. (eds.), *IAU Colloq. 138: Peculiar versus Normal Phenomena in A-Type and Related Stars*. Astronomical Society of the Pacific Conference Series, vol. 44.

Lacey, C. G. and Fall, S. M. 1983. Kinematical and Chemical Evolution of the Galactic Disc. *MNRAS*, **204**(August), 791–810.

Laporte, C. F. P., Johnston, K. V., Gómez, F. A., Garavito-Camargo, N. and Besla, G. 2017. The Influence of Sagittarius and the Large Magellanic Cloud on the Milky Way Galaxy. *ArXiv e-prints*, October

Lattanzio, J., Frost, C., Cannon, R. and Wood, P. R. 1996. Hot Bottom Burning in Intermediate Mass Stars. *Mem. Soc. Astron. Italiana*, **67**, 729

Leaman, R. 2012. Insights into Pre-Enrichment of Star Clusters and Self-Enrichment of Dwarf Galaxies from Their Intrinsic Metallicity Dispersions. *AJ*, **144**(December), 183

Lee, Y. S., Beers, T. C., An, D., et al. 2011. Formation and Evolution of the Disk System of the Milky Way: [α/Fe] Ratios and Kinematics of the SEGUE G-Dwarf Sample. *ApJ*, **738**(September), 187

Lee, Y. S., Beers, T. C., Masseron, T., et al. 2013. Carbon-Enhanced Metal-Poor Stars in SDSS/SEGUE. I. Carbon Abundance Estimation and Frequency of CEMP Stars. *AJ*, **146**(November), 132

Lee, Y.-W. 1992. Evidence for an Old Galactic Bulge from RR Lyrae Stars in Baade's Window: Implications for the Formation of the Galaxy and the Age of the Universe. *AJ*, **104**(November), 1780–1789

Leenaarts, J., Carlsson, M. and Rouppe van der Voort, L. 2012. The Formation of the Hα Line in the Solar Chromosphere. *ApJ*, **749**(April), 136

Lind, K., Amarsi, A. M., Asplund, M., et al. 2017. Non-LTE Line Formation of Fe in Late-Type Stars: IV. Modelling of the Solar Centre-to-Limb Variation in 3D. *MNRAS*, **468**(July), 4311–4322

Linsky, J. L. 2017. Stellar Model Chromospheres and Spectroscopic Diagnostics. *ARA&A*, **55**(August), 159–211

Lodders, K., Palme, H., and Gail, H.-P. 2009. Abundances of the Elements in the Solar System. Landolt Börnstein 4B, 712

Ludwig, H.-G. 2006. Hydrodynamical Simulations of Convection-Related Stellar Microvariability. I. Statistical Relations for Photometric and Photocentric Variability. *A&A*, **445** (January), 661–671

Madore, B. F. and Freedman, W. L. 1998. Calibration of the Extragalactic Distance Scale. Page 263 of: Aparicio, A., Herrero, A. and Sánchez, F. (eds.), *Stellar Astrophysics for the Local Group: VIII Canary Islands Winter School of Astrophysics*

Maeder, A. 1990. Tables for Massive Star Evolution at Various Metallicities. *Astronomy and Astrophysics Supplement Series*, **84**(July), 139–177

Mannucci, F., Della Valle, M. and Panagia, N. 2006. Two Populations of Progenitors for Type Ia Supernovae? *MNRAS*, **370**(August), 773–783

Marconi, M. and Minniti, D. 2018. Gauging the Helium Abundance of the Galactic Bulge RR Lyrae Stars. *ApJ*, **853**(February), L20

Martig, M., Fouesneau, M., Rix, H.-W., et al. 2016. Red Giant Masses and Ages Derived from Carbon and Nitrogen Abundances. *MNRAS*, **456**(March), 3655–3670

Martin, D., Perego, A., Arcones, A., Thielemann, F.-K., Korobkin, O. and Rosswog, S. 2015. Neutrino-Driven Winds in the Aftermath of a Neutron Star Merger: Nucleosynthesis and Electromagnetic Transients. *ApJ*, **813**(November), 2

Mashonkina, L., Christlieb, N. and Eriksson, K. 2014. The Hamburg/ESO r-Process Enhanced Star Survey (HERES). X. HE 2252–4225, One More r-Process Enhanced and Actinide-Boost Halo Star. *A&A*, **569**(September), A43

Matteucci, F. and Calura, F. 2005. Early Chemical Enrichment of the Universe and the Role of Very Massive Population III Stars. *MNRAS*, **360**(June), 447–452

Matteucci, F. and Greggio, L. 1986. Relative Roles of Type I and II Supernovae in the Chemical Enrichment of the Interstellar Gas. *A&A*, **154**(January), 279–287

McCarthy, I. G., Font, A. S., Crain, R. A., Deason, A. J., Schaye, J. and Theuns, T. 2012. Global Structure and Kinematics of Stellar Haloes in Cosmological Hydrodynamic Simulations. *MNRAS*, **420**(March), 2245–2262

Merrill, P. W. 1952. Spectroscopic Observations of Stars of Class. *ApJ*, **116**(July), 21

Metzger, B. D., Martínez-Pinedo, G., Darbha, S., et al. 2010. Electromagnetic Counterparts of Compact Object Mergers Powered by the Radioactive Decay of r-Process Nuclei. *MNRAS*, **406**(August), 2650–2662

Mikolaitis, Š., Hill, V., Recio-Blanco, A., et al. 2014. The Gaia-ESO Survey: The Chemical Structure of the Galactic Discs from the First Internal Data Release. *A&A*, **572**(December), A33

Mösta, P., Roberts, L. F., Halevi, G., et al. 2018. r-Process Nucleosynthesis from Three-Dimensional Magnetorotational Core-Collapse Supernovae. *ApJ*, **864**(September), 171

Nardetto, N., Gieren, W., Kervella, P., et al. 2009. High-Resolution Spectroscopy for Cepheids Distance Determination. V. Impact of the Cross-Correlation Method on the p-Factor and the γ-Velocities. *A&A*, **502**(August), 951–956

Ness, M., Hogg, D. W., Rix, H.-W., Martig, M., Pinsonneault, M. H. and Ho, A. Y. Q. 2016. Spectroscopic Determination of Masses (and Implied Ages) for Red Giants. *ApJ*, **823**(June), 114

Neyskens, P., van Eck, S., Jorissen, A., Goriely, S., Siess, L. and Plez, B. 2015. The Temperature and Chronology of Heavy-Element Synthesis in Low-Mass Stars. *Nature*, **517**(January), 174–176

Nishimura, N., Sawai, H., Takiwaki, T., Yamada, S. and Thielemann, F.-K. 2017. The Intermediate r-Process in Core-Collapse Supernovae Driven by the Magneto-Rotational Instability. *ApJ*, **836**(February), L21

Nissen, P. E. and Schuster, W. J. 1997. Chemical Composition of Halo and Disk Stars with Overlapping Metallicities. *A&A*, **326**(October), 751–762

Nissen, P. E. and Schuster, W. J. 2010. Two Distinct Halo Populations in the Solar Neighborhood: Evidence from Stellar Abundance Ratios and Kinematics. *A&A*, **511**(February), L10

Nomoto, K., Tominaga, N., Umeda, H. and Kobayashi, C. 2005. Yields of Population III Supernovae and the Abundance Patterns of Extremely Metal-Poor Stars. Pages 287–296 of: Hill, V., Francois, P. and Primas, F. (eds.), *From Lithium to Uranium: Elemental Tracers of Early Cosmic Evolution*. IAU Symposium, vol. 228

Nordlander, T., Amarsi, A. M., Lind, K., et al. 2017. 3D NLTE Analysis of the Most Iron-Deficient Star, SMSS0313–6708. *A&A*, **597**(January), A6

Nordlund, Å., Stein, R. F. and Asplund, M. 2009. Solar Surface Convection. *Living Reviews in Solar Physics*, **6**(April), 2

Osorio, Y. and Barklem, P. S. 2016. Mg Line Formation in Late-Type Stellar Atmospheres. II. Calculations in a Grid of 1D Models. *A&A*, **586**(February), A120

Pace, G., Castro, M., Meléndez, J., Théado, S. and do Nascimento, J.-D. 2012. Lithium in M 67: From the Main Sequence to the Red Giant Branch. *A&A*, **541**, A150

Parenago, P. P. 1945. Some Work on the Structure of the Galaxy. *Popular Astronomy*, **53**(December), 441

Pian, E., D'Avanzo, P., Benetti, S., et al. 2017. Spectroscopic Identification of r-Process Nucleosynthesis in a Double Neutron–Star Merger. *Nature*, **551**(November), 67–70

Placco, V. M., Frebel, A., Beers, T. C. and Stancliffe, R. J. 2014. Carbon-Enhanced Metal-Poor Star Frequencies in the Galaxy: Corrections for the Effect of Evolutionary Status on Carbon Abundances. *ApJ*, **797**(December), 21

Placco, V. M., Holmbeck, E. M., Frebel, A., et al. 2017. RAVE J203843.2-002333: The First Highly r-Process-Enhanced Star Identified in the RAVE Survey. *ApJ*, **844**(July), 18

Pont, F., Mayor, M. and Burki, G. 1994. New Radial Velocities for Classical Cepheids. Local Galactic Rotation Revisited. *A&A*, **285**(May), 415–439

Ramírez, I., Allende Prieto, C. and Lambert, D. L. 2008. Granulation in K-Type Dwarf Stars. I. Spectroscopic Observations. *A&A*, **492**(December), 841–855

Riess, A. G., Macri, L., Casertano, S., et al. 2011. A 3% Solution: Determination of the Hubble Constant with the Hubble Space Telescope and Wide Field Camera 3. *ApJ*, **730**(April), 119

Ruchti, G. R., Bergemann, M., Serenelli, A., Casagrande, L. and Lind, K. 2013. Unveiling Systematic Biases in the 1D LTE Excitation-Ionization Balance of Fe for FGK Stars: A Novel Approach to Determination of Stellar Parameters. *MNRAS*, **429**(February), 126–134

Rybizki, J., Just, A. and Rix, H.-W. 2017. Chempy: A Flexible Chemical Evolution Model for Abundance Fitting. Do the Sun's Abundances Alone Constrain Chemical Evolution Models? *A&A*, **605**(September), A59

Salaris, M. 2012. Distance Indicators from Colour-Magnitude-Diagrams: Main Sequence, Red Clump and Tip of the RGB. *Ap&SS*, **341**(September), 65–75

Sbordone, L., Bonifacio, P., Caffau, E., et al. 2010. The Metal-Poor End of the Spite Plateau. I. Stellar Parameters, Metallicities, and Lithium Abundances. *A&A*, **522**(November), A26

Schönrich, R., Asplund, M. and Casagrande, L. 2011. On the Alleged Duality of the Galactic Halo. *MNRAS*, **415**(August), 3807–3823

Schönrich, R. and Binney, J. 2009a. Chemical Evolution with Radial Mixing. *MNRAS*, **396**(June), 203–222

Schönrich, R. and Binney, J. 2009b. Origin and Structure of the Galactic Disc(s). *MNRAS*, **399**(November), 1145–1156

Schönrich, R. and McMillan, P. J. 2017. Understanding Inverse Metallicity Gradients in Galactic Discs as a Consequence of Inside-Out Formation. *MNRAS*, **467**(May), 1154–1174

Schönrich, R., Asplund, M. and Casagrande, L. 2014. Does SEGUE/SDSS Indicate a Dual Galactic Halo? *ApJ*, **782**(May), 35.

Schwarzschild, M. 1975. On the Scale of Photospheric Convection in Red Giants and Supergiants. *ApJ*, **195**(January), 137–144

Searle, L. and Zinn, R. 1978. Composition of Halo Clusters and the Formation of the Galactic Halo. *ApJ*, **225**(October), 357–379

Seitenzahl, I. R. and Townsley, D. M. 2017. Nucleosynthesis in Thermonuclear Supernovae. *ArXiv e-prints*, April

Seitenzahl, I. R., Cescutti, G., Röpke, F. K., Ruiter, A. J. and Pakmor, R. 2013. Solar Abundance of Manganese: A Case for Near Chandrasekhar-Mass Type Ia Supernova Progenitors. *A&A*, **559**(November), L5

Sellwood, J. A. and Binney, J. J. 2002. Radial Mixing in Galactic Discs. *MNRAS*, **336**(November), 785–796

Serenelli, A., Weiss, A., Cassisi, S., Salaris, M. and Pietrinferni, A. 2017. The Brightness of the Red Giant Branch Tip: Theoretical Framework, a Set of Reference Models, and Predicted Observables. *A&A*, **606**(October), A33

Shappee, B. J., Simon, J. D., Drout, M. R., et al. 2017. Early Spectra of the Gravitational Wave Source GW170817: Evolution of a Neutron Star Merger. *Science*, **358**(Dec.), 1574–1578

Siqueira Mello, C., Spite, M., Barbuy, B., et al. 2013. First Stars. XVI. HST/STIS Abundances of Heavy Elements in the Uranium-Rich Metal-Poor Star CS 31082-001. *A&A*, **550**(February), A122

Smartt, S. J., Chen, T.-W., Jerkstrand, A., et al. 2017. A Kilonova as the Electromagnetic Counterpart to a Gravitational-Wave Source. *Nature*, **551**(November), 75–79

Sneden, C., Cowan, J. J., Lawler, J. E., et al. 2003. The Extremely Metal-Poor, Neutron Capture-Rich Star CS 22892-052: A Comprehensive Abundance Analysis. *ApJ*, **591**(July), 936–953

Soderblom, D. R. 2010. The Ages of Stars. *ARA&A*, **48**(September), 581–629

Soszyński, I., Udalski, A., Szymański, M. K., et al. 2014. Over 38000 RR Lyrae Stars in the OGLE Galactic Bulge Fields. *Acta Astronomica*, **64**(September), 177–196

Spagna, A., Lattanzi, M. G., Re Fiorentin, P. and Smart, R. L. 2010. Evidence of a Thick Disk Rotation-Metallicity Correlation. *A&A*, **510**(February), L4

Spina, L., Meléndez, J., Karakas, A. I., et al. 2016. Nucleosynthetic History of Elements in the Galactic Disk: [X/Fe]-Age Relations from High-Precision Spectroscopy. *A&A*, **593**(October), A125

Spite, M., Caffau, E., Bonifacio, P., et al. 2013. Carbon-Enhanced Metal-Poor Stars: The Most Pristine Objects? *A&A*, **552**(April), A107

Spite, M., Cayrel, R., Plez, B., et al. 2005. First Stars VI Abundances of C, N, O, Li, and Mixing in Extremely Metal-Poor Giants. Galactic Evolution of the Light Elements. *A&A*, **430**(February), 655–668

Spite, M., Peterson, R. C., Gallagher, A. J., Barbuy, B. and Spite, F. 2017. Abundances of the Light Elements from UV (HST) and Red (ESO) Spectra in the Very Old Star HD 84937. *A&A*, **600**(April), A26

Starkenburg, E., Aguado, D. S., Bonifacio, P., et al. 2018. The Pristine Survey IV: Approaching the Galactic Metallicity Floor with the Discovery of an Ultra-Metal-Poor Star. *MNRAS*, **481**(December), 3838–3852

Starkenburg, E., Hill, V., Tolstoy, E., et al. 2013. The Extremely Low-Metallicity Tail of the Sculptor Dwarf Spheroidal Galaxy. *A&A*, **549**(January), A88

Starkenburg, E., Oman, K. A., Navarro, J. F., et al. 2017. The Oldest and Most Metal-Poor Stars in the APOSTLE Local Group Simulations. *MNRAS*, **465**(February), 2212–2224

Steigman, G. 2007. Primordial Nucleosynthesis in the Precision Cosmology Era. *Annual Review of Nuclear and Particle Science*, **57**(November), 463–491

Stibbs, D. W. N. 1956. On the Differential Galactic Rotation of the System of Cepheid Variable Stars. *MNRAS*, **116**, 453.

Tajitsu, A., Sadakane, K., Naito, H., Arai, A. and Aoki, W. 2015. Explosive Lithium Production in the Classical Nova V339 Del (Nova Delphini 2013). *Nature*, **518**(February), 381–384

Tanvir, N. R., Levan, A. J., Fruchter, A. S., et al. 2013. A 'Kilonova' Associated with the Short-Duration γ-Ray Burst GRB 130603B. *Nature*, **500**(August), 547–549

Thielemann, F.-K., Eichler, M., Panov, I. V. and Wehmeyer, B. 2017. Neutron Star Mergers and Nucleosynthesis of Heavy Elements. *Annual Review of Nuclear and Particle Science*, **67**(October), 253–274

Thielemann, F.-K., Fröhlich, C., Hirschi, R., et al. 2007. Production of Intermediate-Mass and Heavy Nuclei. *Progress in Particle and Nuclear Physics*, **59**(July), 74–93

Thompson, B. B., Few, C. G., Bergemann, M., et al. 2018. The Gaia-ESO Survey: Matching Chemodynamical Simulations to Observations of the Milky Way. *MNRAS*, **473**(January), 185–197

Tinsley, B. M. 1979. Stellar Lifetimes and Abundance Ratios in Chemical Evolution. *ApJ*, **229**(May), 1046–1056

Tissera, P. B., Beers, T. C., Carollo, D. and Scannapieco, C. 2014. Stellar Haloes in Milky Way Mass Galaxies: From the Inner to the Outer Haloes. *MNRAS*, **439**(April), 3128–3138

Tissera, P. B., Machado, R. E. G., Carollo, D., et al. 2018. The Central Spheroids of Milky Way Mass-Sized Galaxies. *MNRAS*, **473**(January), 1656–1666

Tissera, P. B., Scannapieco, C., Beers, T. C. and Carollo, D. 2013. Stellar Haloes of Simulated Milky-Way-Like Galaxies: Chemical and Kinematic Properties. *MNRAS*, **432**(July), 3391–3400

Tolstoy, E., Hill, V. and Tosi, M. 2009. Star-Formation Histories, Abundances, and Kinematics of Dwarf Galaxies in the Local Group. *ARA&A*, **47**(September), 371–425

Trampedach, R., Asplund, M., Collet, R., Nordlund, Å. and Stein, R. F. 2013. A Grid of Three-Dimensional Stellar Atmosphere Models of Solar Metallicity. I. General Properties, Granulation, and Atmospheric Expansion. *ApJ*, **769**(May), 18

Travaglio, C., Rauscher, T., Heger, A., Pignatari, M. and West, C. 2018. Role of Core-Collapse Supernovae in Explaining Solar System Abundances of p Nuclides. *ApJ*, **854**(February), 18

Tremblay, P.-E., Ludwig, H.-G., Freytag, B., Steffen, M. and Caffau, E. 2013. Granulation Properties of Giants, Dwarfs, and White Dwarfs from the CIFIST 3D Model Atmosphere Grid. *A&A*, **557**(September), A7

Vasilyev, V., Ludwig, H.-G., Freytag, B., Lemasle, B. and Marconi, M. 2017. Spectroscopic Properties of a Two-Dimensional Time-Dependent Cepheid Model. I. Description and Validation of the Model. *A&A*, **606**(October), A140

Viallet, M. and Baraffe, I. 2012. Scenarios to Explain Extreme Be Depletion in Solar-Like Stars: Accretion or Rotation Effects? *A&A*, **546**(October), A113

Wallerstein, G. 2002. The Cepheids of Population II and Related Stars. *PASP*, **114**(July), 689–699

Wanajo, S., Janka, H.-T. and Müller, B. 2011. Electron-Capture Supernovae as the Origin of Elements beyond Iron. *ApJ*, **726**(January), L15

White, S. D. M. and Springel, V. 2000. Where Are the First Stars Now? Page 327 of: Weiss, A., Abel, T. G. and Hill, V. (eds.), *The First Stars*. Springer-Verlag

White, S. D. M. and Frenk, Carlos S. 1991. Galaxy Formation through Hierarchical Clustering. *ApJ*, **379**(September), 52

Wittkowski, M. 2004. Fundamental Stellar Parameters. *arXiv Astrophysics e-prints*, December

Yoon, J., Beers, T. C., Placco, V. M., et al. 2016. Observational Constraints on First-Star Nucleosynthesis. I. Evidence for Multiple Progenitors of CEMP-No Stars. *ApJ*, **833**(December), 20

Yoon, J., Beers, T. C., Dietz, S., et al. 2018. Galactic Archeology with the AEGIS Survey: The Evolution of Carbon and Iron in the Galactic Halo. *ApJ*, **861**(July), 146

Zhang, B. R., Childress, M. J., Davis, T. M., et al. 2017. A Blinded Determination of H_0 from Low-Redshift Type Ia Supernovae, Calibrated by Cepheid Variables. *MNRAS*, **471**(October), 2254–2285

7. Modelling the Atmospheres of Ultracool Dwarfs and Extrasolar Planets

MARK S. MARLEY

Abstract

By absorbing and scattering both incident and emergent radiation, an atmosphere regulates a planet's thermal, chemical and cloud structure, and cooling through time. The photons transmitted through or scattered by an atmosphere provide one of our primary sources of information about planetary composition. Therefore, any effort to fully characterize an extrasolar planet must incorporate atmospheric models that attempt to fully describe the relevant processes and thereby predict a planet's reflected and emitted spectra. Brown dwarfs, ultracool substellar objects with atmospheric composition similar to those of many gas giant planets, provide a tractable training ground to test our ideas and models about atmospheric processes under conditions more exotic than found in the Solar System. This chapter aims to concisely summarize the various ingredients that must be included in any model and the overall process of atmospheric model creation for ultracool dwarfs and extrasolar planets. These considerations include the basic atmospheric structure equations, radiative transfer, atmospheric chemistry, clouds and various disequilibrium processes. Each of these topics is worthy of in-depth treatments, and pointers to appropriate review articles are provided for those wishing to understand each component in more detail.

7.1 Introduction

The last 20 years have witnessed the detection of thousands of extrasolar planets and brown dwarfs. Characterizing and understanding the cool atmospheres of these objects presents a host of challenges in addition to those encountered in the traditional stellar atmospheres problem. The abundance of molecules with a complex forest of rotation-vibration transition lines, the presence of condensates and numerous other issues complicates the modelling process, but these characteristics also mean that the spectra of these objects are very information rich, if we can interpret the data.

In this chapter, I will briefly outline some of the foremost challenges in modelling these types of atmospheres and point to references that further elucidate these issues in more detail. For a more in-depth general overview of cool atmosphere modelling, see Marley and Robinson (2015). More rigorous and complete treatments of the stellar atmospheres problem can be found in Chapter 5 of this work by Puls and in the textbooks by Mihalas (1978) and Hubený and Mihalas (2014).

7.2 Why Do We Need Atmosphere Models?

A one-dimensional radiative-convective model describes the mean profile with altitude of temperature, composition and cloud properties in an atmosphere given a bulk elemental composition and incident and internal energy fluxes. Computing the equilibrium profile requires iteratively finding a single temperature profile that conserves energy and satisfies all chemical equilibrium and other relevant constraints. Figure 7.1 shows an example, in this case a set of temperature profiles for an irradiated planet at progressively closer distances to its primary star.

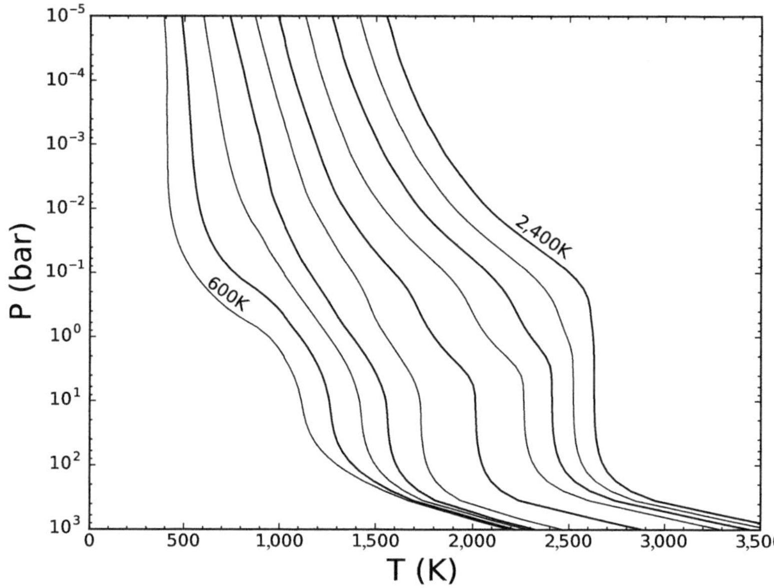

FIGURE 7.1. Model pressure-temperature profiles for planets at various distances from a 6,000 K G0 main-sequence star. Such models give insight into how various physical parameters, in this case stellar insolation, affect atmospheric structure. The models go from $T_{\rm eff}$ of 2,400 K to 600 K in 200 K increments. All models have a Jupiter-like surface gravity of $25\,{\rm m\,s^{-2}}$. Figure courtesy J. Fortney.

Understanding the atmospheric structure is interesting in and of itself, but the greater purpose of computing such models is to aid the interpretation of the reflected light and thermal emission from a planet or brown dwarf. The detailed thermal and chemical structure controls how and where incident light is scattered out of the atmosphere and thermal photons are emitted. The combination of these two radiative fluxes comprises the total spectrum of an object, complete with cloud and scattering and gaseous absorption spectral features. An example showing the best-fitting model to a spectrum of a directly imaged extrasolar planet is shown in Figure 7.2. In addition, transit spectra, obtained as an extrasolar planet transits its primary star, record the imprint of the absorption of incident flux by atmospheric gasses, clouds and hazes. The interpretation of transit spectra likewise requires a model of the whole atmosphere to predict the atmospheric scale height and composition, both of which influence the atmospheric structure.

In addition to controlling the spectrum of an object, the atmospheric thermal profile also controls the rate at which planets and brown dwarfs can lose energy and cool over time. Imagine two gas giant planets with identical internal temperature profiles as a function of pressure, $T(P)$. At depth, both atmospheres are convective, and convection transports energy from the deep interior of the planets. One planet has an optically thin atmosphere that allows flux to radiate away to space at a pressure level where the temperature is 1,000 K. Since the planet is radiating at this temperature, the effective temperature $T_{\rm eff}$ is 1,000 K. The second planet has a more opaque atmosphere, requiring convection to transport energy upwards to lower pressures and temperatures before photons can escape from a level where the temperature is 900 K. Since the total luminosity is proportional to $\sigma T_{\rm eff}^4$, this world will cool $(900/1,000)^4 = 0.656$ times as rapidly as the first planet. The first planet will cool off and evolve to lower temperatures faster than

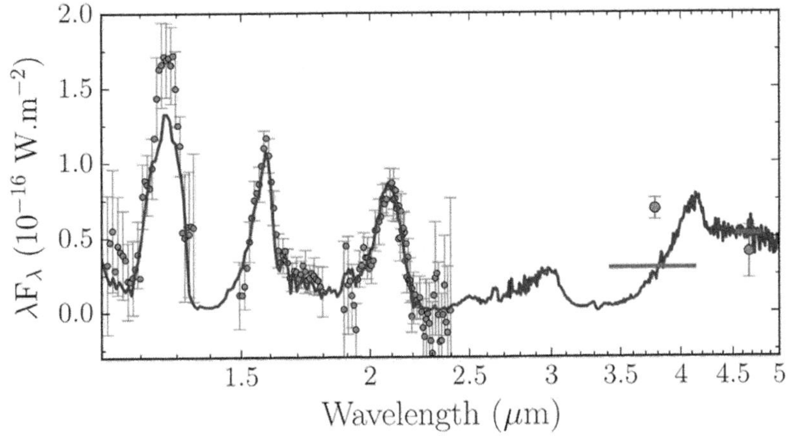

FIGURE 7.2. Example of using models to fit data and constrain the properties of an extrasolar planet. Here the data are for the young giant planet 51 Eri b. Comparison of a variety of models to this spectral and photometric dataset (Rajan et al., 2017) led to this best-fitting model (curve), which has an effective temperature of 900 K, $\log g = 3.25$ and silicate clouds. Figure modified from Rajan et al. (2017). © American Astronomical Society (AAS). Reproduced with permission.

the second planet, all else being equal. As this example shows, atmosphere models are required to model the evolution through time of any planet or brown dwarf.

7.3 Ingredients in a Model

The most basic inputs into an atmosphere model are the physical constraints governing atmospheric structure. Foremost is the equation of hydrostatic equilibrium, essentially a statement that the overlying weight of the atmosphere on a slab of vertical thickness dz is balanced by the pressure gradient, dP, across the slab, or

$$\frac{dP}{dz} = -\rho g. \tag{7.1}$$

In typical brown dwarf and extrasolar giant planet atmospheres, the vertical extent of the visible atmosphere is much less than the radius of the object, thus

$$g = \frac{GM}{R^2}, \tag{7.2}$$

where M and R are the planetary mass and radius. The atmospheric pressure scale height is then

$$H = \frac{kT}{mg} \tag{7.3}$$

leading to the pressure-altitude relation

$$P(z) = P(z_0)e^{-(z-z_0)/H}, \tag{7.4}$$

where m is the mean molecular weight, typically about 2.2 (in atomic mass units) for solar composition giant planet atmospheres.

In addition to these requirements for static stability of the atmosphere, a model structure must conserve energy and the chemical composition must follow some imposed

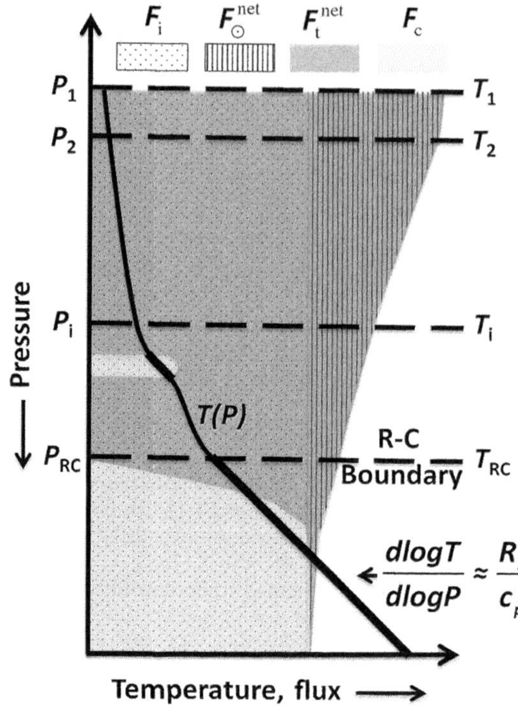

FIGURE 7.3. Sketch of an atmospheric thermal structure model. Pressure is shown on the vertical axis, increasing downwards, while the horizontal axis represents temperature and energy flux. A few model levels are shown by horizontal dashed lines, while the solid line is the temperature profile. Bolded segments indicate a convective region. Level pressures and temperatures are indicated with associated subscripted symbols, and R-C indicates the radiative-convective boundary. For a model in radiative-convective equilibrium, the net thermal flux (F_t^{net}, dark grey) and the convective flux (F_c, light grey) must sum to the internal heat flux (F_i, dotted) and, for an irradiated object, the net absorbed stellar flux ($F_\odot^{net}$, striped). Note that the internal heat flux is constant throughout the atmosphere, whereas the net absorbed stellar flux decreases with increasing pressure, eventually reaching zero in the deep atmosphere. At depth, convection carries the vast majority of the summed internal and stellar fluxes but is a smaller component in detached convective regions (upper light-grey region). Figure modified from Marley and Robinson (2015).

set of rules, such as being in chemical equilibrium. Energy conservation is a requirement that at any level in an atmosphere the energy transported through that level be equal to the net incident flux absorbed below that level plus the internal heat flux, F_i (Figure 7.3). The energy conservation relation can then be cast as

$$F_t^{net}(P) + F_c(P) + F_\odot^{net}(P) = F_i, \tag{7.5}$$

where F_t^{net} is the net thermal flux through a level at pressure P, F_c is the convective flux and $F_\odot^{net}$ is the net stellar flux. The planetary atmosphere problem is to find a self-consistent solution for this equation throughout a one-dimensional mean column of atmosphere.

Introductory reviews by Burrows and Orton (2010) and Seager and Deming (2010) cover the important fundamentals of atmospheric modelling for extrasolar planets. A review by Burrows et al. (2001) discusses the atmospheric and evolution modelling of young gas giant planets and brown dwarfs. In the remainder of this chapter, I briefly highlight the considerations that influence finding a solution of (7.5).

7.4 Opacities

The key distinguishing characteristic of cool brown dwarf and extrasolar giant planet atmospheres from hotter OBAFGK stellar atmospheres is that the radiative transfer is dominated by the opacity of molecules and condensates rather than electrons and atoms. This complicates the problem because a single molecule can have millions to hundreds of millions to billions of individual rotovibrational transitions. Accounting for all of the molecular lines that comprise the total opacity of a gas mixture is consequently a major task of cool atmosphere modelling.

An individual spectral line is described by three key parameters: the line position, the line strength and the line shape function. Large databases, usually referred to as line lists, compile the necessary information for computing absorption line spectra. These databases are based on either lab measurements or quantum chemistry simulations. Commonly used line lists include the HITRAN (Rothman et al., 1987, 2013), HITEMP (Rothman et al., 2010) and ExoMol (Tennyson and Yurchenko, 2012) databases; HITEMP is most appropriate for the range of temperatures encountered in hot exoplanet and brown dwarf atmospheres but only includes data for a limited set of molecules. Sources of line lists for more exotic species and discussion of how to implement these databases are discussed in Sharp and Burrows (2007) and Freedman et al. (2008, 2014).

While the necessary theory to use these line lists to compute opacities is fairly straightforward, issues arise owing to the scope of the problem. Line lists can contain more than 10^9 or even 10^{10} transitions, which makes assembling opacities computationally expensive. Efficiency can be gained by omitting weak lines from opacity calculations, which is most effective when gas concentrations are known (or can be estimated) a priori. A number of approaches for dealing with these large line lists are discussed in the literature. Much current effort is going in to exploring various computational shortcuts for properly accounting for gas opacities and, as always, care must be taken when comparing model approaches to understand how the opacities have been handled.

7.5 Chemistry

Any given atmosphere model calculation starts with some assumption as to the elemental composition of the atmosphere. As with stellar atmospheres, abundances are typically referenced to those of the Sun. Asplund et al. (2009) review the challenges in defining such an abundance set. Unfortunately, the solar C and O abundances are uncertain, and the generally accepted values have varied with time. For this reason, any model comparison between different modelling groups should first begin with a comparison of the assumed elemental abundances.

7.5.1 *Equilibrium Chemistry*

Given the abundances of relevant elements and appropriate thermodynamic data, the abundances of gaseous species at any pressure and temperature can be computed assuming that the system is in chemical equilibrium. Most groups use one of two methods for solving for the gas abundances in chemical equilibrium, commonly termed mass action and Gibbs minimization. The mass action approach utilizes the equilibrium constants for all relevant chemical reactions, as well as mass conservation, to find the abundance of each molecule and condensate of interest at a given pressure and temperature.

Gibbs minimization solves for the mixture of species that has the lowest Gibbs free energy, given the pressure, temperature and assumed elemental abundances. van Zeggeren and Storey (1970) offer the best source for learning about the details of chemical equilibrium calculations.

Sharp and Burrows (2007) explain and offer examples of the free-energy minimization procedure. The mass action approach is described by Fegley and Lodders (1994) in the context of Jupiter's atmosphere. The method has been used to study the chemistry of the major elements in brown dwarf and exoplanet atmospheres by Lodders and Fegley (2002) and Visscher et al. (2006). With identical thermodynamic data, the two approaches should give identical results. However, in practice, typically because of computational limitations, Gibbs minimization methods do not always converge with complete mass conservation. Computational considerations also often limit the number of elements and compounds that can be included in this approach.

A crucial decision must be made when computing equilibrium mixtures regarding how to handle species that condense out of the gas (Lewis, 1972). In one case, the solid grain or liquid drop continues to chemically interact with the surrounding gas to arbitrarily low temperatures, a description of 'true' chemical equilibrium. Burrows and Sharp (1999) explore this assumption, which is implicit in the COND and DUSTY brown dwarf models (Chabrier et al., 2000; Allard et al., 2001). At the other extreme, the condensate can be imagined to rain out of the atmosphere, precluding further reactions with the neighbouring gas, a limit sometimes called rainout chemistry (Burrows and Sharp, 1999; Lodders, 1999, 2004; Lodders and Fegley, 2006). The two approaches give different predictions for some gas species, and care must always be exercised to understand which method has been employed by any given calculation. See Marley and Robinson (2015) for a more in-depth discussion.

7.5.2 *Disequilibrium Processes*

Pure chemical equilibrium, even rainout equilibrium, represents an ideal system that is generally not achieved in real atmospheres. Other processes, including photochemistry owing to incident ultraviolet light from a parent star, or mixing driven by convection or even global wind patterns, can prevent an atmosphere from achieving equilibrium. Blumenthal et al. (2018) consider in detail the impact of such processes on the James Webb Space Telescope (JWST) transit spectra of extrasolar planets. Here we briefly summarize a few important considerations.

Photochemistry

Incident flux, particularly at UV and shorter wavelengths, can dissociate molecules in a planetary atmosphere, leading to a cascade of chemical reactions that produce disequilibrium species that would not otherwise be expected. Ozone, O_3, in Earth's atmosphere being a famous example.

Photochemical processes in exoplanet atmospheres have not yet been comprehensively studied, but some important potential reaction pathways have been identified. In one example relevant to gas giants somewhat warmer than Jupiter, UV-driven dissociation of H_2O produces atomic H (Zahnle et al., 2009), which then reacts with H_2S via the reaction

$$H_2S + H \rightarrow HS + H_2.$$

The HS radical reacts with H to free S,

$$HS + H \rightarrow S + H_2$$

and atomic S reacts with HS to make S_2, leading ultimately to S_8 which, if the atmosphere is cool enough, can condense to form a haze (Zahnle et al. 2016). The impact of such a haze on the reflected light spectrum of a warm giant planet is shown in Gao et al. (2017).

There are likewise reactions among C-bearing species that produce the organic hazes seen in Solar System giant planet atmospheres and Titan, not to mention polluted cities. The potential impact of such photochemical processes must be considered evaluated, as the impact of such hazes can be very large. In solar system giant planets, O- and S-bearing species are trapped beneath optically thick clouds and are not major players in atmospheric photochemistry. This will in general not be true for exoplanet atmospheres, providing a complex set of compositions to participate in photochemically driven reactions. Laboratory experiments (Hörst et al., 2018) are now starting to shed some light on such mixtures, pointing to a complex diversity of outcomes depending on atmospheric composition and incident flux.

Relatively featureless transit spectra seen in some observations of transiting planets (e.g., Sing et al., 2016) have been attributed to the presence of high-altitude photochemical hazes (e.g., Morley et al., 2015). There is unquestionably much more laboratory and theoretical work needed to understand the production and nature of such hazes.

Mixing

A second important disequilibrium process can arise simply from vertical mixing in an atmosphere. If the timescale for a gas parcel to be advected vertically through the atmosphere is much less than the timescale for chemical reactions to bring that parcel to chemical equilbrium, $\tau_{\rm mix} < \tau_{\rm chem}$, then mixing can deliver trace species to the observable atmosphere that would not otherwise be expected.

The usual example of this process is the presence of excess CO in cool, solar composition atmospheres, over that expected by chemical equilibrium. Because of the strong triple bond in CO, it can be difficult for atmospheric chemical processes to convert the species to methane. Figure 7.4 illustrates the chemical pathways linking CO to CH_4. For this reason, Fegley and Lodders (1996) predicted that CO would be discovered in the atmospheres of

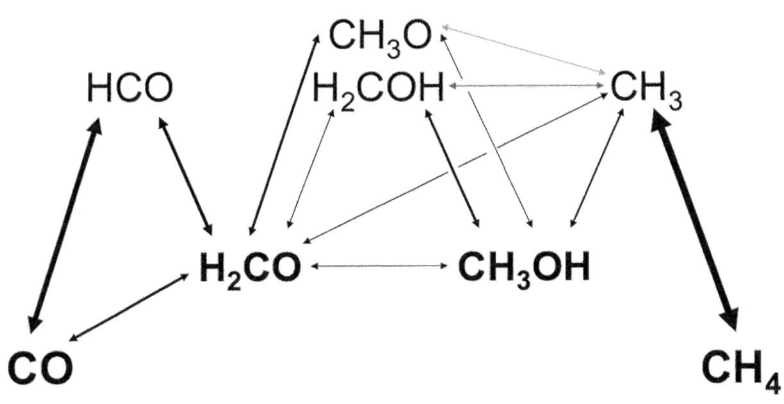

FIGURE 7.4. The key chemical pathways linking CO and CH4 in an H_2-rich atmosphere (from Zahnle and Marley, 2014). Reactions from left to right are with H_2 or H. Key intermediate molecules are shown. The vertical position of individual species gives a rough indication of the energetics. Energy barriers that must be overcome correspond to breaking C-O bonds from triple to double, from double to single and from single to freedom. Relative magnitudes of reaction rates are indicated by arrow thickness. The multiple chemical steps, some of which are quite slow, and the large energy barrier for CO to reach CH_4 account for the long chemical equilibration timescale for this reaction. © AAS. Reproduced with permission.

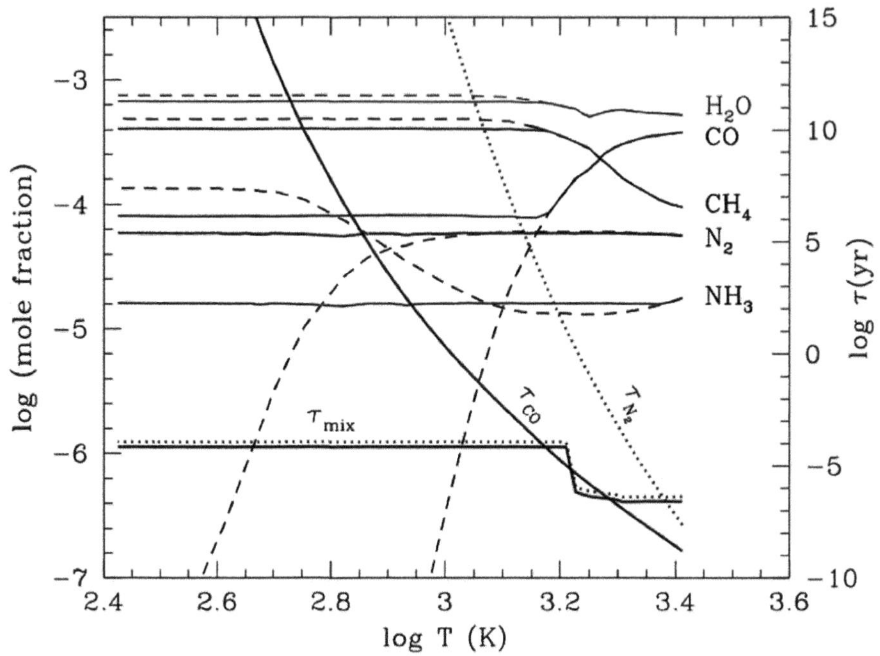

FIGURE 7.5. Chemical and mixing time profiles for a model atmosphere of the brown dwarf Gl 570 D (modified from Geballe et al., 2009). Mole fractions for various species are shown for two assumptions about the atmospheric chemistry as a function of temperature in a model atmosphere (the top of the atmosphere is at the left). Dashed curves show the abundances expected for an atmosphere in pure chemical equilibrium. Solid lines show the abundances in the presence of vertical mixing. The estimated atmospheric mixing timescale is shown as τ_{mix} while the timescales for CO and N_2 to reach equilibrium are shown by the thick solid and dotted lines on the right-hand axis. When the mixing timescale is faster than the chemical equilibrium timescale, the composition is frozen, or quenched, to higher altitudes in the atmosphere. The mixing timescale is nearly discontinuous where the atmosphere becomes convective. © AAS. Reproduced with permission.

what were then termed the methane dwarfs (now the T dwarfs) and, indeed, several observational studies (e.g., Noll et al., 1997; Saumon et al., 2000; Geballe et al., 2009) found excess CO, above what would be predicted by chemical equilibrium, in these objects.

The relative abundance of the disequilibrium to the equilibrium species is set at the depth in the atmosphere where $\tau_{mix} \approx \tau_{chem}$, commonly referred to as the quench level. Figure 7.5 illustrates the expected equilibrium and disequilbrium abundances for several important species in a brown dwarf atmosphere under the presence of vertical mixing and quenching in the CO/CH_4 and the similar N_2/NH_3 systems. In the latter case, N_2 plays the role of a molecule with long equilibration times, in this case to NH_3.

7.5.3 *Clouds*

Clouds and hazes are common features of solar system atmospheres, have been observed in brown dwarf atmospheres and are likely ubiquitous in extrasolar planetary atmospheres as well. They serve as sinks for volatile compounds and influence both the deposition of incident flux and the propagation of emitted thermal radiation. Consequently, they affect the atmospheric thermal profile. General overviews of the importance of clouds in exoplanet and brown dwarfs atmospheres can be found in Marley et al. (2013) and Helling and Casewell (2014).

The terms 'clouds' and 'hazes' are sometimes used interchangeably. In the exoplanet and brown dwarf literature, 'cloud' usually refers to condensates that grow from an atmospheric constituent when the partial pressure of the vapor exceeds its saturation vapor pressure. A general framework for such clouds in planetary atmospheres is provided by Sánchez-Lavega et al. (2004). 'Haze' usually refers to condensates of vapor produced by photochemistry or other nonequilibrium chemical processes. Note that this usage is quite different from that of the terrestrial water cloud microphysics literature where the distinction depends on water droplet size and atmospheric conditions.

Clouds strongly interact with incident and emitted radiative fluxes. The clouds of Earth and Venus increase the planetary Bond albedo (the fraction of all incident flux that is scattered back to space) and consequently decrease the equilibrium temperature. Clouds can also trap infrared radiation and heat the atmosphere. Hazes, in contrast, because of their usually smaller particle sizes can scatter incident light away from a planet but not strongly affect emergent thermal radiation, and thus predominantly result in a net cooling. The hazes of Titan play such an antigreenhouse effect role in the energy balance of the atmosphere. For these reasons, one-dimensional atmosphere models must consider the effects of hazes and clouds.

Because the composition of exoplanetary atmospheres encompasses a very large range of possibilities, a large number of species may be important cloud species. Depending on conditions, clouds in a solar composition atmosphere can include exotic refractory species such as Al_2O_3, $CaTiO_3$, Mg_2SiO_4 and Fe at high temperature and Na_2S, MnS and of course H_2O at lower temperatures. Not all clouds condense directly from the gas to a solid or liquid phase as the same species. For example, in the atmosphere of a gas giant exoplanet, solid MnS cloud particles are expected to form around 1,400 K from the net reaction $H_2S + Mn \rightarrow MnS(s) + H_2$ (Visscher et al., 2006).

The manner in which clouds interact with radiation depends upon the the radiative properties of the constituent particles. The cross sections C_s and C_a with which particles scatter (s) or absorb (a) incident radiation are given by

$$C_s = Q_s \pi r^2$$

and

$$C_a = Q_a \pi r^2$$

for particles with radius r at some wavelength λ. The scattering and absorption efficiencies Q_s and Q_a, and, from them, the extinction efficiency $Q_e = Q_s + Q_a$ are thus defined. The particle single scattering albedo is $\varpi = Q_s/Q_e$. These quantities can be computed by a Mie code, which computes the electromagnetic wave propagation through spherical particles, given a tabulation of the optical properties of the cloud material. Good physically based introductions to radiative properties of particles can be found in van de Hulst (1957), Hansen and Travis (1974) and Liou (2002).

Given this sensitivity of the radiative properties of cloud particles to their size and composition, a cloud model is required to predict cloud species composition, size and vertical extent. There are so many chemical and physical processes that influence these parameters that the number of 'cloud' variables can quickly grow to an unmanageable extent. Thus, relatively simple cloud models are needed for one-dimensional modelling.

For the hydrogen-helium-dominated atmospheres of giant planets, Sánchez-Lavega et al. (2004) review the vapor-pressure condensation framework for cloud formation. Curves tracing the set of pressure and temperature conditions at which a given species condenses assuming equilibrium chemistry ('condensation equilibrium curves') are shown

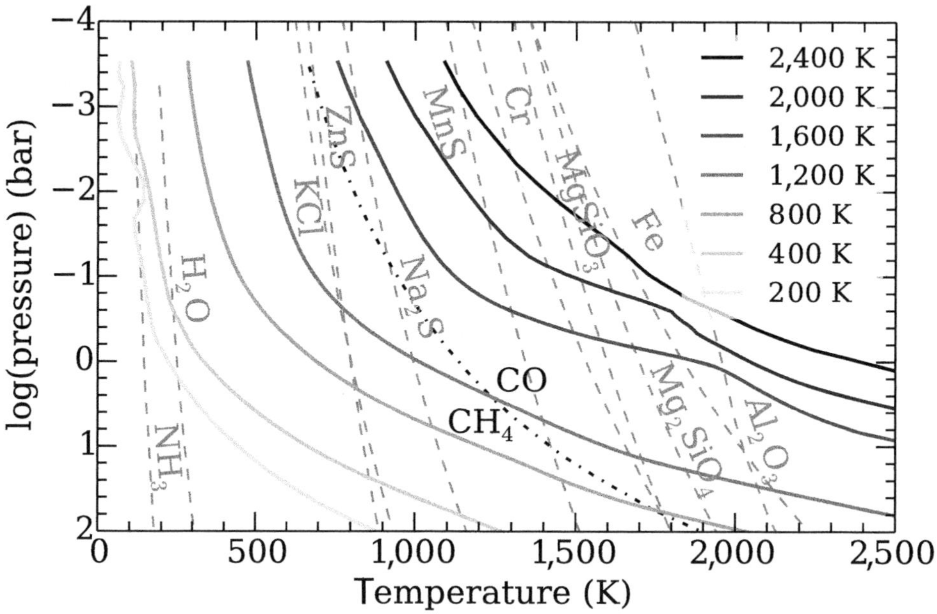

FIGURE 7.6. Model brown dwarf pressure-temperature profiles (modified from Morley, 2016) showing condensation curves for various cloud forming species. Cloud layers are expected to form above the region where a given atmosphere profile crosses the condensation curve. The curves are computed to be the point where the vapor pressure of a given species is equal to the partial pressure of that species in a solar composition gas. © AAS. Reproduced with permission.

for many species in Figure 7.6. A cartoon of the resulting cloud decks is shown in Figure 7.7.

An example of a relatively simple cloud model based on gas condensation and vertical transport that predicts the vertical profile of particle sizes is that of Ackerman and Marley (2001). Albedo spectra of a model giant planet with H_2O clouds and differing cloud thicknesses as predicted by the Ackerman and Marley model are shown in Figure 7.8.

A different approach has been taken in the most extensive body of work on cloud formation in giant exoplanet and brown dwarf atmospheres by Helling and collaborators (Helling et al., 2008, 2016; Witte et al., 2009, 2011), who follow the trajectory of seed particles from the top of their model atmospheres as they sink downwards. The seeds grow and accrete condensate material as they fall, resulting in 'dirty' or compositionally layered grains. To date the datasets on exoplanet, and even brown dwarf, clouds have not yet been sufficient to distinguish between these and other modelling approaches. Improving cloud models and identifying specific observations to distinguish model predictions, particularly particle size and composition and vertical extent, remain important areas for future research (Witte et al., 2011). Higher-quality observations by JWST and the extremely large ground-based telescopes will doubtless provide important new constraints for such efforts.

7.6 Radiative Transfer

Given a draft atmosphere model, specifying temperature, composition and cloud properties through an atmosphere that may be illuminated by a star, we must solve the 1-D, plane-parallel radiative transfer equation,

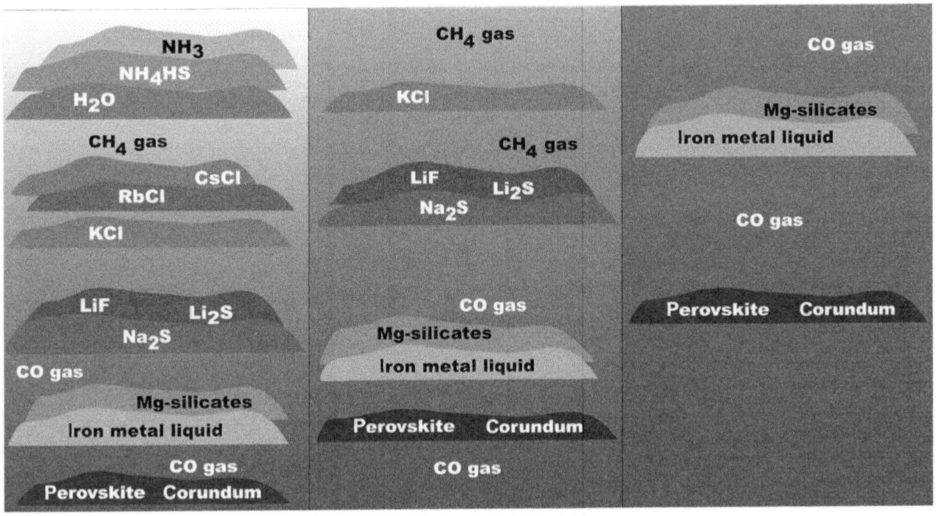

FIGURE 7.7. Schematic illustration (modified from Lodders, 2004) of cloud layers expected in solar composition extrasolar planet atmospheres based on consideration of equilibrium chemistry in the presence of precipitation. The three panels correspond roughly to effective temperatures of approximately 120 K (Jupiter-like, left), to 600 K (middle) to 1,300 K (right). Note that with falling atmospheric temperature, the more refractory clouds form at progressively greater depth in the atmosphere and new clouds composed of more volatile species form near the top of the atmosphere. © AAS. Reproduced with permission.

$$\mu \frac{dI_\nu}{d\tau_\nu} = I_\nu(\tau_\nu, \mu, \varphi) - S_\nu(\tau_\nu, \mu, \varphi),$$

where I_ν is the spectral radiance, τ_ν is the frequency-dependent extinction optical depth (which increases towards higher pressures), μ is the cosine of the zenith angle, φ is the azimuth angle and S_ν is the source function.

For an irradiated atmosphere, the source function is given by

$$S_\nu(\tau_\nu, \mu, \varphi) = \varpi_\nu F_\nu^\odot e^{-\tau_\nu/\mu_\odot} \cdot p_\nu(\tau_\nu, \mu, \varphi, -\mu_\odot, \varphi_\odot)/4\pi + (1 - \varpi_\nu) B_\nu(T(\tau_\nu))$$
$$+ \omega_\nu \int_0^{2\pi} d\varphi' \int_{-1}^{1} d\mu' \cdot I_\nu(\tau_\nu, \mu', \varphi') p_\nu(\tau_\nu, \mu, \varphi, \mu', \varphi')/4\pi,$$

where ϖ_ν is the frequency-dependent single-scattering albedo, $F_\nu^\odot$ is the stellar irradiance impinging at the top of the atmosphere, $\mu_\odot$ is the solar zenith angle, $\varphi_\odot$ is the solar azimuth angle, p_ν is the scattering phase function, B_ν is the Planck function and $T(\tau_\nu)$ is the atmospheric temperature profile. The final term on the right-hand side, which accounts for scattering from directions (μ', φ') into the beam at (μ, φ), complicates radiative transfer calculations, as it couples the source function back to the three-dimensional radiation field, turning the radiative transfer equation into an integro-differential equation. The importance of clouds is apparent by the prominence of the scattering phase function, which is a sensitive function of particle size and composition.

Simplifications and approximations must be brought to bear to rapidly solve the equation of transfer through the atmosphere at a sufficient number of discrete wavelength points to adequately constrain the next upwards and downwards fluxes. Because of the importance of both scattering and molecular gas opacity, with hundreds of millions to billions of discrete lines, many of the approaches commonly applied in the

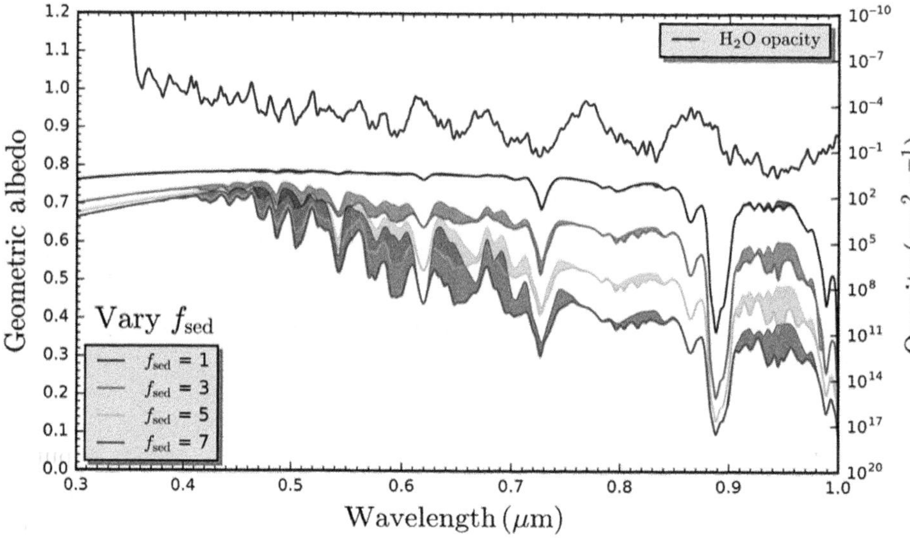

FIGURE 7.8. Geometric albedo spectra for various models of a planet the mass of Saturn, but located at about 1 AU from a solar-type star. Because the planet is relatively warm, the uppermost cloud layer will be composed of H_2O droplets rather than NH_3 ice as in Jupiter and Saturn (see Figure 7.7). Models for five different cloud thicknesses, parameterized here by the f_{sed} parameter of Ackerman and Marley (2001), are shown. Smaller f_{sed} corresponds the vertically thicker clouds, which are brighter. The top curve shows gaseous H_2O opacity, plotted increasing downwards for easier comparison with features in the model spectra. The shaded regions highlight the impact of the water vapour on the albedo spectra. In this case, thinner clouds permit a longer column of H_2O above the cloudtops to absorb and thus exhibit stronger water spectral features. Modified from Macdonald et al. (2018). © AAS. Reproduced with permission.

stellar atmospheres approach for solving the equation of transfer are less applicable to the planetary atmospheres problem. Typically, techniques are divided into either two-stream or multistream solutions, in which a stream refers to a particular azimuth-zenith coordinate through the atmosphere. Solutions in the two-stream category are more computationally efficient than multistream calculations while multistream calculations provide more detailed information about the angular distribution of intensities and, thus, can provide more accurate solutions for radiant fluxes.

The variety of approaches is too extensive to summarize here. Various two-stream approaches commonly used in planetary atmospheres applications are summarized by Meador and Weaver (1980), and Marley and Robinson (2015) give an overview of their application to the exoplanetary atmospheres and brown dwarf atmospheres. In addition, Heng and Marley (2017) give some guidance to choosing among various approaches.

The outcome of a complete radiative-transfer calculation for a given model atmosphere is the upwards and downwards radiative fluxes through each model layer for both the incident stellar and planetary thermal fluxes. These fluxes can then be inserted into (7.5) to evaluate if the given model structure is close to radiative-convective equilibrium.

7.7 Temperature Adjustment

When all of the aforementioned physical considerations have been accounted for, it is time to derive the one-dimensional radiative-convective thermal structure, a task that must be done iteratively, starting from a first guess. There are several approaches that can be followed to determine the equilibrium profile, which satisfies (7.5), and where the temperature and flux profiles are all continuous. The most straightforward technique is to simply time-step the atmosphere to equilibrium; however, more efficient computational approaches have been developed. Such methods are discussed in detail in Hubený and Mihalas (2014). While sophisticated iterative approaches are usually preferred, at times more simplistic Newton–Raphson approaches, such as discussed in Marley and Robinson (2015), can more stably, if less rapidly, converge on a solution.

Regardless of the details of the method, the temperature, composition, cloud and convection profile of the atmosphere must be iteratively modified until a single profile that meets all desired constraints to some accuracy is derived. In some cases, this can be rapid and straightforward, but in cases where cloud opacity is a sensitive function of temperature (e.g., when water clouds are forming), the temperature structure can undergo unstable oscillations as the atmosphere cools, opacity increases, the atmosphere warms in response, clouds evaporate and the process repeats. Various computational tricks are often employed in such cases to press to convergence. Nevertheless, in practice there are occasionally specific volumes of model 'phase space' in which it is difficult to find a preferred model.

7.8 Conclusion

The science we gain from exoplanet spectroscopic observations, as well as from ongoing studies of brown dwarfs, depends our ability to model and understand the atmospheres of these worlds. Although the first two decades of brown dwarf science have seen remarkable advances in the fidelity of atmosphere modelling, there is still much room for improvement. By comparison, exoplanet atmosphere modelling is still in its infancy. Better cloud models, greater exploration of the effect of varying elemental abundances, particularly atmospheric C/O ratios, and greater studies of departures from equilibrium chemistry are all important areas awaiting improvement.

While the number of ingredients in the recipe required for atmospheric modeling may at first seem overwhelming, in practice these can be mastered. As in the kitchen, every chef first learns from the recipe book and then moves on to improvise new dishes. Likewise the next generation of atmospheric modelers will no doubt improve on every aspect of the approaches now commonly followed.

7.9 Acknowledgements

This work benefitted from ongoing discussions about the art and science of atmospheric structure modelling with multiple people, in particular Tyler Robinson, Jeff Cuzzi and Richard Freedman. This work was supported by the NASA Exoplanet Research and Astrophysics Theory Programs. The winter school experience was exceptionally enhanced by the kind hospitality of all of the hosts, and especially that of Dr. Lucio Crivellari.

REFERENCES

Ackerman, A. S. and Marley, M. S. 2001. Precipitating Condensation Clouds in Substellar Atmospheres. *ApJ*, **556**(August), 872–884

Allard, F., Hauschildt, P. H., Alexander, D. R., Tamanai, A. and Schweitzer, A. 2001. The Limiting Effects of Dust in Brown Dwarf Model Atmospheres. *ApJ*, **556**(July), 357–372

Asplund, M., Grevesse, N., Sauval, A. J. and Scott, P. 2009. The Chemical Composition of the Sun. *ARA&A*, **47**(September), 481–522

Blumenthal, S. D., Mandell, A. M., Hébrard, E., et al. 2018. A Comparison of Simulated JWST Observations Derived from Equilibrium and Non-Equilibrium Chemistry Models of Giant Exoplanets. *ApJ*, **853**(February), 138

Burrows, A. and Orton, G. 2010. Giant Planet Atmospheres. Pages 419–440 of: Seager, S. (ed.), *Exoplanet*, University of Arizona Press.

Burrows, A. and Sharp, C. M. 1999. Chemical Equilibrium Abundances in Brown Dwarf and Extrasolar Giant Planet Atmospheres. *ApJ*, **512**(February), 843–863

Burrows, A., Hubbard, W. B., Lunine, J. I. and Liebert, J. 2001. The Theory of Brown Dwarfs and Extrasolar Giant Planets. *Reviews of Modern Physics*, **73**(July), 719–765

Chabrier, G., Baraffe, I., Allard, F. and Hauschildt, P. 2000. Evolutionary Models for Very Low-Mass Stars and Brown Dwarfs with Dusty Atmospheres. *ApJ*, **542**(October), 464–472

Fegley, Jr., B. and Lodders, K. 1994. Chemical Models of the Deep Atmospheres of Jupiter and Saturn. *IAU Circ.*, **110**(July), 117–154

Fegley, Jr., B. and Lodders, K. 1996. Atmospheric Chemistry of the Brown Dwarf Gliese 229B: Thermochemical Equilibrium Predictions. *ApJ*, **472**(November), L37

Freedman, R. S., Lustig-Yaeger, J., Fortney, J. J., Lupu, R. E., Marley, M. S. and Lodders, K. 2014. Gaseous Mean Opacities for Giant Planet and Ultracool Dwarf Atmospheres over a Range of Metallicities and Temperatures. *ApJS*, **214**(October), 25

Freedman, R. S., Marley, M. S. and Lodders, K. 2008. Line and Mean Opacities for Ultracool Dwarfs and Extrasolar Planets. *ApJS*, **174**(February), 504–513

Gao, P., Marley, M. S., Zahnle, K., Robinson, T. D. and Lewis, N. K. 2017. Sulfur Hazes in Giant Exoplanet Atmospheres: Impacts on Reflected Light Spectra. *AJ*, **153**(March), 139

Geballe, T. R., Saumon, D., Golimowski, D. A., Leggett, S. K., Marley, M. S. and Noll, K. S. 2009. Spectroscopic Detection of Carbon Monoxide in Two Late-Type T Dwarfs. *ApJ*, **695**(April), 844–854

Hansen, J. E. and Travis, L. D. 1974. Light Scattering in Planetary Atmospheres. *Space Sci. Rev.*, **16**(October), 527–610

Helling, C. and Casewell, S. 2014. Atmospheres of Brown Dwarfs. *A&A Rev.*, **22**(November), 80

Helling, C., Lee, G., Dobbs-Dixon, I., et al. 2016. The Mineral Clouds on HD 209458b and HD 189733b. *MNRAS*, **460**(July), 855–883

Helling, C., Woitke, P. and Thi, W.-F. 2008. Dust in Brown Dwarfs and Extra-Solar Planets. I. Chemical Composition and Spectral Appearance of Quasi-Static Cloud Layers. *A&A*, **485**(July), 547–560

Heng, K. and Marley, M. S. 2017. Radiative Transfer for Exoplanet Atmospheres. Page 102 in: H. Deeg and Belmonte, J. (eds.), *Handbook of Exoplanets*. Springer.

Hörst, S. M., He, C., Lewis, N. K., et al. 2018. Haze Production in the Atmospheres of Super-Earths and Mini-Neptunes: Insights from the Lab. *ArXiv e-prints*, January.

Hubený, I. and Mihalas, D. 2014. *Theory of Stellar Atmospheres*. Princeton University Press.

Lewis, J. S. 1972. Low Temperature Condensation from the Solar Nebula. *IAU Circ.*, **16**(April), 241–252

Liou, K. N. 2002. *An Introduction to Atmospheric Radiation*. Elsevier

Lodders, K. 1999. Alkali Element Chemistry in Cool Dwarf Atmospheres. *ApJ*, **519**(July), 793–801

Lodders, K. 2004. Jupiter Formed with More Tar than Ice. *ApJ*, **611**(August), 587–597

Lodders, K. and Fegley, B. 2002. Atmospheric Chemistry in Giant Planets, Brown Dwarfs, and Low-Mass Dwarf Stars. I. Carbon, Nitrogen, and Oxygen. *IAU Circ.*, **155**(February), 393–424

Lodders, K. and Fegley, Jr., B. 2006. *Chemistry of Low Mass Substellar Objects*. Praxis Publishing.

Macdonald, R. J., Marley, M. S., Fortney, J. J. and Lewis, N. K. 2018. Exploring H_2O Prominence in Reflection Spectra of Cool Giant Planets. *ArXiv e-prints*, April

Marley, M. S., Ackerman, A. S., Cuzzi, J. N. and Kitzmann, D. 2013. *Clouds and Hazes in Exoplanet Atmospheres*. Pages 367–391 in: Mackwell, S. J., Simon-Miller, A. A.,

Harder, J. W. and Bullock, M. A. (eds.), *Comparative Climatology of Terrestrial Planets.* University of Arizona Press.

Marley, M. S. and Robinson, T. D. 2015. On the Cool Side: Modeling the Atmospheres of Brown Dwarfs and Giant Planets. *ARA&A*, **53**(August), 279–323

Meador, W. E. and Weaver, W. R. 1980. Two-Stream Approximations to Radiative Transfer in Planetary Atmospheres: A Unified Description of Existing Methods and a New Improvement. *Journal of Atmospheric Sciences*, **37**(March), 630–643

Mihalas, D. 1978. *Stellar Atmospheres*, 2nd edition, W. H. Freeman and Co.

Morley, C. 2016. *Clouds and Hazes in Exoplanets and Brown Dwarfs.* PhD Thesis, University of California at Santa Cruz.

Morley, C. V., Fortney, J. J., Marley, M. S., et al. 2015. Thermal Emission and Reflected Light Spectra of Super Earths with Flat Transmission Spectra. *ApJ*, **815**(December), 110

Noll, K. S., Geballe, T. R. and Marley, M. S. 1997. Detection of Abundant Carbon Monoxide in the Brown Dwarf Gliese 229B. *ApJ*, **489**(November), L87–L90

Rajan, A., Rameau, J., De Rosa, R. J., et al. 2017. Characterizing 51 Eri b from 1 to 5 μm: A Partly Cloudy Exoplanet. *AJ*, **154**(July), 10

Rothman, L. S., Gamache, R. R., Goldman, A., et al. 1987. The HITRAN database – 1986 edition. *Appl. Opt.*, **26**(October), 4058–4097

Rothman, L. S., Gordon, I., Babikov, Y., et al. 2013. The HITRAN2012 Molecular Spectroscopic Database. *J. Quant. Spec. & Rad. Tran.*, **130**, 4–50

Rothman, L. S., Gordon, I. E., Barber, R. J., et al. 2010. HITEMP, the High-Temperature Molecular Spectroscopic Database. *J. Quant. Spec. Radiat. Transf.*, **111**(October), 2139–2150

Sánchez-Lavega, A., Pérez-Hoyos, S. and Hueso, R. 2004. Clouds in Planetary Atmospheres: A Useful Application of the Clausius–Clapeyron Equation. *American Journal of Physics*, **72**(June), 767–774

Saumon, D., Geballe, T. R., Leggett, S. K., et al. 2000. Molecular Abundances in the Atmosphere of the T Dwarf GL 229B. *ApJ*, **541**(September), 374–389

Seager, S. and Deming, D. 2010. Exoplanet Atmospheres. *ARA&A*, **48**(September), 631–672

Sharp, C. M. and Burrows, A. 2007. Atomic and Molecular Opacities for Brown Dwarf and Giant Planet Atmospheres. *ApJS*, **168**(Janurary), 140–166

Sing, D. K., Fortney, J. J., Nikolov, N., et al. 2016. A Continuum from Clear to Cloudy Hot-Jupiter Exoplanets without Primordial Water Depletion. *Nature*, **529**(January), 59–62

Tennyson, J. and Yurchenko, S. N. 2012. ExoMol: Molecular Line Lists for Exoplanet and Other Atmospheres. *MNRAS*, **425**(September), 21–33

van de Hulst, H. C. 1957. *Light Scattering by Small Particles.* Wiley and Sons

van Zeggeren, F. and Storey, S. H. 1970. *The Computation of Chemical Equilibria.* Cambridge University Press

Visscher, C., Lodders, K. and Fegley, Jr., B. 2006. Atmospheric Chemistry in Giant Planets, Brown Dwarfs, and Low-Mass Dwarf Stars. II. Sulfur and Phosphorus. *ApJ*, **648**(September), 1181–1195

Witte, S., Helling, C. and Hauschildt, P. H. 2009. Dust in Brown Dwarfs and Extra-Solar Planets. II. Cloud Formation for Cosmologically Evolving Abundances. *A&A*, **506**(November), 1367–1380

Witte, S., Helling, C., Barman, T., Heidrich, N. and Hauschildt, P. H. 2011. Dust in Brown Dwarfs and Extra-Solar Planets. III. Testing Synthetic Spectra on Observations. *A&A*, **529**(May), A44

Zahnle, K. J. and Marley, M. S. 2014. Methane, Carbon Monoxide, and Ammonia in Brown Dwarfs and Self-Luminous Giant Planets. *ApJ*, **797**(December), 41

Zahnle, K., Marley, M. S., Freedman, R. S., Lodders, K. and Fortney, J. J. 2009. Atmospheric Sulfur Photochemistry on Hot Jupiters. *ApJ*, **701**(August), L20–L24

Zahnle, K., Marley, M. S., Morley, C. V. and Moses, J. I. 2016. Photolytic Hazes in the Atmosphere of 51 Eri b. *ApJ*, **824**, 137